Second IASTED International Symposium

ROBOTICS AND AUTOMATION

June 22-24, 1983 Lugano, Switzerland

IASTED

ISBN 0-88986-049-1

ROBOTICS AND AUTOMATION

Editor: M. H. Hamza

A PUBLICATION OF
THE INTERNATIONAL ASSOCIATION OF SCIENCE AND
TECHNOLOGY FOR DEVELOPMENT — IASTED

ACTA PRESS
ANAHEIM * CALGARY * ZURICH

Proceedings of the Second IASTED International Symposium ROBOTICS AND AUTOMATION, Lugano, Switzerland, June 22-24, 1983.

EDITOR: M.H. Hamza Canada

ACTA PRESS
P.O. Box 2481
Anaheim, CA
92804, U.S.A.

ACTA PRESS
P.O. Box 3243, Stn. B
Calgary, Alberta
Canada T2M 4L8

ACTA PRESS
P.O. Box 354
CH-8053 Zurich
Switzerland

ACTA PRESS BOOK CODE: 55

TABLE OF CONTENTS

	PAGE
A Hierarchical Kernel Image Processor, TIP-3 - C. Cherne, B. Dunbar, D. Fronek, A. Hammerquist, D. Hollingsworth, T. Powell, R. Moore, Y. Sugamura, S. Wang	1
An Automatic Complex Assembly With Two Compliant Arms, Force Feedback and Vision - R. Zapata, B. Shariat, B. Jouvencel, P. Dauchez, P. Coiffet	6
The Organization of Information Interchanging in the Systems of Digital Transmission and Switching - G. Bafas	10
A Microprocessor Based Controller for Time Optimal Control - L. Moreno, I. Serra, E. Luque	14
Application of Modern Control Concepts to the Design of a Distillation Column Regulator - N.N. Sorial	17
A Numerical Approach for the Universal Solution of the Inverse Kinematic Problem - E. Ersue, D. Nungesser	20
Modified Minimum Variance Selftuning Control of a Robotic Manipulator R.M. DeSantis, R. Hurteau, S. Lefebvre	24
On Optimal Control of Transients in Not-Very-Large Electrical Power Systems - V.A. Venikov, M.A.H. Tawfik	28
Commande Adaptative en Boucle Fermée d'un Robot Dans un Espace de Travail Local Doté d'un Système de Vision - M. Dufaut, R. Husson	32
Oligarchy: A Control Scheme for Flexible Manufacturing Systems - P.R. Piche, M. Krieger, N. Santoro	35
A Synthesis of Controllers for Sampled-Data Systems - J. Rozinoer	40
A Prototype Robot Workstation for the Motif Application Process Within the Garment Manufacturing Industry - D.R. Kemp, P.M. Taylor, G.E. Taylor	44
Software and Hardware Aspects of a Flexible Workstation for Assembly Tasks Using Sensory Robots - P.M. Taylor, C.A. Stubbings	48
Etablissement d'une Commande Performante pour Actionneur Pas à Pas Appliquée à la Robotique à Partir d'un Modèle Dynamique Complet - C. Rowdo, J.P. Louis, R. Husson	52
Contour and Region Filling Methods - C. Gambaro, P. Sinigaglia	56
Automated Development of Program Modules in Robot Control Operating Systems - M.K. Avramova, B. Manahilov, D. Karastojanov	60
A Robust MRAC for Industrial Robots - F. Nicolo, J. Katende	64
An Adaptive Seam-Tracking Subsystem for Arc-Welding Robots - D. Karastojanov, G. Nachev, B. Manahilov	68
Stability of Non-Linear Continuous Systems - F. Rotella, J.P. Richard, I. Zambettakis	72
Efficient Algorithms for Fast Features Detection in Binary Pictures - N.G. Bourbakis, C.A. Alexopoulos, D. Serpanos	77
A Processor Program for a Float Point Arithmetic Procedures Realizing in a Microprocessing System - B.E. Manahilov, M. Avramova, D. Karastojanov	81
Application de la Régression Orthogonale à l'Identification d'une Chaîne de Mesures Complexe de Type Cascade - M. Robert, J.L. Noizette, J. Ragot, T. Disano	85
Absolute Location of a Mobile Robot With a Scanning Laser Range Finder - L. Marce, M. Julliere, H. Perrichot, G. Utjes, C.J. Zhao	90
Asservissement de Vitesse d'un Moteur Synchrone Diphase Actionnant un Axe de Robot - C. Bergmann, T. Marilier, C. Sol	94

	PAGE
Discrimation des Orientations de Pièces pour l'Alimentation Automatisée Flexible - S. Jacqmart	97
Computer Aided Design of Robust Control Algorithms for Robot Manipulators - R. Dumas M. Leborgne J.J. Borelly B. Espiau	101
Analyse d'Images 2 D. en Vue de la Description d'Objects 3 D. - N. Kerkeni M. Leroi M. Bourton	105
Simultaneous Path-Force-Torque-Teach-In for Robots - The Sensor Programming Concept - J. Heindl G. Hirzinger	109
Tracking of an Underwater Inspection Module. Suivi d'Un Module Sous-Marin d'Inspection - A. Faure P. Lerolland P. Poupel	113
Design of High-Performance Controllers for Remote Manipulator Systems with Explicit Actuator Dynamics - B. Porter	117
Robot Programming by Development Tools - U. De Carlini A. Esposito	120
A Simple Pattern Recognition Scheme Based on Neural Network Structures - N. Dimopoulos R.V. Patel	125
An Outline of a New Approach to Pattern Recognition - L. Goldfarb	129
Generation of Straight Line Motion for Robots - W. Khalil	133
The Development of Modular Simulation Procedures for the Design of Task Independent Industrial Robots - R.C. Michelini R. Molfino G. Acaccia	136
Performance Evaluation of an Automation System I - Methodology - A. Aimo H. N. Koivo A. Halikka	140
Use of a Measurement and Control Processor in Groundwood Mill Control - J. Mustapaa H.N. Koivo J. Kalliokoski	144
Main de Robot à Deux Degrés de Liberté Actionée par Moteur à Courant Continu Geré par Microprocesseur - J.C. Demorise H. Cunha Jr. J.P. L'Huillier	148

	PAGE
Performance Evaluation of Seven Degrees of Freedom Robots - R.G. Fenton B. Benhabib A. Goldenberg	152
Cartesian Path Planning by General Polynomial Interpolation - H. Bruhm E. Ersue	156
An Approach via Robust Linear Multivariable System Design Methods to Nonlinear Manipulator Control - E. Ersue W. Neddermeyer	160
Modular Robot Systems - P.R. Moore R.H. Weston T.W. Thatcher J.D. Gascoigne	165
Design of a Grab Manipulator System for Peat Handling - J. Haeggman H.N. Koivo P. Ranta	170

"A Hierarchical Kernel Image Processor, TIP-3"

C. Cherne, B. Dunbar, D. Fronek, A. Hammerquist, D. Hollingsworth, T. Powell, R. Moore, Y. Sugamura, S. Wang

Image Processing Laboratory
University of Nevada-Reno
Reno, Nevada, U.S.A.

Abstract

The design of recent image processing architectures has largely been driven by VLSI/VHSIC technology and the need for a very high speed data throughput. Future sensors will be highly parallel and will produce data rates that will require special non-Von Neuman architectures. In highly parallel systems, each pixel element is typically assigned an active processor. As larger and larger images are processed in parallel, the interconnection problem and large numbers of active processors becomes a significant if not an impossible restriction. The TIP-3, "The Image Processor version three," provides a reasonable solution to a parallel hierarchical processor without a significant VLSI requirement. The prototype designed so far has shown much promise for real time image processing. The purpose of this paper is to present an overview of the TIP-3 architecture, and a limited amount of data produced by the TIP-3 simulator.

Introduction

Image processing has posed many interesting problems since late 1958 when Unger (1) proposed one of the first binary computer structures to examine simple images. The fabrication of specialized processor architectures was hampered by the lack of large amounts of high speed memory and the absence of LSI Technology. Attention was placed on processor structure, interprocessor communication, and image processing algorithms in an attempt to model certain aspects of the human eye. Uhr (2), for example, proceeded to investigate an "eye-like recognition cone" that proposed layered networks -- actually a series of hierarchical processing planes in a cone-like structure whose apex is a single cell. Image processing was broken down into at least three areas: preprocessing, classification, and description. During the early 1970's there was a tremendous interest in these three processing areas. Processing centered around the "neighborhood function" with a surprising amount of processing done using a weighted average technique.

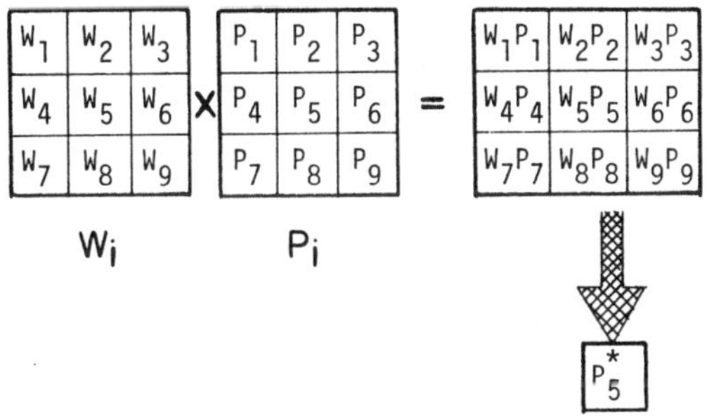

Figure I — Weighted Average Calculation

The processed image is then composed of an array of processed pixels, P_5^* (P_5^* is assumed to be the center of the processed neighborhood).

$$P_5^* = \sum_{i=1}^{n} \frac{W_i P_i}{D} \qquad (1)$$

As the 3 x 3 neighborhood "moves over" the image to be processed (typically 256 x 256, 512 x 512, etc.) the new image is composed of P_5^*'s and has 2-less columns and 2-less rows due to the "edge effect" perimeter on the original image. A typical neighborhood architecture is composed of only one active processor which must process a 3 x 3 window (rows - 2)(columns - 2) times. For a 256 x 256 image array (65536 pixels), the processor repeats equation (1) 64516 times. For an N x N array, the processor is said to process the image in order N^2 times. Depending upon the processing time of equation (1), with 9 multiplications, 8 additions, and 1 division, the entire process could require many minutes to complete. The obvious problem at this point is one of processing speed -- especially at real time rates (33 milliseconds per processed image). An increase in the number of processors was seen in the early 1970's.

Kruse (3) proposed a highly parallel Picture Processing Machine, PPM, capable of neighborhood processing using an array of "modules" or Processing Elements, PE's, shown in Figure (2).

With exception of the perimeter, all pixels are assigned an active processor. This "array processor" is very fast, requires (row - 2) (columns - 2) processors, and has a very complicated interconnection scheme. However, a neighborhood operation can be performed on the entire image in just one time period (the time it takes to execute equation (1)). Certain processes such as global pixel average, max/min pixel, and certain edge detections have been shown (Tanimoto

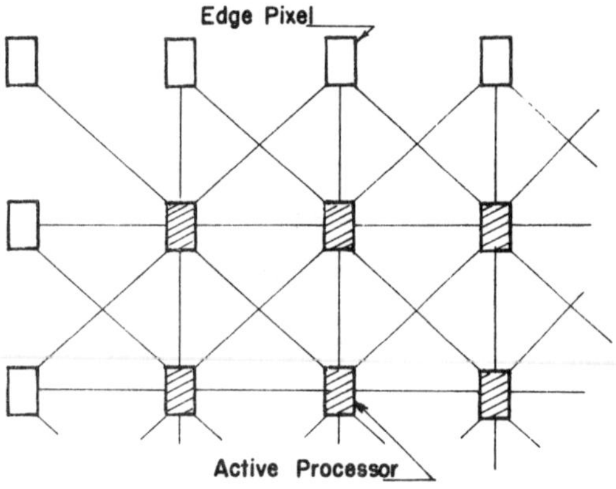

Figure 2—Flat Plane Array Processor

11, 8, 6) to execute in Order n time -- n times faster than the single processor.

Tanimoto (4) proposed a pyramid structure composed of successively smaller planes -- each plane containing an array of parallel processing elements. The processor interconnections not only allowed the nearest neighbors on each plane to communicate but also communicate to certain processors on the plane below (called son pixels) and to a single processor on the plane above (called the father pixel). In this way images can be sent up or down the pyramid structure depending upon the processing algorithm. This structure is shown in figure (3)

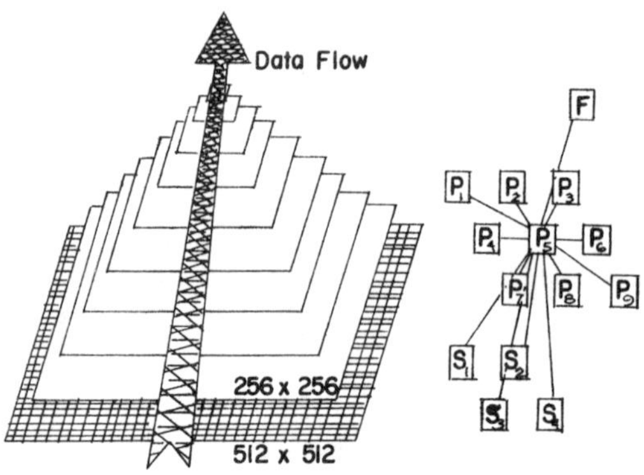

Figure 3—Hierarchical Processor Structure

Algorithms producing a global average, min/max pixel value, special forms of convolution, and many binary operations have been shown to run in Order log n time. For various values of n, (64, 128, 256, 512, 1024) the speed-up is quite evident using the hierarchical processor structure.

Hierarchical Processor	Array Processor	Single Processor	$Gain_1$	$Gain_2$
Log n	n	n^2	$\frac{n}{\text{Log } n}$	$\frac{n^2}{\text{Log } n}$
6	64	4096	10.67	682
7	128	16384	18.29	2340
8	256	65536	32.00	8192
9	512	262144	56.89	29127
10	1024	1048576	102.40	104857

Table 1 -- Comparing Processing Time

The gain in speed of the hierarchical processor is accompanied by a modest 33% increase in hardware over conventional flat plane array architectures. This price is quite easy to justify since a gain in speed of 32 (256 x 256 image) over the array processor is typical and even higher gains are obtainable for very large images. The problem with large images is the brute number of active processors. VLSI/VHSIC technologies provide some solutions to this problem, and some chip manufacturers (Texas Instruments, Fairchild and Harris) are investigating "wafer integration" of array sensors and active preprocessing.

There have been several approaches proposed to parallel processing without excessively large numbers of active processors. Most schemes assume that available sensors do not yet produce a large number of parallel data paths and typically produce a single serial stream of data. The load time, then, becomes a significant portion of the total processing time. While sensor data is being loaded, processing time is partitioned and executed in a time-multiplexed parallel fashion. Large images are partitioned into "tiles" with each tile having a single row of active processors at the top of the tile as shown in figure (4).

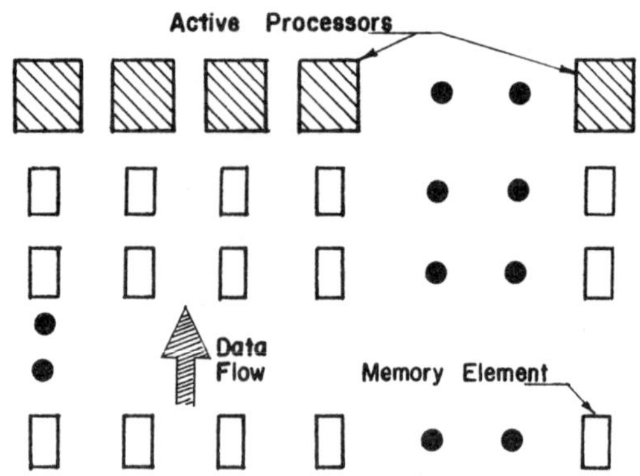

Figure 4—Systolic Array Processor

The columns under each active processor are simple memory elements that contain pixel data. This data is systollically "pushed" through the top row of active processors.

Since each tile must communicate with neighboring tiles (except of the edge of the image frame), the interconnection and timing problem between tiles is no easy matter. Boeing Aerospace (9) is currently fabricating a 4 x 16 processor systolic tile with all adjacent communications and timing on a single chip. When these chips are placed together, large images, in multiples of 4 columns and 16 rows, can be processed in sixteen machine cycles.

TIP-3 Architecture

Several excellent processor architectures have been developed including the CLIP(5), Base (6,8), Pyramid(7), PPM(3) and the VHSIC image processors. Each architecture gives consideration to:

1. Neighborhood pixel communication
2. The arithmetic and logic unit
3. Individual pixel processor memory
4. Serial/parallel arithmetic processing
5. Machine instruction and control.

To a large extent, the image processing application will significatnly impact the architecture design. Typically, the original image is processed for enhancement (equalized histogram, noise removal, edge detection, thresholds, etc.) and attemps are made to generate a "level of abstraction" as soon as possible. The preprocessing step requires critical consideration in specifying an image processing architecture.

TIP-3 circuitry external to the processor deals with the eight nearest neighbors ($P_1 \ldots P_9$) excluding P_5 (center pixel) in much the same way as the pyramid machine parallel architecture. Father and son communication is also provided with an additional communication channel for global memory (read and write). TIP-3 is a three dimensional 4-plane hierarchical kernel with 8 x 8, 4 x 4, 2 x 2, 1 x 1 planes and side inputs for neighbor communications. This kernel is totally parallel and requires 85 active processors. To process a 128 x 128 image, for example, the hierarchical kernel is placed on the bottom plane in a "corner" position. Neighborhood processing, including hierarchical neighborhoods (father/sons), occurs almost instantaneously owing only to arithmetic and logical gate delays within the kernel. Even though the controller (6809/68000) is synchronous, processed data is ready for storage at each plane in the structure within a controller clock cycle. For the 128 x 128 image, 120 x 120 processing cycles are required at the bottom plane ((128-8)(128-8)). The apex of the hierarchical kernel produces an 8 x 8 array after this step and is temporarily stored. It takes 15 kernel positions on the bottom plane (128 x 128) to map one pixel at the 8 x 8 level. Figure 5 illustrates the kernel scanning operation.

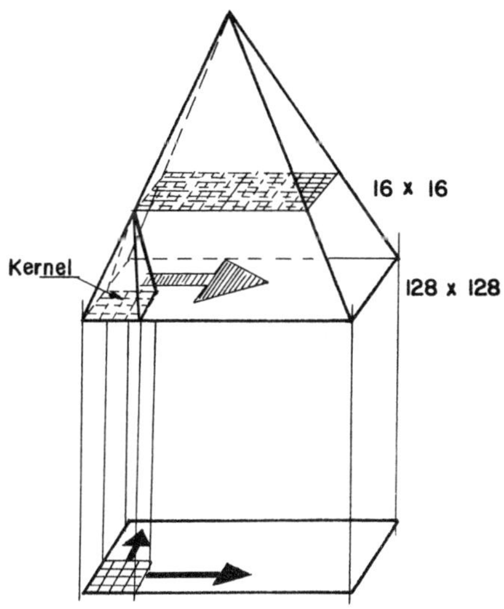

Figure 5— Kernel Scanning

The processor, shown in figure 6, is essentially the controller peripheral in the sense that it must be driven by a host computer (6809/68000). The controller writes an opcode for the desired operation to the instruction register of the TIP-3. No handshake protocol is necessary since each processor will operate asynchronously in parallel on its own data. Each processor is identical and communicates to neighboring processors identically except for those processors at the "edges" of the plane. These edge pixels receive data from programmable memory elements. The problem of interprocessor communication becomes somewhat clumsy but is straightforward. Each processor will receive input data from its eight surrounding neighbors, from four sons, and a father pixel. This same processor must also provide data to these 13 neighboring processor cells.

The processor consists of the neighborhood function, binary preprocessors, an input multiplexer, a 4-input ALU/Boolean processor, and memory. The neighborhood function selects any of the pixels ($P_1 \ldots P_9$), any of the four sons ($S_1 \ldots S_4$), the father pixel, F, or an auxiliary memory input from the base image plane. This selection process is based upon 4-bits of the TIP-3 instruction word. The functions implemented by the Boolean preprocessors are neighborhood contraction/dilation, and sons contraction/dilation. These functions are realized as combinational logic and are quite fast. Any ALU function used regularly in a particular process is considered as a permanent member of the instruction set.

The multiplexer selects one of the five function sources to be propagated to the ALU. This selection is based on three bits of the instruction word shown in figure 7. The ALU is the heart of the processor and a critical component from a speed standpoint. This ALU has 4-inputs, a multiplexer output, a B-memory output, a A-memory output, and C-memory output (for serial carrys). The three inputs (A,B,C) yield 256 possible functions of three variables -- a number much too large for a random logic implementation. A fourth input, T, is provided to operate on incoming pixel data and pixel data contained in the individual processor memory (A-memory). An EPROM is used to realize this function due to its flexibility in the prototyping stage. This version of the TIP uses 50 of the more commonly used ALU/Boolean functions. Up to 12 address lines are available in the instruction word for A and B memory addressing. The lower address lines are split providing A and B to have a small offset. Two bits of the instruction word are used for A-memory read/write, and B-memory read/write. A third bit is used to control the clock of the C-memory. Six more bits of the instruction word are used to decode one of the 50 ALU/Boolean processes. Most of the ALU functions use only one output but a second output is necessary for any function producing a borrow or a carry.

The speed of the TIP-3 is currently limited by the controller and the ALU-ROM. The processing ROM speed is in the 50-500 nanosecond range depending on the technology used. Some advantages of the TIP-3 are:

1. ROM-based 4-input/2-output ALU providing a function and its complement
2. Bit serial processing for hierarchical flexibility
3. Four basic instruction types: Transfer, Compare, Boolean, Special
4. Multiple instructions done in one pass
5. Separate input and output memory.

Hierarchical Algorithms

The impact of "the way in which problems are solved" is a major impact on the design and execution of processing images. Clearly, a "good" algorithm must provide an efficient tradeoff between the procedures employed and the

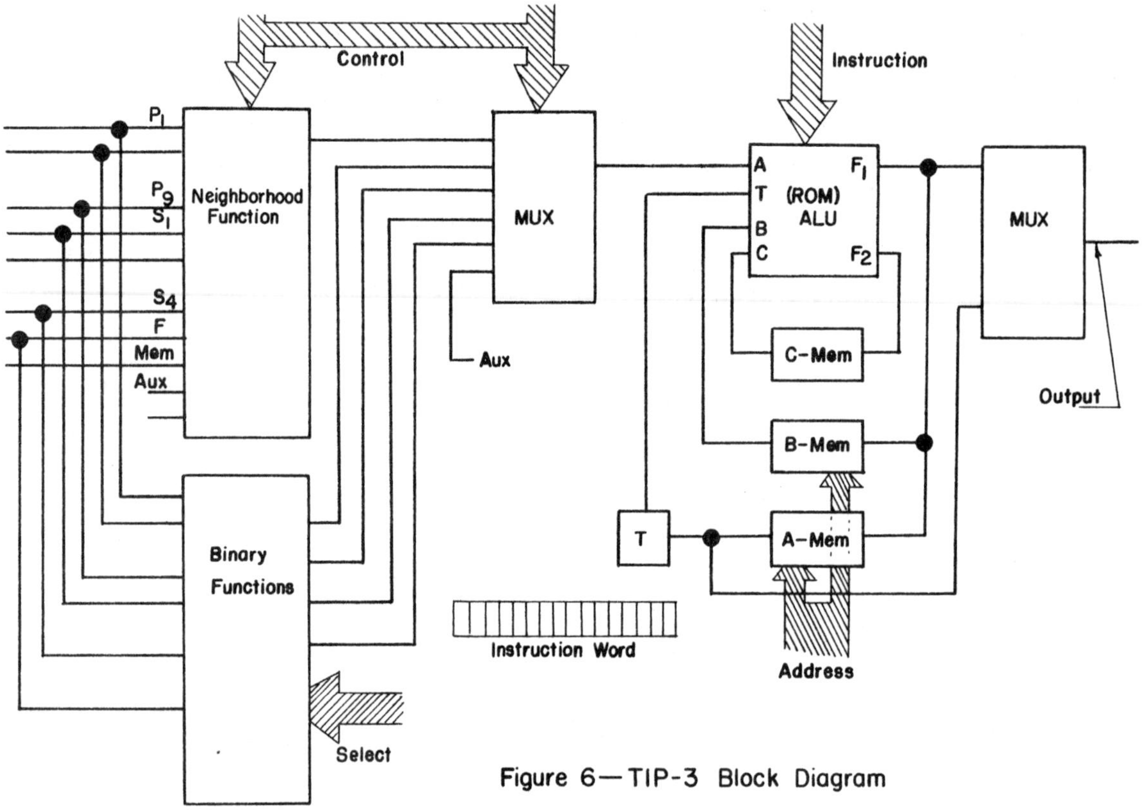

Figure 6—TIP-3 Block Diagram

hardware available within the processing system. There is no direct way to provide efficient algorithms without the knowledge of the processor architecture, sensor data format, and display requirements with respect to processing speed. And, even with extensive knowledge of these attributes, the task is not straightforward at all and usually requires extensive interaction between hardware and software capabilities.

To further appreciate the data structure and parallel hierarchical processing, consider the following algorithm: The average of all pixels such as might be used to determine a gray scale threshold. For this example, the bottom plane contains an array of 8-bit pixels. After each addition instruction, the result is placed in the next higher plane. Since two 8-bit numbers add together with a potential 9-bit result, variable length pixel size and addressing control is needed. As the addition continues upward in the structure, additional bits are required for processed data. For a bottom array of 256 x 256 only 8 addition instructions are required (log 256) to calculate the sum of the entire plane. The division necessary for averaging is done by shifting the apex result 8 times to the right -- now an 8-bit pixel again. The shifting is done simply by addressing and causes no delay after the addition instructions.

As was noted earlier, the hierarchical processor requires more hardware than does the flat plane processor by 33% owing to the vertical stacking of additional planes each one successively smaller. The additional 1/3 in hardware complexity is well worth the throughput gain in speed and execution time -- Order Log N for the hierarchical processing vs. Order N time for the flat plane processor. The TIP-3 processor is capable of the following algorithms:

a. Addition
b. Subtraction
c. Compare
d. Maximum Pixel
e. Minimum Pixel
f. Average Pixel (Local, Global)
g. Contraction
h. Dilation
i. Top Down Search
j. Image Perimeter
k. Image Area
l. Convolution
m. Multiply
n. Histogramming
o. Special Functions (template driven)

It should be noted that the flat plane processor is a proper subset of hierarchical processors since hierarchical processors also process data on individual planes. The additional ability of selecting the "plane of sufficient resolution" is a strong argument for increased throughput. Currently, algorithms are running with the following machine cycles (calculated).

	Machine Cycles
Dilation/Contraction (Binary)	2
Add two 4-bit pixels	15
Add two 8-bit pixels	29
Multiply two 4-bit pixels	33
Multiply two 8-bit pixels	57
Load 8-bit pixel	51
Max/Min Pixel	280
Ave 2 x 2 Matrix 4-bit pixels	376
8-bit pixels	722

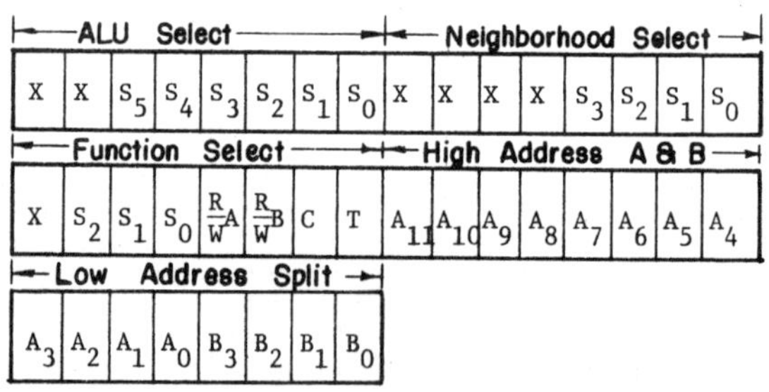

Figure 7- TIP-3 Instruction Word

Summary

Real time image processing (defined as the processing time to display a standard television raster--33 milliseconds) presents a variety of challenges dependent upon the:

Sensor data format
Processor architecture
Processing algorithms.

Architectures like the Pyramid (11,12,13,16) TIP-3 (17), CLIP (15), BASE-8(8), CEDAR (19), TRAC (19) and several others have considered the data rates produced by highly parallel sensors that will be available in the late 1980's. As the demand for processing throughput continues to increase, non von Neuman architectures like the TIP-3 will provide the real time processing required.

High speed processing is proceeding in at least three directions (19):

1. Fast single processors--Josephson junctions, gallium arsenide, etc.
2. Low degree of parallel processors
3. High degree of parallel processors with several hundred processors on the same wafer.

Highly parallel machines are primarily being designed at many universities. Architectures like PRINGLE, and PASM (1024--68000 CPU's Purdue) TOPSTAR and EDDY (Japan), STARLET (West Germany), INMOS (Britain) and a huge additional list of clever approaches to parallel computing. In addition to the hardware, a major thrust is occuring in software development to support these various architectures.

Industry is taking an "I'll believe it when I see it" (19) approach to the highly parallel architecture and considers many of the highly parallel machines as "very narrow application areas".

Just in the last two years have actual models been constructed and used in evaluating both hardware and software. Many of the processors mentioned here will be modified or constructed within the next two years. TIP-3 for example, has undergone extensive upgrade and simulation of the pyramid-like architecture. A 2 x 2 prototype has been fabricated and work is proceeding on a 4 x 4 and 8 x 8 planes to complete the TIP-3 kernel.

Much work has been done with algorithms running in real time on the TIP-3 simulator with excellent results. These results are not presented in this paper but will be available in September 1983.

References

1. Unger, "A Computer Oriented Toward Spacial Problems", Proceedings IRE, p. 1744, October 1958.

2. Uhr, "Layered Recognition Cone Networks", IEEE Transactions -- Computers, July 1972, p. 758.

3. Kruse, "A Parallel Picture Processing Machine", IEEE Trans -- Computer, December 1973, p. 1057.

4. Tanimoto, "A Hierarchical Data Structure for Picture Processing", CGIP, June 1975.

5. Duff, "CLIP-4: A Large Scale Integrated Circuit Array Parallel Processor", 3rd Int. Conf. on Pattern Recognition, 1976, p. 728.

6. Reeves, Anthony, "A Systematically Designed Binary Array Processor", IEEE Transactions on Computers, April 1980, p. 278.

7. Tanimoto, "An Image Processor Based on an Array of Pipelines", Computer Architecture and Image Professor Conference Proceedings, Hot Springs, Virginia, November 1981.

8. Reeves, "The BASE-8 Binary Arrey Processor", Proc. Pattern Recognition and Image Processing Conf., Chicago, August 1979, p. 250.

9. Fronek, Snapp, "The Boeing Aerospace Pyramid Image Processor -- A summary and Capability Study, June 1982.

10. Wang, Hall, "Sequential Hierarchical Scene Matching, IEEE Trans -- Computer, April 1978.

11. Steiglitz, "Operations on Images Using Quad Trees", IEEE Transactions, PAMI, April 1979.

12. Tanimoto, "Image Data Structures in Structured Computer Vision", Academic Press.

13. PMI Company, "Report of Algorithms, Simulations, and Architecture Updates", Report Submitted to Boeing Aerospace Company, December 1981.

14. Kelley, "Edge Detection by Computer Using Planning", Mach. Int., Vol. 6, 1971.

15. Duff, "A Parallel Computer for Array Processing", Info. Proc., North Holland, 1974, p. 94.

16. Dyer, "A VLSI Pyramid Machine for Hierarchical Parallel Image Processing", IEEE Transactions -- Computer, August 1981.

17. Cherne, Dunbar, Fronek, Hammerquist, Hollingsworth, Powell, Moore, Sugamura, Wang, "A Hierarchical Image Processor, TIP-3, for Polarimetric Radar Signals", Second Workshop on Polarimetric Radar Technology, Huntsville, Alabama, Conference Proceedings, May 1983.

18. Masatsugu Kidode, "Image Processing Machines in Japan", Toshiba Corporation, Computer, January 1983, p. 68.

19. Manuel, "Advanced Parallel Architectures get Attention as Way to Faster Computing", Electronics, June 1983, p. 105.

An automatic complex assembly with two compliant arms, force feedback and vision.

R. Zapata, B. Shariat, B. Jouvencel, P. Dauchez, P. Coiffet.

Laboratoire d'Automatique et de Microélectronique de Montpellier,
U.S.T.L., Place Eugène Bataillon
34060 Montpellier Cedex, FRANCE.

ABSTRACT

In this paper, we describe an assembly experiment carried out by two compliant arms (MA-23, Société La Calhène), controlled by computer and using force feedback and vision. The aim of this experiment is to assemble a vice of 8 parts initially located under a C.C.D. camera (100x100 pixels).

A vision processor provides informations about position and orientation of each part, and an host computer defines their grasping position. A planning algorithm is then used to compute the arms trajectories in order to avoid any collision between the both arms, or between the arms and fixed obstacles (worktable, camera,...).

To build the vice, eight assemblies are needed : four of them are performed by a single arm, the four others require the use of both manipulators. Because of inaccuracy of the robots, and lack of specialized assembly tools, the forces and torques acting on the manipulators are measured to compute correcting movements making the assemblies possible.

I - INTRODUCTION

This paper deals with the problem of the use of robots acting in true cooperation in assembly tasks ; this kind of problem may be essential in the future, for the industrial robotics.

The first part of this paper describes the experimental situ.

The second part consists in a description of the algorithms and the assembly 's strategies used in this experiment :

1) the recognition : pictures acquisition and treatments in order to compute the grasping positions of the parts

2) the displacements : a safe trajectories algorithm to avoid any collisions between the arms

3) the assemblies : the assembly parts, chosen in order to carry out various mechanical joinings (prismatic, cylindrical, multi-insertion, screwing,...), and the assembly strategies used to overcome the lack of precision of the manipulators (no possible precise position assemblies)

II - THE EXPERIMENTAL SITU

The experimental situ (Figure 1) is composed of :

1) two manipulators MA-23 /1;2/: each of them is a six degrees of freedom robot, driven by six electric motors.

2) a C.C.D. camera which is used to recognise the assembly parts.

3) a worktable.

4) a moveable plate on which the parts are disposed at the begining of the experiment ; this plate is divided into three zones, because the assembly pieces are too numerous to be seen in a single time by the camera.

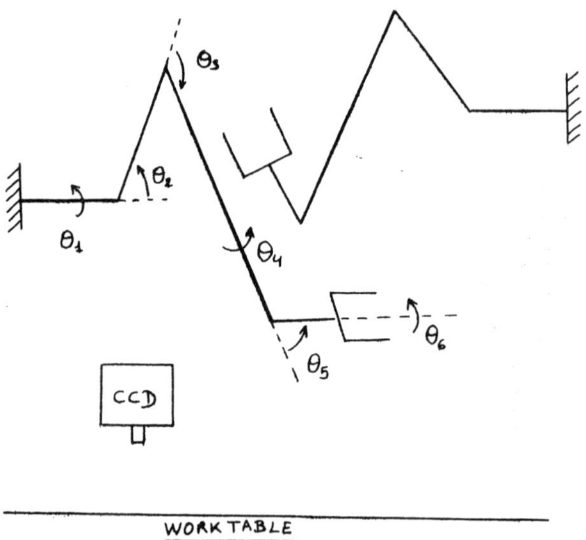

Figure 1

III - DESCRIPTION OF THE EXPERIMENT

1) The recognition /3;4;5/

The picture is obtained by a C.C.D matrix camera with 100x100 pixels. The picture acquisition rate is set by a clock pulse which varies from 0.25 to 5 MHz ; we set this latter one in order to regulate the picture acquisition time which is a function of the scene brightness. The image signal is converted by a "flash" A.D.C and then digitalized on four bits (16 gray levels). All the materials and the treatments can be adapted to any kind of matrix cameras. One input multiplexer allows to manage up to 3 cameras. The pattern treating and recognition programs are in 8085 Intel assembly language. The block diagram of the vision system is shown in Figure 2.

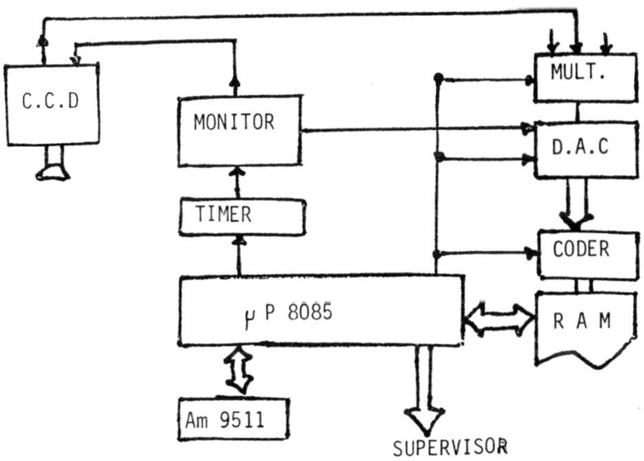

Figure 2

Each part is disposed in bulk without overlapping. We use a sufficient contrast between pieces and bottom of the scene. We bind tracking with gradient operator to find sequentially every edge. We induce a point belongs to edge by comparison with a local threshold and we consider the proximities between points. We calculate in line partial sums allowing resolve total characteristics : perimeter P, maximum radius, third combination of invariant moments and the number of holes. Those parameters are calculated for every face of stability of parts. They are caracterised by trapezoidal probability law. We calculate the class for which the conditionnal probability is upper than threshold. The order of introduction of parameters is determined with the theory of information. We research the parameters that gives an average maximum information.

Position and orientation of a part are procured by :

* the known face

* the knowledge of three points : gravity center, points corresponding to maximum and minimum radii. So wo make a system of axis that permits to seize the piece for orientate the manipulators grippers.

2) The displacements /6;7/

During the grasping and carrying stages of the assembly parts, the two MA-23 have to move along large trajectories while avoiding any collision between themselves and with the work table. In these stages, the robot needs to be controlled by a safe trajectories algorithm which will be introduced in the experiment procedure.

a) The safe trajectories planner

The inputs of this subroutine are the initial robot configuration, and the final one to be reached. The output is a safe trajectory between these two configurations. This planner uses a graph search algorithm. The nodes are the arms configurations and the links are the elementary displacements of the robot. (Depth First Search Algorithm) /8/.

b) The environment models

The checking of the collisions relations needs models of the robot and of the robot environment. These models must be as simple as possible to permit a fast trajectory planning. We have chosen to modelise each forearm by four spheres, and to modelise each terminal device by one sphere, as it is shown in Figure 3. The other arms links do not need to be modelised, because they cannot collide any object in the environment. The camera is also modelised by a sphere, and the table by a plan.

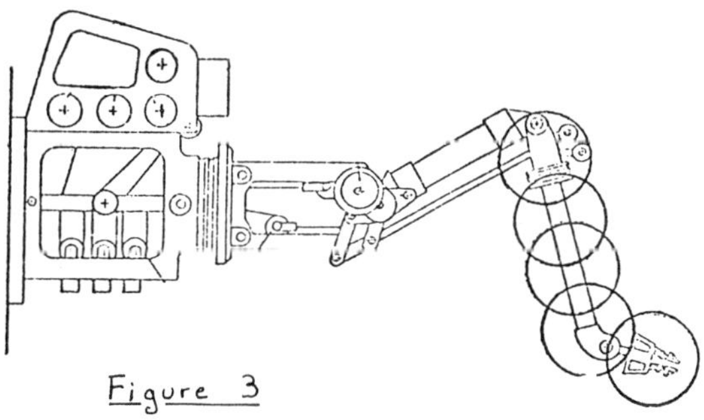

Figure 3

c) The graph generating function

Each node in the graph has 5 sons defined by :

1°son : $Ts(q)=\eta\left(F(q)-T(q)\right) / \left(\sum_{q=1}^{12}\left(F(q)-T(q)\right)^2\right)^{1/2}$

2°son : $Ts(3)= T(3)-\eta$

3°son : $Ts(9)= T(9)-\eta$

4°son : $Ts(2)= T(2)-\eta$

5°son : $Ts(8)= T(8)-\eta$

where : $T(q)$ is the qth component of the node configuration

$Ts(q)$ is the qth component of its son configuration

$F(q)$ is the qth component of the final configuration

η is a sampling parameter

3) Carrying out of the assemblies/9;10/

a) The parts

The aim of the experiment is to assemble a vice composed of eight parts and a base fixed on the worktable (Figure 4).

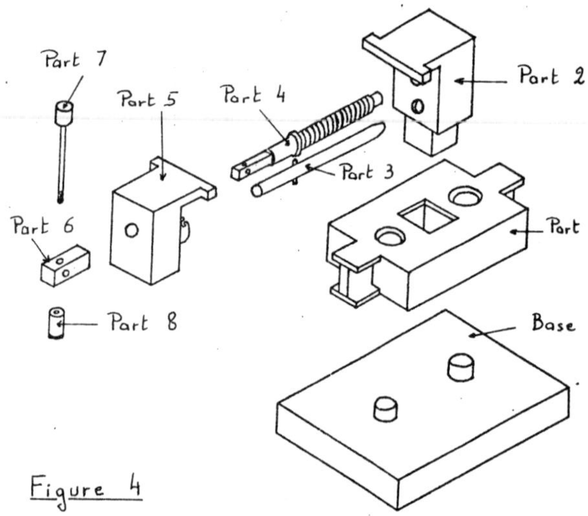

Figure 4

b) The assembly procedure

The complete experiment can be divided into 9 stages:

stage 1 : Grasping and assembly of part 1 on the base by two arms.
stage 2 : Grasping and assembly of part 2 on part 1 by one arm
stage 3 : Grasping of parts 4 and 5 and assembly in the air of these two parts.
stage 4 : One of the arms holds the block (part 4-part 5) while the other arm grasps part 3 and carries out a second assembly in the air.
stage 5 : Exchange of the block (parts 3, 4 and 5) between the two arms in order to bring it to an orientation compatible with the following stage.
stage 6 : Assembly by one arm of the block (parts 3, 4 and 5) on part 2.
stage 7 : Grasping and assembly by one arm of part 6 on part 4.
stage 8 : Grasping and assembly by one arm of parts 6 and 7.
stage 9 : Grasping by one arm of part 8 and assembly by two arms of parts 7 and 8.

To do a displacement Δx during an assembly, two methods are possible :

1) Δx is computed from the forces acting on the joints /11/.

2) Δx is chosen from a priori informations : the inaccuracy of the manipulators causes lateral and angular errors which should be compensated. The lateral error is compensated by the receptacle 's chamfer. To compensate the angular error, a strategy is adopted. It consists in doing a rotation around the assembly axis before the boarding phase. This allows to set an error in orientation which masks the error inherent in the robot and which will be then gradually corrected to do the hooking of the parts.

Moreover, to lessen jamming during the assembly phase, we execute little back and forth movements Δx in a plan perpendicular to the assembly axis. Each assembly of the complete experiment requires the use of these methods or some of them.

V - CONCLUSION

Using two manipulators is convenient for many tasks and allows to do tasks very hard or impossible by using only one (for example to carry a heavy or large part).

The use of two robots in the described experiment has allowed to carry out complete assemblies which cover many mechanical joinings. The adaptability of the robots is due to the use of vision and force feedback. The robots inaccuracy (flexibility) is compensated by special "intelligent" algorithma that ensure the success of the assembly tasks. The strategies proposed are sufficiently general to be used in other experimental conditions.

Nevertheless, using two robots needs to solve some hard problems :

- to avoid considerable effort on a part carried by two manipulators
- to avoid collisions between moving robots : we have proposed a solution based on a spherical modelisation of the manipulators, and on the use of a graph search algorithm to generate co-ordinated displacements of the robots.

In spite of these difficulties, we are interested in developing cooperation and coordination of two (or more) robots in various tasks, because we think it will be in the future a very important area of industrial robotics /12/.

REFERENCES

/1/ W.KHALIL,
"Modélisation et commande par calculateur du manipulateur MA- 23. Extension à la conception par ordinateur des manipulateurs",
Thèse de Docteur Ingénieur, U.S.T.L, Montpellier 1976

/2/ J. VERTUT, P. COIFFET,
"Bilateral Servomanipulation MA-23 in direct mode and via optimized computer control"
M.M.T. Journal IFTOMM special issue, 1976.

/3/ B. JOUVENCEL, P. COIFFET, J.POT,
"Un système de vision artificielle pour robot-manipulateur"
OPTO 82 ESI Publications

/4/ J. POT,
"Localisation et reconnaissance des pièces pour un système de vision autonome"
Thèse de Docteur-Ingénieur,17 dec. 1982, U.S.T.L., Montpellier

/5/ P. COIFFET,
"Les Robots - Interaction avec l'environnement"
Tome 2, Hermès France, 1981.

/6/ R. ZAPATA, P. COIFFET, A. FOURNIER,
"Trajectory planning for a multi-arms robot in an assembly task"
D.S.I.A., New-York, 1983, (to appear).

/7/ R. ZAPATA,
"Le problème de l'évitement des collisions dans la commande des robots-manipulateurs"
Thèse de 3ème cycle, 13 juin 1983, U.S.T.L., Montpellier.

/8/ P. H. WINSTON,
"Artificial intelligence"
Addison Wesley, 1977.

/9/ B. SHARIAT, P. COIFFET, A. FOURNIER,
"A strategy to achieve an assembly by means of an inaccurate, flexible robot"
D.S.I.A., New-York, 1983, (to appear).

/10/ P. DAUCHEZ, P. COIFFET, A. FOURNIER,
"Cooperation of two manipulators in assembly tasks"
D.S.I.A., New-York, 1983, (to appear).

/11/ P. BORREL,
"Modèle de comportement des manipulateurs. Application à l'analyse de leur performance et à leur commande automatique"
Thèse de 3ème cycle, U.S.T.L., Montpellier, 1979.

/12/ P.DAUCHEZ,
"Etude de la commande de deux robots manipulateurs lors de tâches coordonnées"
Thèse de troisième cycle, 13 juin 1983, U.S.T.L., Montpellier

THE ORGANIZATION OF INFORMATION INTERCHANGING IN THE SYSTEMS OF DIGITAL TRANSMISSION AND SWITCHING

G. Bafas : Teaching and Research Scientist,
National Technical Univ.of Athens.

Abstract

By the development of digital technique's transmission and switching and their introduction to the telecommunication networks of various countries,the serious problem of the adaptation of the new technique to the existing analogue environment,appears.Naturally,the problem of the adaptation as well as of the organization of information interchanging between the electronic exchanges of various types,becomes more difficult in the case of the Integrated Services Digital Network and can only be confronted by the fastest possible international standardization of the protocols of information interchanging and the required switching's criterions. This paper presents a method of adaptation of the digital technique's switching to the existing analogue environment,by applying the pulse code modulation systems,as well as methods for the information interchanging between subscribers that use different terminal types and the information banks through the various exchanges of an Integrated Services Digital Network.

1. Introduction

By the last important achievements of the technological development in the field of electronics and particularly of microelectronics,the construction of digital systems of transmission and switching - widely applied to the existing telecommunication networks - becomes economically possible. This development was so rapid,that in a period of two years only,the construction of a chip which contains the codec,the required filter became possible,and also the chip itself has the ability of codec with μ-law (15 segment characteristic for PCM 24 transmission systems) and A-law (13 segment characteristic for PCM 30 transmission systems).

So,the difficult problem of the adaptation of the North American systems of digital switching to existing analogue European environment,seems to be finaly solved and with relatively very low cost.

Basic difficulty still remains the adaptation of the digital technique to the existing analogue environment and particularly in those cases in which an abundance of existing electromechanic exchanges is noticed,as it is in the case of the Greek Telecommunication network. For the more successful adaptation of the digital technique to the analogue environment through Pulse Code Modulation (PCM) systems,the determination of the interchanging method as much between the systems of digital switching and transmission,as well as of the digital exchanges and the corresponding existing electromechanic ones through PCM connection,is required.

The specific problem becomes very important in the case of the establishment of an Integrated Digital Network (IDN) and the further evolution of it to an Integrated Services Digital Network (ISDN).

Therefore, the attainment of the most economic solution for the creation of the ISDN and the cooperation of the ISDN exchanges with the existing analogue ones,the timely thorough analysis and study of the existing network of a country is required and the fastest possible establishment of the interchanging method of all the systems of the mixed networks.

In the case of the Integrated Services Digital Network,the subscribers of the mixed network, will have the demand for the following access methods:

- Simulataneous access to several independent services,a main channel (usually the existing telephone line)and an auxiliary channel.
- Dedicated access to a particular service, with the aid of terminals of various types.
- Alternate access,possibly on a call-by-call basis to two or more services.

For the attainment of the above-mentioned access possibilities for the subscribers,the mixed switching system must have the ability of the following switching methods :

- Packet switching.
- Circuit switching.
- Message switching.

Basic precondition of course in the case of the exchange of an Integrated Services Digital Network,is to have the possibility of switching of the following non voice facilities |1| :

- Dedicated packet switched data network access for calls originating from start-stop and packet user terminals.
- Dedicated circuit switched data network access for calls originating from start-stop and synchronous user terminals.
- Circuit switched calls for start-stop and synchronous user terminals.
- Packet switched calls for packet user terminals.
- Circuit switched calls for start-stop users accessing packet user terminals.
- Simultaneous circuit and packet switched calls for digital multiservice subscribers.

This paper presents methods for the adaptation of the digital technique's switching in difficult analogue environments,as in the case of the Greek network which has an abundance of electromechanic exchanges and also suggestions are given for the probable methods of interchanging between the various exchange types,in the case of direct creation of an Integrated Digital Network and the evolution of it to an Integrated Services Digital Network.

2. Information interchanging in Pulse Code Modulation systems.

For the adaptation of the digital technique to the analogue environment,the case of the Pulse Code Modulation systems (PCM) of first order with 2.048 Kbit/sec,is researched. The interchanging of these systems and their cooperation with the switching systems must be attained in such a way,that the users will be independent from the each time supplier and the solution with the most possibilities and the lowest cost will be attained.

According to the references of CCITT -which are expressed by G.711,G.712 and G.732 - by the Multiplexing method,30 voice channels are converted to a PCM multiplexing signal on the send side and on the receive side 30 voice channels are reproduced from PCM multiplexing signal.

The voice signals,with 300 till 3400 Hz reach the send side from where after the required amplification and filtering,are led to the multiplexing arrangement,where after the sampling,are converted to pulse amplitude signals (always on the send side) with simultaneous quantizing and codification in continuation.By this way,for each sampling value an 8 bit word is created. The pulse frame of the PCM-signals,is formed by the 30 codified words and the additional information which are necessary for the cooperation of the digital technique in the existing analogue environment,as for example: information for the recognition of the frame,all the required criterions for the connection of the PCM system with two or more analogue exchanges of different types,as well as a telesignalling of the alarm situation of the system. For forwarding the operations of the receive side,the synchronization of the system with the pulse frame of the PCM, signal that is received,is at first required. After that,the system of the receive side,separates the additional information from the codified voice signals and by decoding these signals it converts them to pulse amplitude modulation signals. Then, after the required demultiplexing the voice signals are led to the corresponding voice channels,after they have been converted to signals of 300 till 3400 Hz through the corresponding low-pass filters. For each one of the 30 different channels of the system, 8000 sampling values per second are transmitted to both directions - according to the sampling frequency - under the form of 8 bit PCM words,which means that during the 125 μs , 30 codified words of 8 bit are transmitted to the two directions (receive - send) as well as 8 bit which are necessary for the signalling and other 8 bit which alternately contain a bunched frame alignment signal and a service word. The required synchronization of the receive and the send side of the PCM system,is obtained with the aid of the bunched frame alignment signal of the pulse frame. For this purpose, the bit 2 till 8 of the bunched frame alignment must always have the same place.

In Figure 1 the architecture of the pulse frame of the PCM 30 system is shown.

The necessary signalling criterions of the exchanges are transmitted through the signalling time slot No 16.

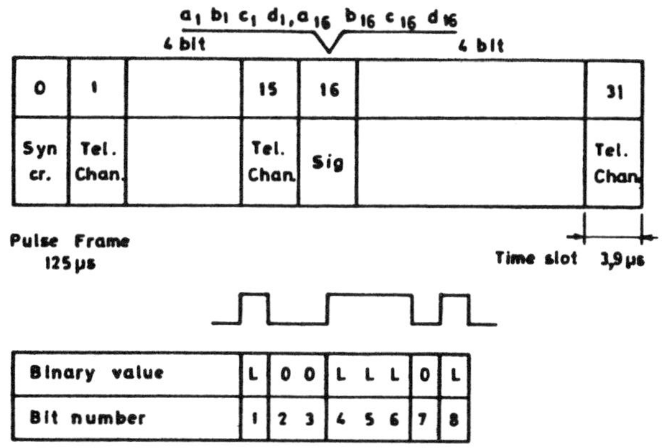

Fig.1 Architecture of the pulse frame of the PCM 30 system

For the case of the Greek network, for the time slot 16 at first only 1 bit had been chosen,but finally it was found out that this solution was not the proper one and 4 bits were at last decided for the intercommunication,as it is shown in figure 1.

3. Information Interchanging in the case of the Integrated Services Digital Network.

Internationally,has not been established yet a uniform method of interchanging for the various exchange types of the Integrated Services Digital Network (ISDN).

The references of ISO 4335 and 3309 for the basic frame format are an initial effort.
According to the last information from CCITT,three evolution phases of the ISDN seem to be prearranged: basic integration,medium integration and full integration.

According to the CEPT references,five different types of information for the access of the subscribers to the ISDN are determined :

- Information f ,for speech signals corresponding to 64 Kbit/sec PCM 30 systems.
- Information d ,for data communication with bit rates of up to 48 Kbit/sec.
- Information s ,for subscriber to network signalling.
- Information t ,for telemetry applications.
- Information d' , for applications of data restricted in bit rate and frame length.

The five above-mentioned different information types,will be used for the following channel types :

- Channel b with a bit rate of 64 Kbit/sec for information f or d .
- Channel b' for low rate data services and therefore,for information d.
- Channel Δ for out-slot signalling between the subscribers and the local exchange and therefore,for information s, t and d'.

The three above-mentioned channel types can be - per case - applied as much with the fixed channel allocation method,as well as with the dynamic channel allocation method,according to which messages are queued according to a defined priority to minimize delays to signalling messages.

Basic role for the cost-effectiveness of the

ISDN plays the method which will be applied to the multiplexing.

For the case of the mixed telecommunication network and the creation of the ISDN,the following possibilities for digital multiplexing exist:

- Fixed multiplex structure according to which the available bandwidth is preallocated.
- Semi-flexible multiplexing,according to which the available bandwidth is preallocated on a semipermanent basis.
- Flexible multiplexing,which permits the available bandwidth to be allocated on demand in a store and forward manner.
- Intelligent multiplexers.
- Frame multiplexers.

Basic precondition for the creation of the Integrated Services Digital Network (ISDN) is the determination of the switching methods of the information and the signalling way for the recognition as much of the subscribers and the terminals which they use,as well as of the kind of the information which they want to get from the specific network.

Basic switching methods which can have direct application to the ISDN are the following three:

- Circuit switching,in which an established connection between two or more terminals has the exclusive use of the transmission circuit until the connection is released.
- Packet switching,in which the information is transmitted by addressed packets and the transmission channel is only occupied during the transmission of a packet and the channel is then available for use by packets being transfered between other terminals.
- Message switching,in which the complete messages are received,stored and retransmitted after appropriate routing to the partial desirable exchanges of the mixed telecommunication network.

In the existing telecommunication networks,basic role in the information transmission plays the signalling method and the speed of the used channels.
In the specific case of the IDN and the ISDN,instead of the term transmission,the term transport is used and so,for the attainment as much of the information' transport from one exchange to the other,as of the realization of the required cooperation between the various types of exchanges and terminals,the determination of the transport method of the imformation is required.

Due to the fact that the case of the creation of Integrated Services Digital Network,internationally seems to interest many scientists,it is considered as expedient a series of thoughts to be given in this paper related to the method of information' transport in mixed telecommunication networks,which involve from thorough analysis of the experience which was obtained in the Greek telecommunication network.

As it is known,today two basic methods of transport have been proposed :Synchronous Time Division and Labeled or Framed Time Division,which are based on the following function principles |1| :
According to the framed time division the time on the bearer is allocated to information of a particular significance on a statistical basis and the information significance is defined an accompanying frame.

The application of the above-mentioned method for the information' transport to a mixed information network involves the following characteristics :

- Special apparatus are not required for the switching and transmission of subscribers with different bit rates.
- The same system can be used for the transport of different capacities.
- One part of the capacity is consumed for the information' transport of the frame itself.
- The method gives possibilities of concentration and therefore dimensioning is on a traffic basis,the criterion being the allowable waiting time.
- For a particular input transit delay depends on the traffic offered to the bearer,the type of protocol used and the priority assigned to the input.
- For the transmission of the frames an intermediate buffering is required.

According to the transport method of Synchronous Time Division the time on the bearer is divided into periods which repeat regularly and if multiplexing is considered are subdivided into slots,which means that the information of a particular significance is always carried in the same slot in each period. It must be explained in the specific case,that the significance of the information is defined by the point in time at which it occurs.

Main characteristics of the method,which are related to the application of it for the transport of information to a mixed network of informatics are :

- constant transit delay
- each channel of the system is dedicated to information of a particular significance and in case that no such information is present the channel remains unused,which means that the channel is not available for other puproses
- one part of the carrier's capacity is completely consumed to carry a synchronization pattern that enables the time slots to be identified at the receiving end.
- basic disadvantage of the method is the relative inflexibility,from the fact that the capacity of individual channels is fixed.

Today. the most of the Telecommunication Organizations,have already established a number of non voice facilities for their subscribers,despite the fact that internationally ther is not yet a uniform line for the confrontation of the problem. Due to the fact that the establishment of new facilities of the informatics for the subscribers will lead ,without any doubt to the creation of a mixed Integrated Digital Network, every effort must be made for the best further development of this system to an ISDN.

After an analysis of the Greek telecommunication network,which as it is known has an astonishing abundance of exchanges of conventional technique, the connection for information interchanging in Integrated Services Digital Network of fig.2 is proposed for more reasonable maximizing of transmission efficiency.

The communication between the various terminals of the customers is attained by the protocol, sending commands and information between terminals and exchanges for call handling as well as between terminals for end to end signalling and information transfer.

The existing networks of informatics use the method of the framed format of fixed or variable

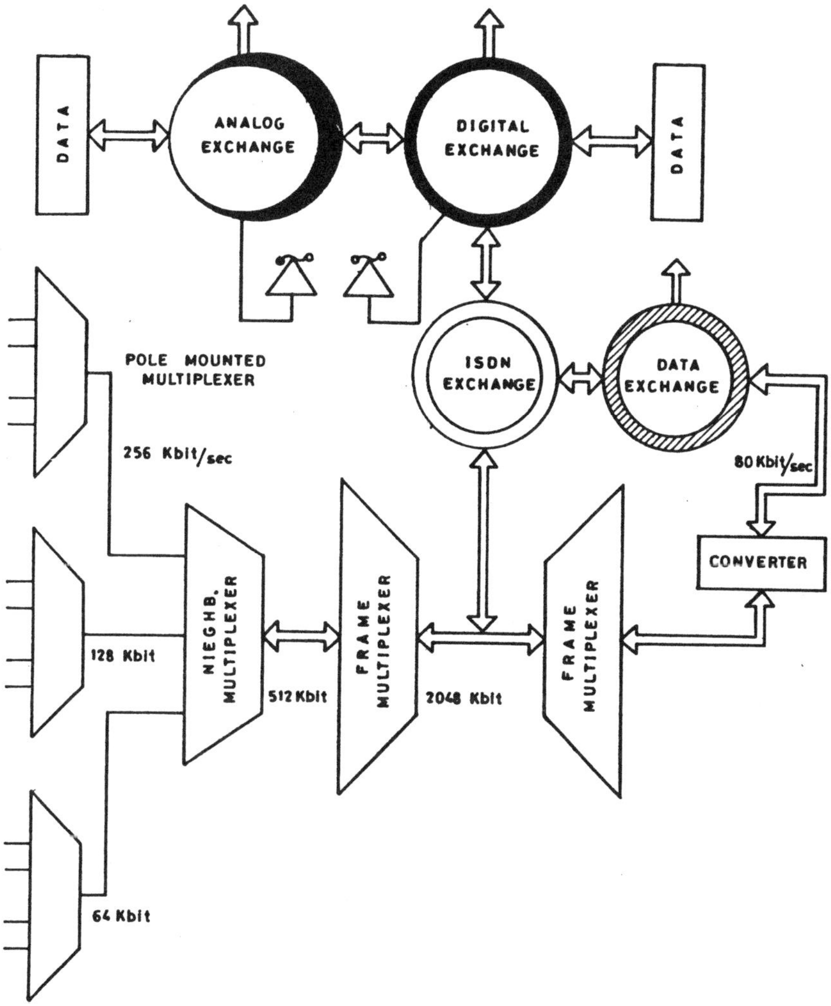

Fig. 2 Connection for information interchanging in ISDN.

length. The customer access procedures should be structured on a functional level basis similar to that provisionaly defined in the International Standards Organization ISO/T97/SC 16 N227 working document.

The packet switching systems and the common channel signalling have as basis the suggested method of the High Level Data Link Control (HDLC).

4. References

|1| R.C. Slatter,W.A.G. Walsh,F. Kaderali, P. Schmidt : Customer access to ISDN, Electrical communication N1 , 1983.

|2| G. Bafas:A simulator for the functional test of Digital Exchanges.IASTED, Modelling and Simulation '83, Lugano.

|3| G. Bafas :Digital telephony and the Integrated Services Digital Network. IASTED, Mini and Microcomputers and their Applications,Lugano 1983

|4| G. Bafas :A universal signalling converter for pulse code modulation systems. Melecon 83, IEEE conference in Athens.

|5| G. Bafas :Digital telephony and the space-time switch. AMSE - computer Bermuda conference 1983.

|6| G. Bafas: The introduction of the digital technique to the Greek telecommunication network. MECO '83, Athens.

A MICROPROCESSOR BASED CONTROLLER
FOR TIME OPTIMAL CONTROL

L. Moreno, I. Serra, E. Luque
Dpt. Electricity and Electronic
Universidad Autónoma de Barcelona
Bellaterra. Barcelona. (Spain)

ABSTRACT

A microprocessor based time-optimal control algorithm is based on the bang-bang approach which incorporates an asymmetrical control strategy to allow load disturbance. The implementation is carried out in such a way that all kinds of computations in real-time can be avoided. A relationship involving the sampling period, the asymmetrical value and the amplitude disturbance has been developed. The time response deterioration as a consequence of the asymmetrical control signal is also evaluated. The method has been applied to a second order process and experimental results are presented as well.

INTRODUCTION

Since the late 1970's the microprocessor has been incorporated to many types of control processes. This incorporation is possible if several conditions on the control algorithm as well as crucial decisions on the design process for a microprocessor based controller are fulfilled {1},{2}. Since there, a large number of works have reduced the gap existing between the control theory and its applications. This paper deals with a time optimal control process. An approach to this problem is achieved by recalculating the control sequence each sampling period {3}{4}. The bang-bang method has been already applied by the authors and properly works in absence of disturbance {5}. The purpose of this paper is the improvement of the bang-bang approach by redesigning it specifically for load disturbances. The principle underlying the method is to avoid, in real-time, all kinds of computations. This is carried-out by separating the control strategie in two parts. In the first stage the method evaluates the disered trajectory (the switching curve) and the second part consists in the real-time control and is based on the following characteristics: The system state is obliged to remain on the desired trajectory previously computed, and the command signal has asymmetrical bounds, in order to direct the system state towards the trajectory, even in the case of disturbances could occur.

SYSTEM CONFIGURATION

The control system configuration used in real-time control is shown in Fig. 1. The process to be controlled was simulated on the analog computer EAI 180 and the controller was implemented in the microcomputer AIM 65 based on the microprocessor Rockwell 6502 {6}. A Burr-Brown analog I/o system MP 7400, bus compatible with Rockwell 6502 has been conected to microcomputer AIM 65. This analog I/o system performs the following functions: Analog data acquisition and Analog data output. The analog data input includes up to 32 channel differential (64 channel single-ended), and A/D converter with a resolution of 8 bits and 44 μs of conversion time. The analog data output consists of two D/A converters with a resolution of 8 bits and settling time of 5 μs and providing 5 mA of output current.

The load disturbance is simulated by means of a square signal which is added to the command signal. This signal is produced by a function generator. The off-line programs that are necessary in order to carry out the method have been implemented in Apple II Europlus System.

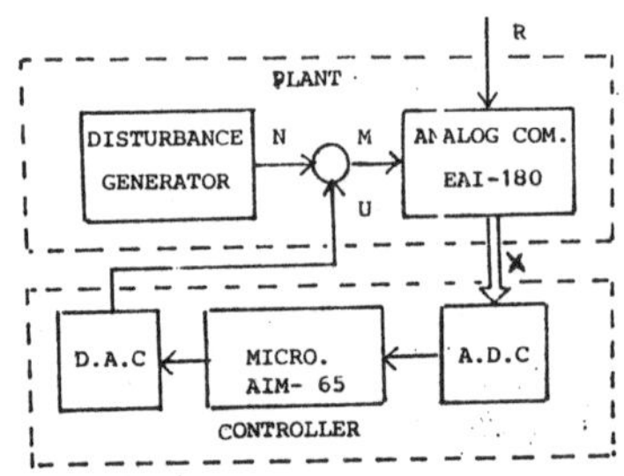

Fig.1.- System configuration

METHOD DESCRIPTION

It is well know that in order to obtain a time optimal zero error response of a second-order system it is necessary to realize three switchings, where the manipulated variable switches between its limiting values ($\pm$um). In order to avoid the presence of limit cycles in the response signal the third switching consists in changing to linear behavior (dual mode) {7}.

The strategy belongs to the theorical case, where disturbances are not present. If a disturbance appears it could occur that the system state never reaches the origin of the phase-plane, thus, it is necessary to dispose a strategy of asymmetrical bound {8} which reduces the minimum control value from $u = -um$ to $u = -\lambda \cdot um$, where $\lambda < 1$.

Let us see what happens when a disturbance appears. We can consider two different possibilities:

1). The disturbance appears before the system state reaches the curve Λ relative to the command value $u = -\lambda \cdot um$ (Fig. 2).

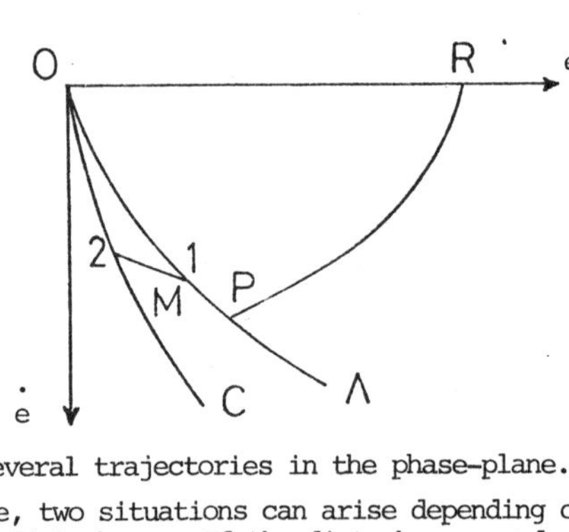

Fig.2.- Several trajectories in the phase-plane.

In this case, two situations can arise depending on the sign of the disturbance. If the disturbance produces the acceleration of the system, the response time will be enhanced. On the other hand, if the disturbance slows the dynamic system, the suitable strategy consists in to increment the control command.

Considering these facts we have adopted the solution consisting in applying the maximum control value u = +um until the system state reaches the curve Λ at which time the controller produces the second switching $u = -\lambda \cdot um$.

2). The disturbance appears after the system state reaches the curve Λ. As the controller disposes of an additional command $\Delta U = (1-\lambda)um$, it will produce the command signal value $u = -\lambda \cdot um \pm \Delta U$ accorgind to whether the sign of disturbance is positive or negative in order to direct the system state towards the switching curve Λ.

RELATIONS INVOLVING DESIGN PARAMETERS

In order to ensure the reachability of the phase-plane origin, a relationship involving the maximum sampling period, the amplitude disturbance and the asymmetrical parameter value λ, has been developed.

Considering the system state moving along the curve Λ and that, at some instant, a disturbance appears (point 1 Fig. 2), this disturbance will produce a deviation in trajectory M. The perturbed trajectory M, which corresponds to an input process $u = -\mu \cdot um$, will reach the curve C ($u = -um$) at point 2. This point defines the last point of the trajectory M, which must not be surpassed in order to ensure the attainability of the origin.

In the following, we shall consider a second-order process $G(S) = 1/S(1+\tau S)$. For this process the curve Λ is given by:

$$e_1 = -\tau \dot{e}_1 - \tau \cdot \lambda \cdot um \ln \frac{\lambda . um - \dot{e}_1}{\lambda \cdot um} \qquad (1)$$

where $(e_1, \dot{e}_1)$ is the state relative to point 1.

Then the equation of the branche 1,2 can be written as:

$$e_2 = e_1 + \mu \cdot um\, t + \tau (e_1 - \mu . um)(1 - e^{-t/\tau})$$
$$\dot{e}_2 = (\dot{e}_1 - \mu \cdot um)\, e^{-t/\tau} + \mu \cdot um \qquad (2)$$

Substituting Eq. (2) into the equation of curve C (obtained from (1) with $\lambda = 1$) we have:

$$um \cdot e^{-t/\tau} = p \cdot um\,(1-\mu) - p \cdot q\, e^{-t/\tau} \qquad (3)$$

where $p = e^{(e_1 + \tau \dot{e}_1)\tau \cdot um}$, $q = \dot{e}_1 - \mu \cdot um$

In Eq. (3), t is the time needed to move the state from 1 to 2 and, consequently, defines the time limiting value for the controller to detect the deviation and apply the necessary correction.

Note that the input process, when a disturbance n is present, is given by $m = -\lambda \cdot um + n = -\mu \cdot um$. Thus, the parameter is $\mu = \lambda - (n/um)$.

Now we consider the time response increase as a result of applying an asymmetrical control on the command signal. To obtain this increase we have to find the time required to reach the origin when the system state moves along the phase trajectory RPO, (Fig. 2).

The first branche of the trajectory RP corresponds to the command signal $u = +um$ and the second one to $u = -\lambda \cdot um$. From the equation trajectories we obtain the following relations

$$e^{-R/\tau \cdot um} = e^{\lambda t_3/\tau}\,(1 + \lambda(1 - e^{t_3/\tau})) \qquad (4)$$

$$e^{-R/\tau \cdot um} = \left[\frac{1 + \lambda - e^{-t_4/\tau}}{\lambda}\right]^{\lambda} \cdot e^{-t_4/\tau} \qquad (5)$$

t_3 and t_4 being the times required to go from P to o and R to P respectively.

The minimum time response which corresponds to symmetrical control $u = \pm um$ is given by

$$t_{OP} = \tau . \ln \frac{1+\alpha}{1-\alpha} \quad \text{where} \quad \alpha = (1 - e^{-R/\tau \cdot um})^{\frac{1}{2}}$$

Finally, the increase in the time response will be found introducing the values t_3, t_4, t_{OP} in the expression

$$F = t_3 + t_4 - t_{OP} \qquad (6)$$

IMPLEMENTATION OF THE METHOD

In the following discussion we shall consider the control algorithm which implements the asymmetrical bang-bang strategie. The steps involved in this algorithm according to the sections mentioned above are:

1. Obtainment of parameters

   We have developed an off-line program which solves the equations (3),(4) and (5) according to the following statement: given a process subject to load disturbances n, find the values of the time t, λ, and the function F which satisfy such equations. According to equation (3), these values depend on the point $(e_1, \dot{e}_1)$ of the curve Λ. Taking into account that the sampling period must be equal or smaller than the time t, we have selected this point $(e_1, \dot{e}_1)$ in the vecinity of the origin to ensure the control action in every point of the curve Λ.

2. Obtainment of switching curve Λ

   Fixed a suitable couple of values (λ, T) according to the specifications, a second off-line program solves the equation (1) and gives the set of points which defines the curve Λ. The number of points is imposed by the A/D converter resolution.

3. Real-time control

   The set of points is incorporated as a table to a real time program which properly realizes the control of the process. Now suppose the system state is completely available. The program scans the system state and produces the suitable command actions to direct the system state to the phase-plane origin. This is carried out by comparing between the measured system state and the corresponding state stored in the memory of the microprocessor.

The real-time program uses the interrupt system of the microprocessor through the programation of sampling period on a timer. It consists of three parts:

1. It scans continously the error signal to detect the reference value. When a change is detected, the controller produces a control signal $u = +um$.

2. It reads the system state $(e_R, \dot{e}_R)$ and obtains the corresponding derivative error $\dot{e}_T$ relative to the curve Λ, through an indexed addressing from a starting address for each e_R. Finally, if $|\dot{e}_R - \dot{e}_T| < \delta$ it generates the second switching $u = -\lambda \cdot um$.

3. As it was mentioned in section 3 the controller disposes of an additional command $u = -\lambda\, um \pm \Delta U$ that allows to correct the effect of disturbances. In this step, the role of the controller is to impose the system state to move on the curve Λ. In order to eliminate the presence of oscillations on the command signal as much as possible, the controller will produce the additional command $u = -\lambda \cdot um \pm \Delta U$ if $|\dot{e}_R - \dot{e}_T| > \delta_1$. This value of the command signal remains until the relation $|\dot{e}_R - \dot{e}_T| < \delta_2$ is verified, where $\delta_2 < \delta_1$. At this time the system state intersects the curve C and then the controller will apply the signal $u = -\lambda \cdot um$.

For illustrating the behavior of the method, we will apply it to a second order process given by $G(S) = 1/S(1+5S)$. The disturbance specification is considered to be 20% relative to the command signal. The system state is directly measurable and no restrictions are imposed on sampling period.

The first off-line program yields the parameters λ, T, F where $um = \pm 10$ volts and $\dot{e}_1 = -0.3$ volts. For a reference value change of 4 volts $t_{OP} = 2.84$ sec. The results are plotted in Fig. 3. As there is not restriction on sampling period T, we can adopt the smaller possible value T compatible with the microprocessor and A/D converter. However, for illustration purpose, we have adopted T = 1 ms, which results in $\lambda = 0.81$ and the degradation control performance F = 12 %.

The second off-line program obtains the points of the curve Λ corresponding to $\lambda = 0.81$. As the controller uses a 8 bit A/D converter, the curve Λ will have 128 points. The values $(e,\dot{e})_T$ are given in hexadecimal code.

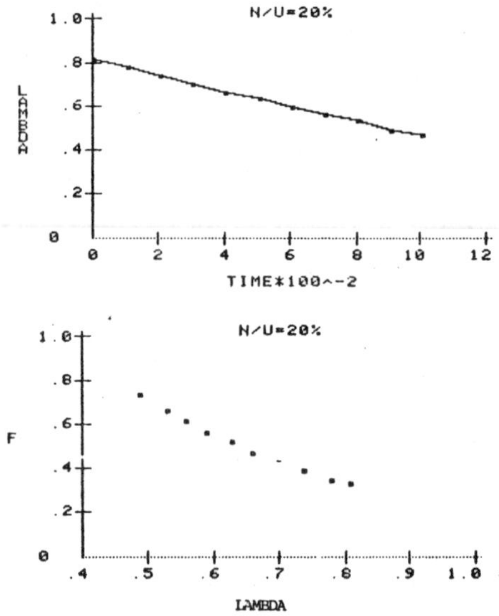

Fig.3.- Asymmetrical parameter λ vs. Sampling period T. Time response increase vs. parameter λ.

Finally, the real-time program realizes the correction applying an additional command $\Delta U = \pm 2$ volts.

The Fig. 4 illustrates the behavior of the input signal of process (command signal + disturbance) and the phase-plane portrait corresponding to steps changes in the reference variable of 4 volts for $\delta_1 = 2$ and $\delta_2 = 5$. In this case, the corrective action occurs immediately after the disturbance vanishes. This is so because of the duration of the disturbance.

Now let us consider some further comments concerning the implementation of the method.

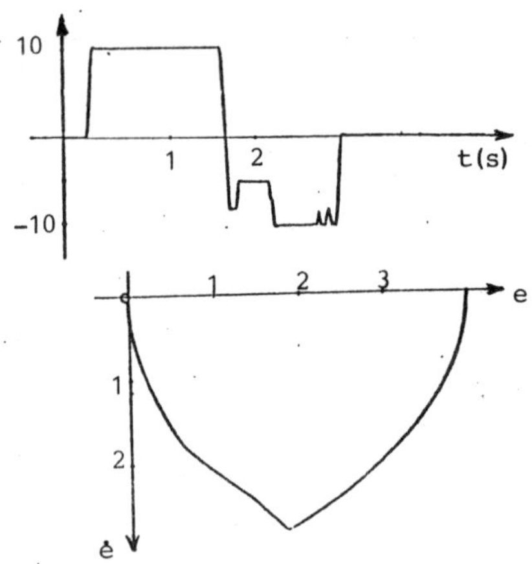

Fig.4.- Input process signal and phase-plane trajectory

1). The method has resulted to be sucesful in the case where the disturbance is considered as a small variation of system parameter τ. Note that round off error due to discretization of switching curve can be considered as a disturbance on the τ parameter. Thus, the asymmetrical control strategy has shown to be useful even in the case where load distutbances are not present.

2). The control performance can be improved by decreasing gradually the values of δ_1 and δ_2 as the system state moves to the origin.

CONCLUSIONS

We have developed an approach that solves the time optimal control through the incorporation of a microprocessor in the feedback control loop. The method allows the presence of disturbances by aplying an asymmetrical control law to the process. The principle of the method consists in evaluating all data which correspond to the desired trajectory. Unlike others methods where computations are needed at each sampling period to recalculate the remaining optimal trajectory, in this method the controller detects deviations produced by disturbances and directs the system state to the switching curve.

We have developed as well a relationship involving the magnitude disturbance, the sampling period and the asymmetrical parameter λ.

Our method has been applied to a second-order system, in presence of disturbances, and works even in those cases where some changes on the parameters system can occur.

REFERENCES

{1}. A:H: Glattfelder, F. Huguenin and W. Schaufelberger. "Design considerations for microprocessor based controllers". Proceedings of the 6th IFAC/IFIP Conference. Dusseldorf. October 1980.

{2}. D.M. Auslander, Y. Takahashi, M. Tomizuka. "Direct digital process control: practice and algorithms for microprocessor application". Proceedings of the IEEE Vol.66 no 2, February 1978.

{3}. J:H: Vlieger, H.B. Verbruggen, P.M. Bruijn. "A time optimal control algorithm for digital computer control". Automatica. Vol.18, No 2, September 1981.

{4}. R. Ortega. "Experimental evaluation of four microprocessor based advances control algorithms". Microprocessing and Microprogramming. The Euromicro Journal. Vol.10, No 4, November 1982.

{5}. E. Luque, I. Serra, L. Moreno. "A time-optimal bang-bang control algorithm for microprocessor control". 1983 American Control Conference. June 1983. San Francisco, California. (Accepted).

{6}. AIM 65 Microcomputer User's Guide. Rockwell International Corporation, 1979.

{7}. Y. Takahashi, M.J. Rabins, D.M. Auslander. "Control and dynamic systems". Addison-Wesley 1970.

{8}. H.B. Verbruggen, P.M. Bruijn. "A comparison of some time (sub) optimal control algorithms for in-line computer control" Proceedings of the 6th IFAC/IFIP Conference. Dusseldorf. October 1980.

APPLICATION OF MODERN CONTROL CONCEPTS TO THE DESIGN OF A DISTILLATION COLUMN REGULATOR

By

Dr.N.N. Sorial

Department of Electrical Engineering
Faculty of Engineering, Alexandria University
Alexandria,Egypt

ABSTRACT

In this paper an improved output feedback controller, based on the "Internal Model Principle", is designed for a distillation column. The resulting controller provides absolute stability for the closed-loop system and, moreover, it maintains the output composition of the overhead product constant without any offset due to step disturbance in the feed flow rate. The system preserves these properties even in the presence of small system parameters variation.

1.INTRODUCTION

Recent developments to design well-posed regulator for a control system that preserves output regulation and system stability in the presence of small parameter changes, have been achieved by many authors[1],[2]. The controller structure incorporates in addition to the invariant term, dynamic model resembling the disturbance signal applied to the system. The design procedure steps are presented in section 2. In section 3, a distillation column model is developed, and then, the controller design is presented in section 4.

2.CONTROLLER DESIGN PROCEDURE

Starting from the standard system state-space model

$$\dot{x} = A_1x + A_3d + B_1u$$

$$\dot{d} = 0$$

$$y = D_1x + D_2d \qquad (1)$$

where the plant state x is of dimension n_1 and the control u is of dimension m. The dimensions of disturbance d and output vector y are n_2 and q respectively.

Assume that: the pair (A_1 , B_1) is controllable, and

$$\text{the pair}\left\{[D_1 , D_2], \begin{bmatrix} A_1 & A_3 \\ 0 & 0 \end{bmatrix}\right\}$$

is obsevable; the design algorithm could be summerized in the following steps:

step 1

The state feedback matrix $v=F_1x+F_2d$ is chosen such that $(A_1+B_1F_1)$ is stable and

$$\begin{bmatrix} A_1+B_1F_1 & B_1 \\ D_1 & 0 \end{bmatrix} \begin{bmatrix} R \\ F_2 \end{bmatrix} = - \begin{bmatrix} A_3 \\ D_2 \end{bmatrix} \qquad (2)$$

step 2

The following matrices are defined

$$A = \begin{bmatrix} A_1 & A_3 \\ 0 & 0 \end{bmatrix} \quad B = \begin{bmatrix} B_1 \\ 0 \end{bmatrix} \quad D = [D_1 \quad D_2] \qquad (3)$$

and a minimal-order observer is designed to obtain the three matrices W, T and K so that T is stable and

$$W(A-KD) = T.W \qquad (4)$$

step 3

The following matrices are defined

$$Q \triangleq \begin{bmatrix} D \\ W \end{bmatrix} \qquad (5)$$

$$T_e \triangleq T+W.B.F.Q^{-1} \begin{bmatrix} 0 \\ I_{n_1} \end{bmatrix} \qquad (6)$$

It can be shown that there exists a nonsingular matrix P so that

$$P^{-1}T_eP = \begin{bmatrix} T_1 & 0 \\ 0 & 0_{qxq} \end{bmatrix} \qquad (7)$$

step 4

The controller is then synthesised as

$$u = u_e+[C_1+\frac{1}{s}C_3K_3+C_2(sI-T_1)^{-1}K_2]y \qquad (8)$$

where the matrices C_1, C_2, C_3, K_2 and K_3 are computed as follows:

$$F.Q^{-1}=[C_1,\hat{C}_2,\hat{C}_3] \;,\; C_1:=mxq \quad C_2:=mx(n_1-q)$$

$$C_3:=mxq \;,\; [C_2,\, C_3]=[\hat{C}_2 \;,\; \hat{C}_3]P \;,\; \text{and}$$

$$P^{-1}W\left\{K+BFQ^{-1}\begin{bmatrix} I_q \\ 0 \end{bmatrix}\right\}=\begin{bmatrix} K_2 \\ K_3 \end{bmatrix} \;,\; K_2:=(n_1-q)xq,$$

$$K_3:=qxq$$

3.REPRESENTATION OF THE DISTILLATION COLUMN MODEL

The distillation column cnsidered in this paper is used for processing a binary mixture of methanol-water[3]. The accurate mathematical model of the column is nonlinear. Purturbing the plant variables about a specific operating condition, a linearized third order model representing the dynamics of the distillation column is obtained in the form

$$\dot{x} = A_1x + A_3d + B_1u$$

$$y = D_1x + D_2d \qquad (9)$$

where $x = [x_1 \quad x_2 \quad x_3]'$ is the state vector

$u = [H \quad R]'$ is the control vector, H is the heating power, and R is the reflux flow rate.

$y = X_D$ is the output variable where X_D is the concentration of the overhead product.

F = the feed flow rate, considered as disturbance function.

Assuming that the column is to operate with the following steady state values:

$F_{s.s.}$=156 cc/min , $H_{s.s.}$ =2.2 KW , and

$X_{D_{s.s.}}$=97%, then A_1,B_1,A_3,D_1 and D_2 matrices take the values:

$$A_1 = \begin{bmatrix} -0.167 & 0 & 0 \\ 1.667 & -1.667 & 0 \\ 0 & 4.034 & -0.3 \end{bmatrix} \qquad B_1 = \begin{bmatrix} 0.167 & 0 \\ 0 & 0 \\ 0 & -0.3 \end{bmatrix}$$

$$A_3 = \begin{bmatrix} 0 \\ 0 \\ 1.1 \end{bmatrix} \qquad D_1 = [0 \quad -13.23 \quad 0.937] \qquad D_2 = 0.9$$

4.DESIGN OF THE DISTILLATION COLUMN CONTROLLER

The design steps presented in section 2 are followed to derive the closed-loop model. The feedback matrix F_1 is chosen as follows

$$F_1 = \begin{bmatrix} 1 & 0 & 1 \\ 1 & 13.446 & 1 \end{bmatrix}$$

the resulting model $(A_1+B_1F_1)$ will have the eigenvalues -0.1, -0.5, and -1.667. Entries of matrix F_2 are calculated from equation(2) to give

$$F_2 = \begin{bmatrix} -13.159 \\ -23.65 \end{bmatrix}$$

The third order minimal observer is designed to have the eigenvalues -10, -20, and -100.

The numerical values of matrices C_1,C_2,C_3,K_2, K_3, and T_1 composing the controller configuration of equation(8) are given by

$$C_1 = \begin{bmatrix} 0.9812 \\ -0.0085 \end{bmatrix} \qquad C_2 = \begin{bmatrix} -0.833 & -0.047 \\ 0.00718 & -0.0622 \end{bmatrix}$$

$$C_3 = \begin{bmatrix} -0.352 \\ 0.0027 \end{bmatrix} \qquad K_2 \begin{bmatrix} 22.81 \\ -0.671 \end{bmatrix} \qquad K_3 = [0.205]$$

$$T_1 = \begin{bmatrix} -21.44 & 0.7113 \\ -0.0051 & -108.56 \end{bmatrix}$$

Accordingly, the controller is synthesised as follows

$$u = \begin{bmatrix} 0.9812-\frac{0.0722}{s}-\frac{18.97s+2061.66}{(s+108.6)(s+21.5)} \\ -0.0085+\frac{0.00554}{s}+\frac{0.206s+18.68}{(s+108.6)(s+21.5)} \end{bmatrix} y+u_e$$

The step response of the closed-loop system is determined using the transition matrix technique with a sampling interval of 0.005 sec. Fig.(1) shows the response of the output variable X_D due to step change in the disturbance d with zero control signals. The response confirms that the closed-loop response settles down to zero after transience, even in the presence of closed-loop perturbation[fig.(1-c)]. The response of the output variable X_D to 10% step change in the control input H, with zero disturbance, is shown in fig.(2). It could be concluded that the system under the action of the proposed controller achieves the required output even in the presence of external and, or, parameters disturbances.

5.CONCLUSIONS

A controller is designed in this paper for a bi-mixture distillation column. The proposed controller is based on the Internal Model Principle. The compensated system has a stable closed-loop and shows no offsets due to the change of th e feed-flow rate, even in the presence of small parameter variations. Although the analytical design seems complicated, the practical realization of the designed controller is quite simple.

6.REFERENCES

[1] W.M. WONHAM,"Linear Multivariable Control: A Geometric Approach", Springer-Verlag, New-York, 1974, pp.184-226.

[2] O.A. SEBAKHY, and W.M. WONHAM,"A Design Procedure for Multivariable Regulators", Automatica, September 1976.

[3] N.N. SORIAL,"Design of Sensitivity-reducing Output Feedback Controller for A Distillation Column", Presented at the IFAC Conference on System Approach for Development, Rabat, Morocco, 24-27 November,1980.

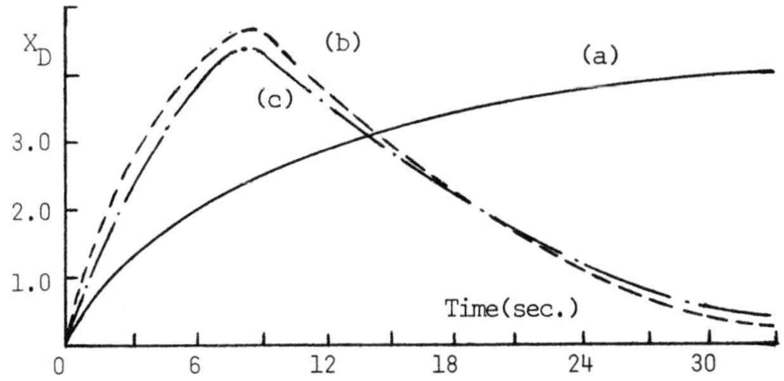

Fig(1) Response to -1.0 step in disturbance d.

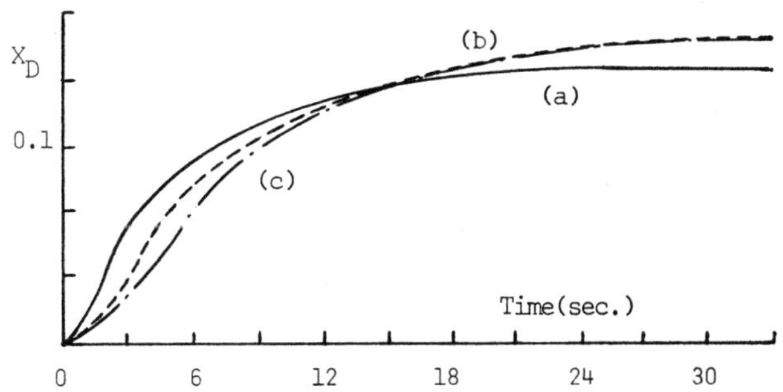

Fig.(2) Response to -10% step in control input H

(a) Response of open-loop system

(b) Response of closed-loop system

(c) Response of perturbed closed-loop system.

A NUMERICAL APPROACH FOR THE UNIVERSAL SOLUTION OF THE INVERSE KINEMATIC PROBLEM

E. Ersü and D. Nungesser

Institute of Control Engineering
Section Control Systems Theory
Technical University of Darmstadt
D-6100 Darmstadt, FRG

Abstract. One basic problem in a robot control system is the mathematical transformation between cartesian coordinates and joint coordinates of the robots's arm. Both the transformation from joint to cartesian coordinates as the Kinematic Problem (KP) and the transformation from cartesian to joint coordinates as the Inverse Kinematic problem (IKP) have to be solved. Due to the geometric properties of the kinematic system described by nonlinear transcendental functions especially the IKP represents a severe problem and in spite of heavy research work there exist only intuitive solutions for simple kinematic structures usually up to six joints.

This paper describes a numerical algorithm based on homogeneous transformations which solves the kinematic equations in both directions for a series of one degree-of-freedom (d.o.f.) joints either revolute or prismatic. The algorithm handles redundant d.o.f. either in joint or in cartesian coordinates satisfying the constraints of the joint workspace and joint velocities. An additional procedure is used to minimize the overall computational errors along a prescribed motion path.
The numerical results and the computation time recommend the method for real-time robot control systems.

I. INTRODUCTION

Cartesian motion control is one of the indispensable properties of modern robot control systems. One of the essential problems in implementing cartesian motion control is the transformation between the different coordinate systems.

The first is to compute the position and orientation of the roboter hand for a given set of joint positions, the kinematic problem, the second is the inverse task i. e. to compute a set of joint positions for a given position and orientation of the hand, the inverse kinematic problem (IKP).

The solution of the KP is always possible and has an unique solution. The more severe problem is the solution of the IKP because of the transcendental character of the kinematic equations and it is impossible to solve it analytically except for simple kinematic structures.

To have compatible control systems for a big class of robots with different kinematic structures there is a need to find an universal solution of KP and IKP independent of the kinematic structure and the geometrical parameters of the roboter arm.

Two main groups of existing solutions can be distinguished. The methods in the first group use the exact non-linear kinematic equations allowing precise solutions. The main disadvantage for an algorithmical solution is the fact, that in the general case an effective handling is possible only if the designer of the system has sufficient experience and intuition.

The methods in the second group are based on linearization of the exact kinematic equations in the neighbourhood of prescribed points. The procedure results in an algebraic equation system whose solution might be possible using algorithms for linear algebraic problems.

Some algorithmical methods have already been developed applicable to certain classes of kinematic structures. Benati [1] proposes an algorithm to solve the IKP partly numerical and partly analytical for anthropomorphic manipulator arms with seven degrees of freedom. It can handle all structures with 3 rotary joints (elbow, shoulder, wrist) each joints with 2 or 3 d.o.f. Lipkin [4] describes an application of the theory of screws for the kinematic analysis of industrial robots resulting in linearized kinematic equations for all six joint manipulators. Solutions of these equations are given in case of three intersecting joint axes or in case of three following prismatic joints. Konstantinov [3] develops a more common approach to a numerical solution of the IKP for n d.o.f. arms using linearization of the kinematic equations. A look at the number of mathematic operations shows that an enormous computational effort (approximately about 2500 multiplications and 2000 additions for a six joint arm) is necessary.

The proposed method consists of an algorithm which gives an universal solution for the KP using the exact kinematic equations and an universal solution for the IKP using linearized kinematic equations. Both parts can handle arms consisting of a series of n one d.o.f. joints either revolute or prismatic.

II. KINEMATIC EQUATIONS

The kinematic description of serial link manipulators in this paper is based upon homogeneous transformation matrices. It is one of the possible ways to handle kinematic structures and it is well developed for the application to robotics (Uicker, et.al. [8], Denavit, Hartenberg [2]).
A very useful feature of this mathematical tool is the comfortable combination of position and orientation vectors in one matrix. The homogenous transformation matrices as introduced by [2] and used in robotics is a (4x4) matrix

$$T = \begin{pmatrix} \underline{n} & \underline{o} & \underline{a} & \underline{d} \\ & & & \\ 0 & 0 & 0 & 1 \end{pmatrix} \qquad (II.1.)$$

with a (3x3) orthonormal upper partial matrix indicating the rotation of a transformed c ordinate system (c.s.) C (see figure 1) with respect to w.r.t.) the reference coordinate system R and the vector $\underline{d}$ indicating the translation of the origin of C w.r.t. the origin of R. In this sense T represents a series of three successive rotations about the three principal axes of the reference coordinate system and a translation:

$$T=Rot(x,\psi)\cdot Rot(y,\theta)\cdot Rot(z,\phi)\cdot Trans(x,y,z) \qquad (II.2.)$$

with $Rot(x,\psi) \hat{=}$ Rotation about the x-axis by the angle ψ, $Trans(x,y,z) \hat{=}$ Translation with vector $\underline{d}=(x,y,z)^T$.

Thus the transformation in Fig. 1 can be expressed by a product

$$R \xrightarrow{T} C \qquad (II.3.)$$

The application to kinematics of a roboter arm takes place by attaching coordinate systems to each link of the arm in a well defined manner. The coordinate system aligned with the gripper of the roboter hand is ($\underline{n},\underline{o},\underline{a}$) in Fig. 1. The definition which aligns coordinate systems to each links and the relative transformation matrix between following links was also developed in [2] (Fig. 2 and 3) for a series of links connected by revolute or prismatic joints with one d.o.f. each.
The origin of the coordinate system lies on the joint axis, the z-axis coincides with the joint axis. There are four basic transformations necessary to transform c.s. of link i-1 into c.s. of link i.

1. rotation about z_{i-1} with angle θ_i
2. translation along z_{i-1} with a distance d_i
3. translation along new $x_{i-1}=x_i$ with a distance a_i
4. rotation about x_i with angle α_i

The product of the four corresponding transformation matrices results in the matrix

$$A_i = Rot(z_{i-1},\theta_i)\cdot Trans(0,0,d_i)\cdot Trans(a_i,0,0)\cdot Rot(x_i,\alpha_i) \qquad (II.4a)$$

$$A_i = \begin{pmatrix} \cos\theta_i & -\sin\theta_i\cdot\cos\alpha_i & \sin\theta_i\cdot\sin\alpha_i & a_i\cdot\cos\theta_i \\ \sin\theta_i & \cos\theta_i\cdot\cos\alpha_i & -\cos\theta_i\cdot\sin\alpha_i & a_i\cdot\sin\theta_i \\ 0 & \sin\alpha_i & \cos\alpha_i & d_i \\ 0 & 0 & 0 & 1 \end{pmatrix} \qquad (II.4b)$$

For a revolute joint the joint variable q_i is the angle θ_i, for a prismatic joint the distance d_i. The other parameters are constants. All n joint variables are summarized in the (nx1) vector $\underline{q}$.

The matrix T describing hand position and orientation w.r.t. the base of the arm is then calculated by the product

$$T_n = \prod_{i=1}^{n} A_i = A_1\cdot A_2\cdot \ldots\ldots \cdot A_n \qquad (II.5.)$$

which contains n joint variables q_i. Setting II.5. equal to II.2.

$$T_n = T \qquad (II.6.)$$

a position vector

$$\underline{p}^T = (x,y,z,\psi,\theta,\phi) \qquad (II.7.)$$

can be calculated for the hand position (x,y,z) and its orientation (ψ,θ,ϕ).

Equation (II.6.) can be solved for $\underline{p}$ for any series of one d.o.f. joints.

In the case of IKP the eq. (II.6.) is to be solved for the joint variables q_i by giving the hand position and orientation $\underline{p}$. Due to the nonlinear trigonometric functions in T_n this is possible only for simple kinematic arm configurations [6], [7].

Differentiating (II.6) in fixed points in $\underline{p}$-space or $\underline{q}$-space results in linearized equations for the elements of a differential matrix dT of the hand located at T_n.

Neglecting higher order terms dT can be calculated as

$$dT = \Delta \cdot T \qquad (II.7.)$$

with

$$\Delta = \begin{pmatrix} 0 & -d_\phi & d_\theta & dx \\ d_\phi & 0 & -d_\psi & dy \\ -d_\theta & d_\psi & 0 & dz \\ 0 & 0 & 0 & 0 \end{pmatrix} \qquad (II.8.)$$

consisting of the six elements of a differential change vector

$$\underline{dp} = (dx, dy, dz, d\psi, d\theta, d\phi)^T. \qquad (II.9.)$$

Differential changes w.r.t. different reference coordinate systems are transformed by

$$dT = {}^B\Delta \cdot T = T \cdot {}^T\Delta \qquad (II.10.)$$

with ${}^B\Delta$, ${}^T\Delta$: differential change matrices w.r.t. base (B) or hand (T) coordinates.

In the same way differential changes of each joint variable q_i can be transformed from the corresponding coordinate system i to a common system, for example to base coordinates by

$${}^B\Delta_i \cdot (A_1\cdot A_2\cdot\ldots\cdot A_{i-1}) = (A_1\cdot A_2\cdot\ldots\cdot A_{i-1})\cdot {}^{i-1}\Delta_i \qquad (II.11.)$$

The impact of differential changes in each joint variable dq_i to changes in handposition $\underline{dp}$ w.r.t. base coordinates are summarized by the equation system

$${}^B\underline{dp} = ({}^B\underline{dp}_1, {}^B\underline{dp}_2, \ldots, {}^B\underline{dp}_n)\cdot\underline{dq} \qquad (II.12)$$

with

$${}^BJ = ({}^B\underline{dp}_1, {}^B\underline{dp}_2, \ldots, {}^B\underline{dp}_n) \qquad (II.13.)$$

${}^BJ$ is called the Jacobian and can be calculated analytically [5].

III. ALGORITHM FOR NUMERICAL SOLUTION

The adaptation of the universal algorithm to a special roboter needs only an interactive dialog to specify the constant geometrical parameters of the link configuration:

- number n of joints
- type of joints: revolute or prismatic
- link parameters: α_i, θ_i, a_i, d_i
- mechanically determined workspace of the joint variables θ_i or d_i.

The transformation from joint to cartesian coordinates can be solved in a straight forward way by eq.(II.5.) to compute the T_n matrix of the hand and by eq. (II.6.) to compute the corresponding $\underline{p}$ vector.

This transformation has an unique solution, i.e. each arm configuration results in just one hand position and orientation. The solution is possible for any number n of joints.

The transformation from cartesian to joint coordinates is calculated through linearized equations by getting solutions of eq.(II.12.) for differential changes of the joint variables $\underline{dq}$ with given $\underline{dp}$. In each step, during execution of the motion both solutions, the KP and the IKP, are necessary, the former to transform actual joint position in cartesian as a feedback, the latter to command the next joint position.

During calculation of the KP-solution the columns of the actual Jacobian are evaluated by e.(II.11.) for i=1,...,n and the Jacobian is then prepared for the ongoing IKP-solution. The IKP-solution, i.e. the evoluation of $\underline{dq}$ by eq. (II.12.) for given $\underline{dp}$ is even theoretically a nontrivial problem because of the diversity of solutions. The trivial solution

$$\underline{dq} = J^{-1}\,\underline{dp} \qquad \text{(III.1.)}$$

is only possible if J is quadratic and nonsingular. In general there are three cases to distinguish.

1. Eq.(II.12.) has an unique solution, if
   $\text{rank}(J) = \text{rank}(J\ \underline{dp}) = n$ (III.2.)
2. Eq.(II.12.) has an unlimited numer of solutions,
   if $\text{rank}(J) = \text{rank}(J\ \underline{dp}) \neq n$ (III.3.)
3. Eq.(II.12.) has no solution, if
   $\text{rank}(J) \neq \text{rank}(J\ \underline{dp})$ (III.4.)

All three cases are handled together by modified versions of linear programming algorithms which transform eq.(II.12.) in a moderate form to decide upon the kind of solution and to compute it [5]. The following remarks have to be made:

To 1.) A fundamental advantage of kinematic computation with linearized equations is the fact that a unique solution is determined for configurations in which manifold solutions may exist because of the ambiguity of the trigonometric functions. This unique solution is the one nearest to the actual configuration of the arm. If this solution is outside the possible workspace according to

$$q_{i\ min} \leq q_i \leq q_{i\ max} \qquad \text{(III.5.)}$$

the commanded motion cannot be executed and an error signal occurs.

To 2.) Arms with redundant degrees of freedom (n>m) can reach in the interior of the bounded workspace every desired handposition with an unlimited number of different arm configurations. The algorithm is able to choose one of these configurations at the same time satisfying the restrictions of the workspace (eq.(III.5.)) and constraints on the maximal joint velocity corresponding to a maximal increment

$$|dq_i| \leq dq_{i\ max}. \qquad \text{(III.6.)}$$

To 3.) If the equation system contains linear dependent equations, there is no solution $\underline{dq}$ possible satisfying all equations. Only a reduced system can be solved according to the rank J. In this case the arm is degenerated i.e. because of the special relation of the different joint axes the manipulator loses possible degrees of freedom. The solution of the reduced system commands the arm to a position near to the commanded position.

Any position error can be eliminated by a procedure described in the following. The linearization errors in each step would run up along the motion path to non-negligible values. To cope with these errors the difference between commanded and actual cartesian position is superponed to the next commanded step. So the overall error along the path is reduced to the size of a one-step-error which is a function of the special configuration of the arm and of the commanded step length [5].

If the manipulator has redundant degrees of freedom, i.e.

$$n > r$$

with r = rank (J), (n-r) joint variables can be chosen arbitrarily. Before computation of $\underline{dq}$ a desired ordering of the joints eq. due to their energy consumption can be selected. Thus mainly these joints are involved in the arm motion.

The position error of each KP- and IKP-solution is checked if it exceeds the maximal error limit. In this case the actual solution is iterated until the error bound is satisfied.

If these iterations don't converge to the desired position or if the computed joint velocities exceed the maximum velocities the solution procedure can be repeated with reduced step length to fulfill the error limits.

The algorithm accepts the cartesian description of the next commanded step both as absolute or differential. These inputs may be in form of the T_n-matrix or $\underline{p}$-vector. In each case the input representation is transformed to the coordinate system of the roboter base before the Jacobian is evaluated and the IKP is solved.

To keep maximum flexibility another criterion is implemented by which the joints are made as far as possible, to operate in the middle of their workspace. Defining

$$r_i = \left| \frac{q_i - \dfrac{q_{i\ max} + q_{i\ min}}{2}}{q_{i\ max} - q_{i\ min}} \right| \qquad \text{(III.7.)}$$

again an ordering of the joints can be selected thus joints with small r_i's are preferably chosen.

IV. RESULTS

Simulation experiments have been made with data of existing industrial robots (6 d.o.f.) as well as with data of robots (6,8 and 12 d.o.f.) whose parameters are chosen arbitrarily.

The number of operations to be executed depends on the number n of all joints and on the number r of revolute joints. For an example of a 6 d.o.f. arm with 6 revolute joints the total number of operations for solution of KP and IKP is:

467 Multiplications + 338 Additions + 23 trigonometric functions

Based on an Intel's 8086/87 μ-processor combination the computation time of this configuration would last about 20 ms. Depending on the special structure of the IKP equation system the computation time for a 12 d.o.f. roboter increases about two to four times of the 6 d.o.f. example.

The numerical results show that the computational errors are several orders smaller than the errors caused by the linearization. A precise evaluation of the linearization errors is hardly possible because their magnitude depends on the special configuration. The second derivations of the kinematic equations determine to which extent the linearized equations deviate from the exact equations in a fixed neighbourhood of the linearization point. These erros can't be predicted without additional computation

The computed deviations between the commanded and actual position of the hand are proportional to the commanded step length. They vary for different configurations with the same step length of 5 cm in all three position axes and 5 degree in all three rotation axes between 1% and 100% of the commanded step length. Without error-compensation this leads to nonacceptable position errors at the end of the path. Using the proposed compensation procedure the overall error remains in the order of a one-step-error. With one or two additional iterations this error can be reduced to an order of the computational accuracy. Computing the path off-line without any restrictions in computation time any desired accuracy for each point on the motion path can be reached.

The algorithm computes solution of the IKP in singular

points for a reduced equation system according to the rank J. The resulting errors which arise in the original system for this solution are treated as normal position errors in the next step by the error compensation procedure. The size of the relative position deviations near and in singular points requires decreased step lengths and increased computation time.

V. CONCLUSIONS

The proposed algorithm and its numerical results demonstrate the universal solution of the kinematic problems. For robot control systems two different fields of application can be taken into consideration, off-line simulation and programming of motions and on-line real-time control of their execution. For the off-line application the algorithm is suited very well because there are no restrictions on the computation time. The cycle time of 20 ms stated above for KP and IKP-solution of a six axis arm is short enough for existing robot control systems compared with their special implemented solutions.

In comparison with the constant cycle time for the solution of exact equations the execution time for a solution with desired accuracy using linearized equations depends on the number of necessary iterations and reductions of the step length. For the application in real-time control it is recommended to test each desired motion path in an off-line computation if the algorithm computes solutions without iterations and step reductions.

VI. REFERENCES

[1] Benati, M., Morasso, P., Tagliasco, V.: The Inverse Kinematic Problem for Anthropomorphic Manipulator Arms. Journal of Dynamic Systems, Measurement and Control. Vol. 104, March 1982, S.110-113

[2] Denavit, J., Hartenberg, R.S.: A Kinematic Notation for Lower-Pair Mechanisms Based on Matrices. ASME Journal of Applied Mechanics 22 (June 1955) - 215-221

[3] Konstantinov, M.S., Patarinski, S.P., Zamanov, V.B., Nenchev, P.N.: A Contribution to the Inverse Kinematic Problem for Industrial Robots. Proceedings 12th International Symposium on Industrial Robots, 6th International Conference on Industrial Robot Technology, June 9-11, 1982, Paris, France, S.459-467

[4] Lipkin, H., Duffy, J.: Analysis of Industrial Robots via the Theory of Screws. Proceedings 12th International Symposium on Industrial Robots, 6th International Conference on Industrial Robot Technology, June 9-11, 1982, Paris, France, S.359-370

[5] Nungesser, D.: Implementierung eines Ansatzes zur Lösung des inversen Kinematischen Problems für Industrieroboter. Diplomarbeit TH Darmstadt, Institut für Regelungstechnik, 1982 (directed by E. Ersü)

[6] Paul, R.P., Shimano, B., Mayer, G.E.: Kinematic Control Equations for Simple Manipulators. IEEE Transactions on Systems, Man and Cybernetics, vol. SMC-11, June 1981, S.449-455

[7] Paul, R.P., Shimano, B., Mayer, G.E.: Differential Kinematic Control Equations for Simple Manipulators. IEEE Transactions on Systems, Man and Cybernetics, vol. SMC-11, June 1981, S.456-460

[8] Uicker, I.J., Denavit, J., Hartenberg, R.S.: An Iterative Method for Displacement Analysis of Spatial Mechanisms. Journal of Applied Mechanics, June 1964, S.309-314

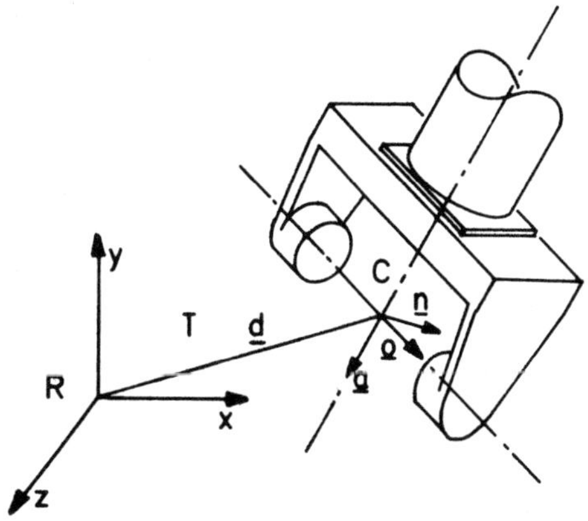

Figure 1 Gripper

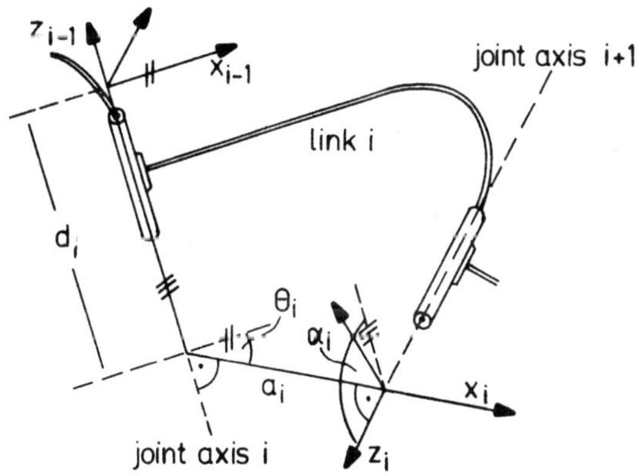

Figure 2 Revolute joint

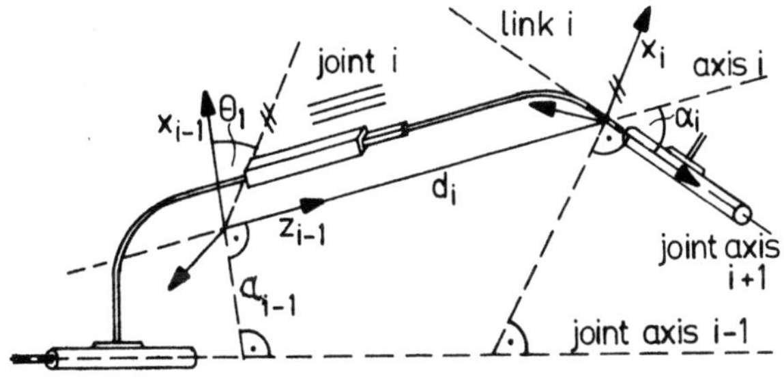

Figure 3 Prismatic joint ($a_i = 0$)

MODIFIED MINIMUM VARIANCE SELFTUNING CONTROL OF A ROBOTIC MANIPULATOR

DeSantis, R.M.
Département de Génie Electrique
Ecole Polytechnique de Montréal
Montréal, Québec
CANADA

Hurteau, R.
Département de Génie Electrique
Ecole Polytechnique de Montréal
Montréal, Québec
CANADA

Lefebvre, S.
Institut de Recherches
de l'Hydro-Québec
Varennes, Québec
CANADA

SUMMARY

The present paper considers the adaptive control of a leight weight flexible robotic manipulator. In line with the classical selftuning approach the controller is made of a regulator/observer block, a plant parameters estimator and a gain tuning module. The regulator gains are continuously adjusted so as to minimize a minimum variance criterion of the type proposed by Clark and Gauthrop.

The novelty is that the time horizon intervening in this criterion is somehow itself adjusted to the estimated values of the plant parameters. Advantages and salient features of such a strategy are illustrated on the basis of simulated experiments.

THE SYSTEM

We consider a microcomputer based adaptive control of the angular position of the horizontal joint of a modified version of Rhino XR1, a robotic manipulator by Sandhu Mfg Co [1]. Controlled plant aside, the experimental set-up is identical to that considered in [2] (see figure 1). A discrete simplified model of the plant is given in figure 2. During the operation of the plant the values of the parameters appearing in this model vary in a way which is a priori unknown to the controller.

The controller consists of a state regulator, a state observer and a plant parameters estimator (see figure 3). The structure of the state regulator is as indicated in figure 4. The plant parameters estimator uses a least square algorithm of the type described in [3] and can be represented in terms of the dynamical model of figure 5. The state observer responds to the simple scheme of figure 6.

The objective of the controller is to provide a control function so as to force the angular position of the horizontal joint of Rhino XR1 to follow a desired reference as accurately and as rapidly as possible. This is done by measuring the actual position, x1, by comparing it with the desired reference, REF, and by developing the required control, u. The control law is

$$u = -k1 * (x1 - REF) - k2 * x2$$

where x2=x1. The problem is to generate the values of k1 and k2 as a function of the estimated values of the plant parameters.

THE MINIMUM VARIANCE

Given the system

$$\begin{bmatrix} x1(i+1) \\ x2(i+1) \end{bmatrix} = \begin{bmatrix} 1 & F(1,2) \\ 0 & F(2,2) \end{bmatrix} \begin{bmatrix} x1(i) \\ x2(i) \end{bmatrix} - \begin{bmatrix} B1 \\ B2 \end{bmatrix} u(i)$$

the idea at the basis of the minimum variance criterion (see Clark and Gauthrop [4]) is to compute the feedback gains k1 and k2 so that the issuing control minimizes the functional

$$j(k1,k2)=[x1(i+1)-REF]\wedge 2+R*u(i)\wedge 2$$

The ponderation coefficient R is to be choosen so as to attain a reasonable compromise between accuracy and control level requirements. For R=0 one obtains the classical minimum variance approach [5]. Some of the typical difficulties associated with this appraoch as reflected by our case study are illustrated by the results reported in figure 7.

These results are relevant to the case where the plant parameters values are

$$TT = .1 \text{ sec} \qquad K = 12 \text{ rad/volt*sec} \qquad T = .005 \text{ sec}$$

where TT, K and T are respectively the plant time constant, the dc motor static gain and the sampling period. Figure 7 represents the output response and the control applied to the plant in correspondence with a step input of .25 rads. The gains k1 and k2 are computed according to the classical minimum variance criterion. Figure 7a is associated with the ideal case where the parameters of the plant are perfectly known and no saturation is supposed to exist in the controller. The exceptionally fast and accurate ouptut response is obtained at the price of an unrealistically high control level. Figure 7b illustrates the effect of the saturation imposed by the power amplifier (SL = 5 volt). One notes that the output response is now far from being ideal and the fast frequency components in the control will probably cause undesirable higher mode vibrations. Figure 7c illustrates the further deterioration brought about by the combined effect of saturation level (again SL = 5 volt) and imperfect parameters knowledge (k1 and k2 are computed by using the estimate $\hat{TT}$ = .2 sec and $\hat{K}$ = 10 rad/volt*sec).

The literature suggests to overcome these difficulties by conveniently choosing a nonzero value of R [6] [7]. How one would proceed about doing this in the case of our case study is, however, far from being clear. The crux of the problem is illustrated by the results represented in figure 8. These results are relevant to the same setting of conditions as that considered in the experiment of figure 7a except that k1 and k2 are now determined in correspondence with various tentative values of R. One notices that if R is too small (figure 8a) the transient response of the system is excessively oscillarory; if it is too high the transient response becomes excessively slow (figure 8b); for intermediate values the response is consistently characterized by an excessive over shoot (figure 8c). In addition to this, one finds that a value of R which might possibly be acceptable in correspondence with a given experimental setting does typically result into being totally unacceptable in correspondence with an alternative setting.

THE PROPOSED MODIFICATION

In spite of the negative aspects discussed in the aboved section one notices that in the context of our case study the minimum variance criterion does work satisfactorily when the sampling period T is not too small in comparison with the time constant TT. This fact plus a natural extension of the classical reciding horizon approach sugtest the following variation. The gains k1 and k2 are chosen so that the issuing control

u(i) = -k1*(x1(i)-REF)-k2*x2(i)

has the property that by posing

u(j) = u(i)   for   i< = j< = i+N-1

one minimizes the functional

j(k1,k2) = [x1(i+N)-REF]∧2+R*u(i)∧2

where

N = TT/B*T

The acceleration factor B is determined experimentally.

SIMULATED EXPERIMENTS

The salient features of such an approach are illustrated by the results in figures 9 and 10. Figure 9 illustrates the behavior of the system under nonadaptive control. The proposed criterion for selecting the gains is apparently satisfactory over a wide range of parameters values. Figure 10 illustrates the behavior of the modified minimum variance under an adaptive strategy. In this experiment the reference position is represented by a square wave with an amplitude oscillating between +-.25 rad and a period equal to .125 sec. The DC motor static gain is equal to K=12 rad/volt*sec and the time constant TT is equal to .1 sec. The initial expected values for these constants are $\hat{K}$=24 and $\hat{TT}$=.6 ; .03. Additional details concerning the precise setting of the experiment are given in table 1. The improvement offered by the proposed modification is quite clear. The gains can be conveniently adapted to suit a wide range of parameters values and lead to an accurate and reasonably fast response. The control is quite smooth and is characterized by a reasonable level. As a second term of comparison figures 11 and 12 illustrate the results corresponding to the behavior under identical conditions of the classical and the Clark and Gauthrop minimum variance criteria.

REFERENCES

[1] Sandhu, H., Hands-On Introduction to Robotics: The Rhino XR1, Sandhu Machines D. Inc., 1981.

[2] DeSantis, R.M., Hurteau, R., Authié, G., A Real Time Computer Control Demonstrator for Use in a University Laboratory, IEEE Trans. on Education, Vol. E-25, pp. 18-27, Feb. 1982.

[3] Hsia, T.C., System Identification: Least Squares Methods, Lesington Books, DC Heath and Co., 1977.

[4] Clarke, D.W., Gauthrop, P.J., Self-Tuning Controllers, Proc. IEEE, Control and Science, Sept. 1975.

[5] Astrom, K.J., Self-Tuning Regulators: Design Principles and Applications, Applications of Adaptive Control, Academic Press, 1980.

[6] Koivo, A.J., Guo, T.H., Control of a Robotic Manipulator with Adaptive Controller, Proc. 20th IEEE Control and Decision Conference, pp. 271-276, San Diego, Dec. 1981.

[7] Landau, Y.D., Adaptive Control: The Model Reference Approach, Dekker 1979.

[8] Dubowsky, S., Desforges, D.T., The Application of Model Referenced Adaptive Control to Robotic Manipulators, Journal of Dynamic Systems Measurement and Control, ASME Sept. 1979.

FIGURES LIST

1. The test bench
2. Simplified model of the plant
3. The adaptive scheme
4. State regulator structure
5. Plant parameters estimator
6. Elementary state observer
7. Minimum variance behaviour (nonadaptive control)
   TT = .1 sec   K = 12 rad/volt*sec   T = .005 sec
   a) ideal case: no saturation and perfect parameters knowledge
   b) perfect parameters knowledge but saturation is taken into effect (SL=5 volt)
   c) combined effect of saturation and imperfect parameters knowledge:
   $\hat{TT}$ = .2 sec $\hat{K}$ = 10 rad/volt*sec   SL = 5 volt
8. Clark and Gauthrop output response (non adaptive; perfect parameters knowledge)
   TT = .1 sec   K = 12 rad/volt*sec   T = .005 sec
   a) R = 10∧-4   b) R = 10∧-3   c) R = 2*10∧-3
9. Modified minimum variance (nonadaptive control)
   K = 12 rad/volt*sec   SL = 5 volt   T = .005
   a) TT = .1 sec (perfect parameters knowledge)
   b) TT = .03 sec (perfect parameters knowledge)
   c) TT = .33 sec (imperfect parameters knowledge ; $\hat{TT}$ = .03 sec)
10. Modified minimum variance (adaptive control)
    Basic experimental setting as specified in table 1
    a) $\hat{TT}$ = .6 sec $\hat{K}$ = 24 rad/volt*sec
    b) $\hat{TT}$ = .03 sec $\hat{K}$ = 24 rad/volt*sec
11. Classical minimum variance (adaptive control)
    Basic experimental setting as specified in table 1
    a) $\hat{TT}$ = .6 volt $\hat{K}$ = 24 rad/volt*sec
    b) $\hat{TT}$ = 0.3 sec $\hat{K}$ = 24 rad/volt*sec
12. Clark and Gauthrop minimum variance (adaptive control)
    Basic experimental setting as specified in table 1
    a) $\hat{TT}$ = .6 sec $\hat{K}$ = 24 rad/volt*sec   R = .001
    b) $\hat{TT}$ = .03 $\hat{K}$ = 24 rad/volt*sec   R = .001

TABLE 1: BASIC EXPERIMENTAL SETTING

DC motor time constant TT = .1 sec
DC motor static gain K = 12 rad/volt*sec
Position transducer gain KP = .02 volt/rad
Speed transducer gain KS = .06 volt*sec/rad
Amplifier input saturation level SL = 5 volt
Initial estimate of TT, TT = .03; .6
Physically admissable upper bound for TT, TT(max)=.6 sec
Physically admissable lower bound for TT, TT(min)=.03 sec
Initial estimate of K, K = 24 rad/volt*sec
Physically admissable upper bound for K,
K(max) = 24 rad/volt*sec
Physically admissable lower bound for K,
K(min) = 6 rad/volt*sec
Covariances associated with initial estimate
P11 = 10, P22 = 1000, P33 = 1000
Estimator forgetting factor F = 0.5
Sampling period T = .005 sec
Acceleration factor, B = 4

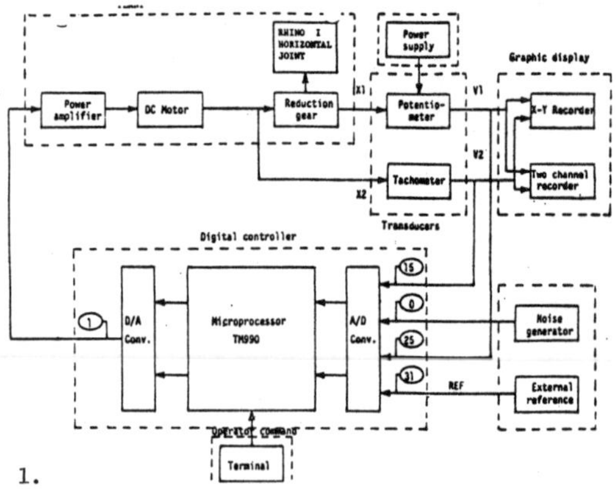
RHINO I HORIZONTAL JOINT
Power supply
Graphic display
Power amplifier
DC Motor
Reduction gear
X1
V1
Potentiometer
X-T Recorder
V2
Tachometer
Two channel recorder
X2
Transducers
Digital controller
D/A Conv.
Microprocessor TM990
A/D Conv.
Noise generator
REF
External reference
Operator command
Terminal

1.

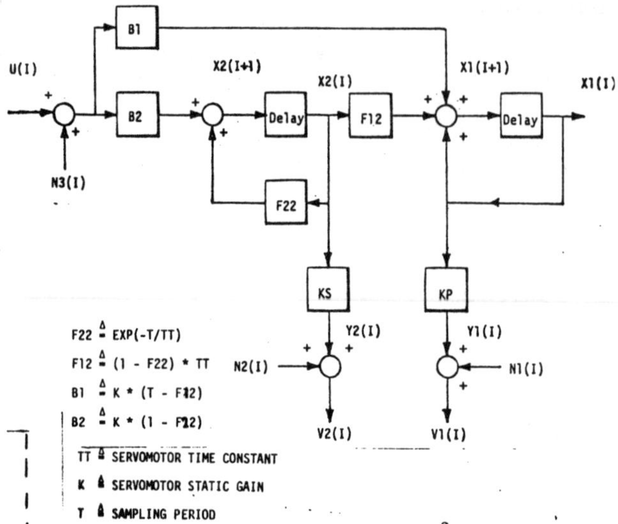
B1
U(I)
X2(I+1)
X1(I+1)
X1(I)
X2(I)
B2
Delay
F12
Delay
N3(I)
F22
KS
KP
Y2(I)
Y1(I)
N2(I)
N1(I)
V2(I)
V1(I)
F22 ≜ EXP(-T/TT)
F12 ≜ (1 - F22) * TT
B1 ≜ K * (T - F12)
B2 ≜ K * (1 - F12)
TT ≜ SERVOMOTOR TIME CONSTANT
K ≜ SERVOMOTOR STATIC GAIN
T ≜ SAMPLING PERIOD

2.

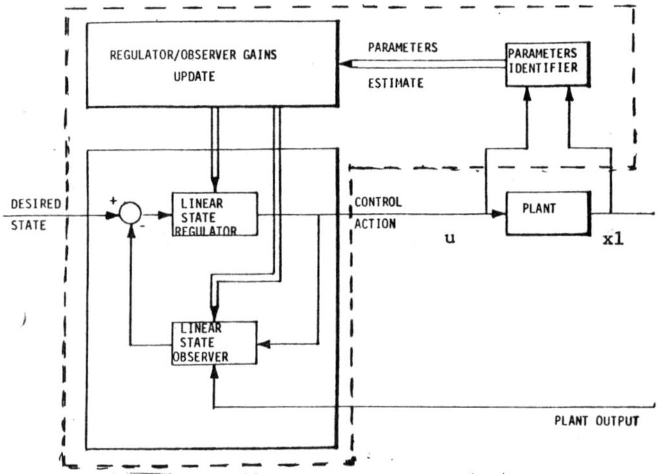
REGULATOR/OBSERVER GAINS UPDATE
PARAMETERS
ESTIMATE
PARAMETERS IDENTIFIER
DESIRED STATE
LINEAR STATE REGULATOR
CONTROL ACTION
u
PLANT
x1
LINEAR STATE OBSERVER
PLANT OUTPUT

3.

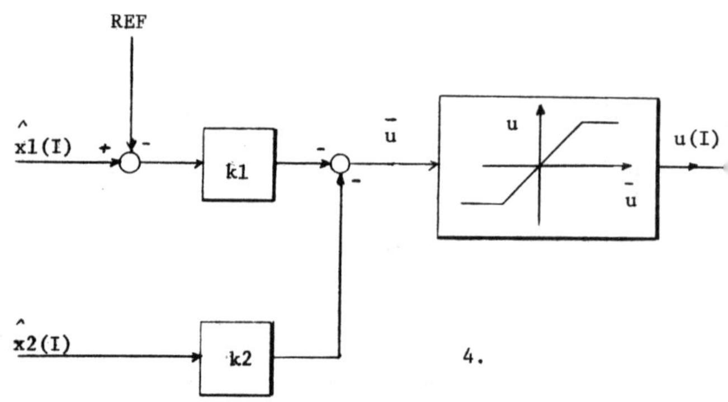
REF
x1(I)
k1
u
u(I)
x2(I)
k2

4.

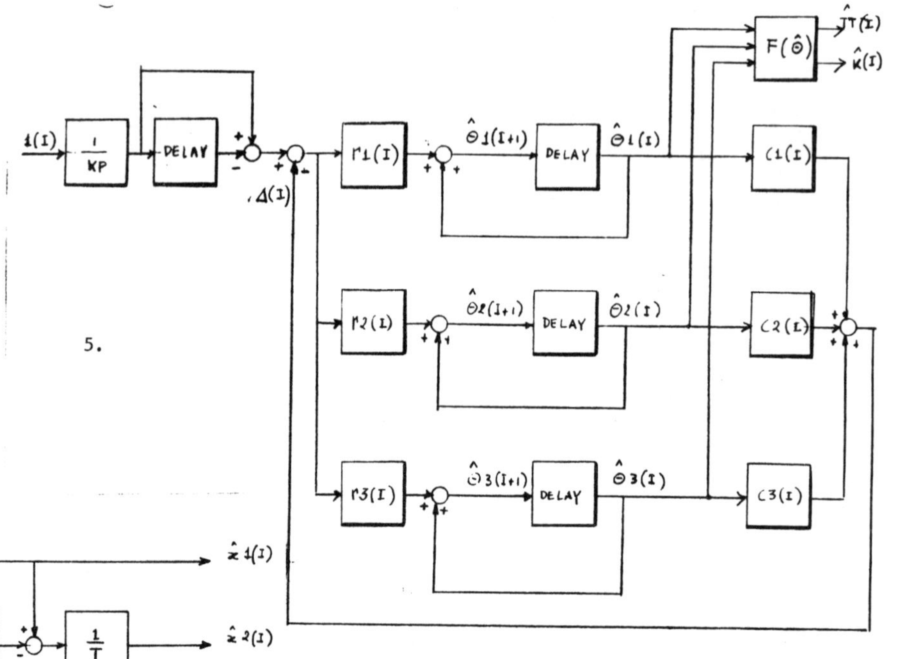
1(I)
1/KP
DELAY
Δ(I)
Γ1(I)
Θ1(I+1)
DELAY
Θ1(I)
F(Θ)
TT(I)
K(I)
C1(I)
Γ2(I)
Θ2(I+1)
DELAY
Θ2(I)
C2(I)
Γ3(I)
Θ3(I+1)
DELAY
Θ3(I)
C3(I)

5.

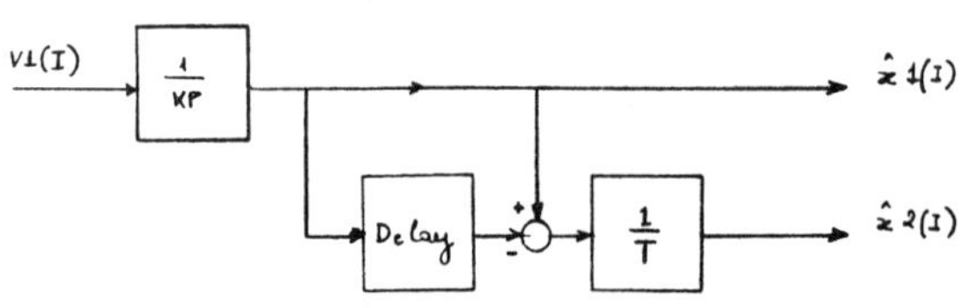
V1(I)
1/KP
x1(I)
Delay
1/T
x2(I)

6.

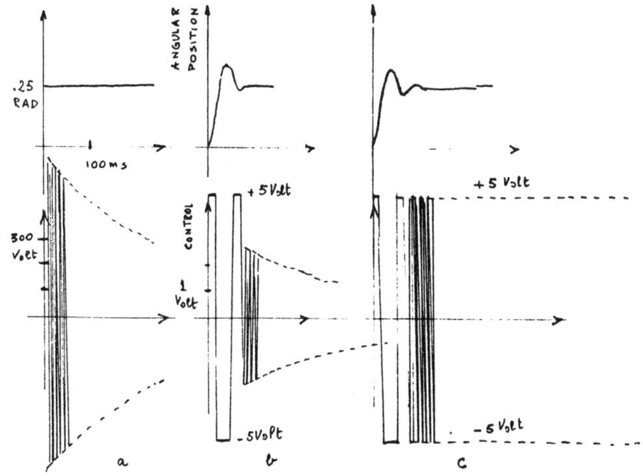
.25
RAD
100 ms
ANGULAR POSITION
300
Volt
CONTROL
1
Volt
+ 5 Volt
- 5 Volt
+ 5 Volt
- 5 Volt
a
b
c

7.

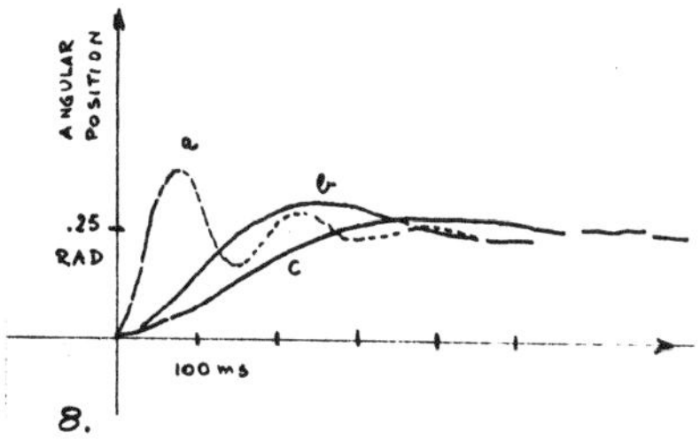
ANGULAR POSITION
a
b
c
.25
RAD
100 ms

8.

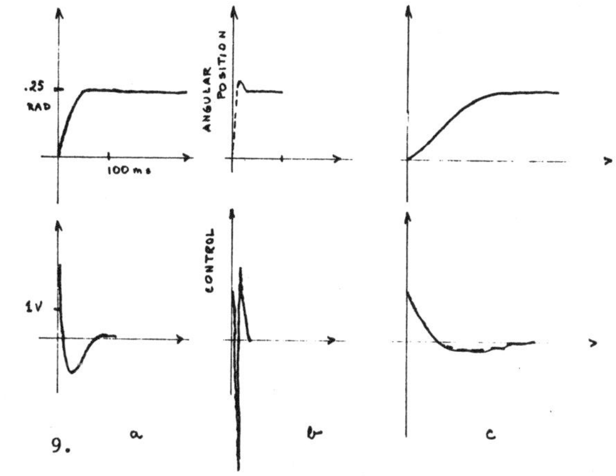
.25
RAD
100 ms
ANGULAR POSITION
1V
CONTROL
a
b
c

9.

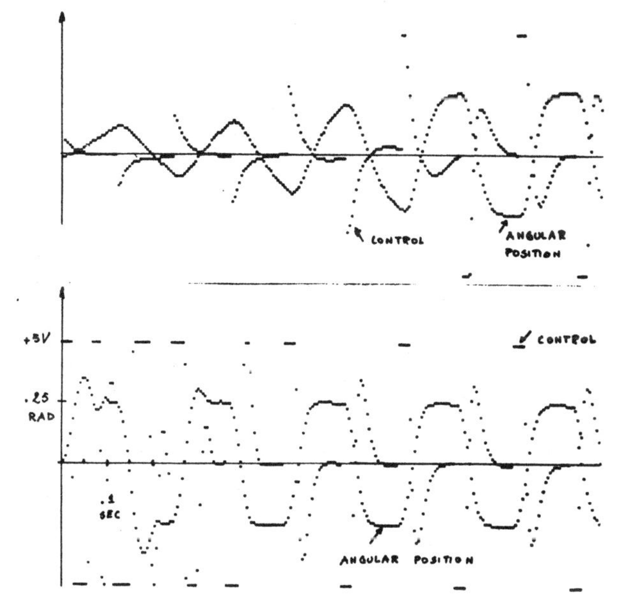
CONTROL
ANGULAR POSITION
+5V
CONTROL
.25
RAD
.1
SEC
ANGULAR POSITION

10.

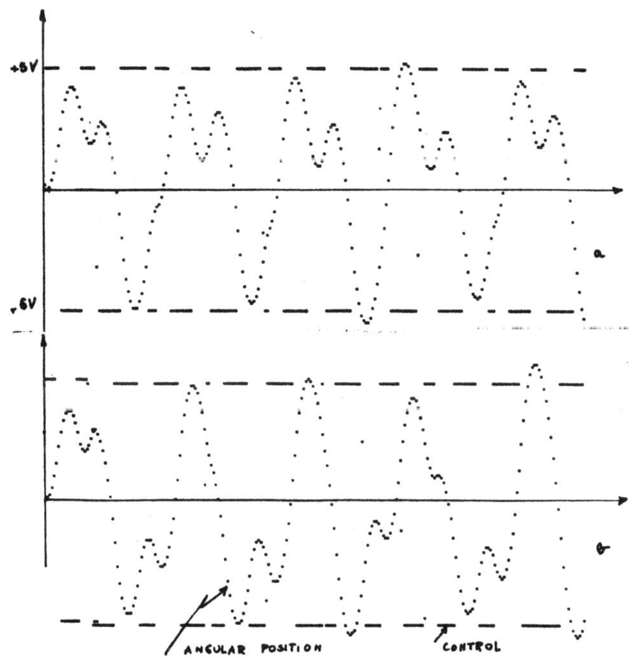
+6V
a
-6V
b
ANGULAR POSITION
CONTROL

11.

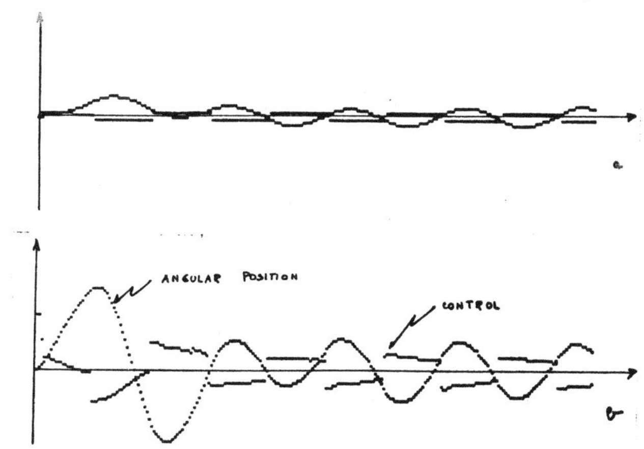
a
ANGULAR POSITION
CONTROL
b

12.

ON OPTIMAL CONTROL OF TRANSIENTS IN NOT-VERY-LARGE ELECTRICAL POWER SYSTEMS

V.A. Venikov,
Moscow Power Institute,
Moscow, USSR

Mohsen A.H. Tawfik,
Ain Shams University,
Cairo, Egypt

Abstract

In this paper the optimal control problem of transients in a 2-machine electrical power system has been formulated through the application of a variational technique, then the resulted two-point boundary value problem has been solved using modified quasilinearisation, gradient, and variation of extremals techniques.

A comprehensive comparative study has been conducted to analyse the differences between the above mentioned techniques concerning the easiness of problem programming, requirements to computer-storage, solution time, convergence and sensitivity to initiall values to start solution.

A conclusion has been made about the choice of augmented techniques, which overcome, to a great extent, the problems of covergence of solution, minimum requirements to computer storage and time of solution.

INTRODUCTION

During the last two decades many studies have been conducted in the field of optimization of transients in electrical power systems. The main concern was to adapt some of the well known optimization techniques to the specific problem of transients in electrical power systems. With the system transients simulated linearily it is relatively easy to solve the problem. But with nonlinear representation of the system transients the problem is rather difficult[5]

It is well known that the capabilities of the optimization techniques used, by now, do not exceed the possibility of obtaining off-line solution **of nonlinear problems, where** the time required to calculate the optimal control is longer than that in which the phenomenon takes place. Off-line calculations are usually used to show the maximum capabilities of any control system and also used as an ultimate for controllers designed using any other technique. But, **for nonlinear problems it is not** easy to get solution. This is mainly due to one or more of the factors: sensitivity of the method to the initially suggested values to start solution, requirements to computer storage and improper choice of the weighting coefficients.

Among the known techniques for the solution of nonlinear dynamic optimization problem the dynamic programing has proved not to be efficient due to its high requirements to the computer memory storage even for small systems.

Nonhierarchical techniques have been proved to be suitable for small and not-very-large systems where the solution is a single-level one.

Gradient, quasilinearisation and variation of extremals techniques can be considered as well established methods for the solution of nonlinear optimization problem. Each of these methods has its own drawbacks which may make it not feasible for use under certain conditions.

The main objective of this paper is to get to suitable augmented algorithms that may avoid the difficulties, while making use of the advantages of the considered technicues.

FORMULATION OF OPTIMAL CONTROL PROBLEM

If the system is described by

$$\dot{X} = F(X,U) \quad (1)$$

where X,U are the vectors of the system states and controls respectively, it will be required to find the control $U \in \mathcal{U}$ that makes the performance index (PI)

$$I = \int_0^T \Phi(X,U)\, dt \quad (2)$$

attains a minimum.

According to the maximum principle of Pontryagin the Hamiltonian H

$$H = \Phi(X,U) + P^t\dot{X} \quad (3)$$

should be formulated and then the state- costate equations

$$\dot{X} = \partial H/\partial P\ ,\ X(0) = X^0 \quad (4)$$

$$\dot{P} = -\partial H/\partial X\ ,\ P(T) = 0 \quad (5)$$

to be solved. The control U(t) can be found from

$$\partial H/\partial U = 0 \quad (6)$$

The solution of the state-costate equations (5,6) is a typical nonlinear two-point boundary value problem (TPBVP) where the given boundaries are the initial conditions of the states and the end conditions of the costates.

In the following, a concise description of each of the gradient, variation of extremals, and modified quasilinearization algorithms is given to show its main features. Full algorithms can be found in the relevant references.

GRADIENT (STEEPEST DESCENT) ALGORITHM [1]

1- suggest the m control trajectories
2- integrate the system state equations forwards.
3- integrate the system costate equations backwards.
4- if the end condition is not satisfied correct the control trajectory and get back to step 2 for a new iteration.

In this algorithm we have to store n trajectories of X and **2m** trajectories of **U** and gradient, and to integrate 2n differential equations in each iteration.

If the computer used is of small storage the above algorithm can be modified to avoid the storage requirements. This can be done when replacing the above steps: 2 & 3 with the following:

2- integrate the system state equations forwards and store the end points only.
3- integrate the system state-costate equations simultaneously backwards.

In this modified procedure we have to store 2m trajectories and to integrate 3n differential equations in each iteration. However, in any gradient algorithm the value of the control step (steepest descent) that leads to a minimum PI should be calculated using a single search method.

THE VARIATION OF EXTREMALS METHOD [2] :

1- suggest the costate initial values.
2- integrate simultaneously the 2n state-costate equations and the 2n x n equations of the influence matrix and store the final points.
3- if the end conditions is not satisfied correct the initial values of the costates and go back to step 2.

In this algorithm there is no need to store any trajectories while it is required to solve 2n(n+1) differential equations in addition to the inversion of an n x n matrix in each iteration.

MODIFIED QUASILINEARISATION ALGORITHM [3]

1- suggest 2n trajectories for the state-costate equations.
2- integrate the 2n particular and 2n x n homogeneous equations of the linearized state-costate set and store the final points.
3- find the integration constants by solving the system costate (algebraic) equations at the final point.
4- integrate the original nonlinear state-costate equations and store their 2n trajectories. Further, if the solution criterion is not satisfied go to step 2 with the constants found in step 3 used as initial conditions for the costate variables.

In this algorithm we have to store 2n trajectories and to integrate 2n(2+n) differential equation in addition to inversion of an n x n matrix in each iteration.

For all the above mentioned methods the solution may be considered obtained and the iteration process can be stopped when the maximum modification in initial guesses obtained after any iteration becomes smaller than a given small positive value.

APPLICATION TO DOUBLE MACHINE SYSTEM:

If each machine is simulated by 3 nonlinear differential equations the system can be represented by the 5th order model [4]:

$$\dot{x}_1 = x_3 - x_2$$
$$\dot{x}_2 = a_1u_1 - c_1x_4^2 - c_2x_4x_5 - c_3x_5^2$$
$$\dot{x}_3 = a_2u_2 - c_4x_4^2 - c_5x_4x_5 - c_6x_5^2$$
$$\dot{x}_4 = a_3u_3 - c_7x_4 - c_8x_5$$
$$\dot{x}_5 = a_4u_4 - c_9x_4 - c_{10}x_5$$

where $x_1 = \delta_{21}$, $x_2 = \omega_1$, $x_3 = \omega_2$, $x_4 = E'_{q_1}$, $x_5 = E'_{q_2}$, $u_1 = P_{t_1}$, $u_2 = P_{t_2}$, $u_3 = E_{q_{e1}}$, $u_4 = E_{q_{e2}}$. The constants a's, c's are as given in [4].

So, if it is required to transfer the system from a disturbed state $X(0)=X^o$ to a final steady state $X(T) = X^f$ in such a way to minimize the PI

$$I = \frac{1}{2}\int_0^T [\sum_{i=1}^{n} W_{x_i}(x_i - x_i^f)^2 + \sum_{\ell=1}^{m} W_{u_\ell}(u_\ell - u_\ell^f)^2]dt$$

where W_x's and W_u's are the state and control weighting coefficients respectively. The Hamiltonian (3) is to be formed and then expressions for control to be found from (6) then to be substituted in (4) and (5). The obtained state-costate 2n differential equations with their initial and final conditions is the TPBVP to be solved according to the algorithm used.

PROBLEM SOLUTION

The double-machine system has been disturbed by a sudden loss of input power P_{t_1} for 0.2 sec. after which the control starts acting. Different alternatives of input power and exitation controls for the first and/or the second machine have been conducted. Many results that show interesting relations between different controls and the machines and system parameters have been obtained. These results may be used for design of actual realizable controllers for such system. But as far as we are concerned, in this paper, with the analysis of digital methods of solutions, we will show the results of one case-study where only excitation control of both machines are used. So, for the same initial conditions and the same weighting coefficients the problem has been solved using the three studied algorithms. Fig.(1) shows the power angle response obtained due to the application of each method.

Discrepencies in the obtained optimal responses may be attributed to the fact that the reached optimal trajectories in each case is a local optimal. Moreover, the effect of accumulated error in each algorithm should be considered. Actually, discrepencies in results may be obtained when the same problem is solved with different calculation precisions (e.g. when using double precision).

RESULTS ANALYSIS

To evaluate the differences between the three studied methods all factors dealing with algorithm programing, requirements to computer memory storage, computational effort, and sensitivity of the algorithm to the suggested

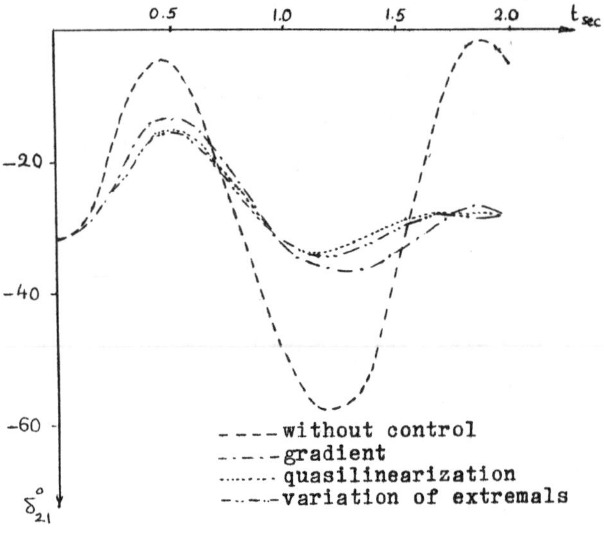

Fig.1. Relative Power angle response

initial conditions and to the weighting coefficients should be considered. In the following each factor will be discussed.

A. EASINESS OF PROBLEM FORMULATION AND PROGRAMING:

Before applying any algorithm it will be assumed that the TPBVP is formulated with the state-costate set written and an expression for control vector is deriven. For the gradient algorithm no additional derivation is required. For the variation of extremal algorithm it is required to formulate 2n x n influence function differential equations which needs the derivation of second partial derivatives of the Hamiltonian w.r.t. the state and costate variables. The quasilinearization algorithm requires the formulation of the n x n Jacobian matrix and then the homogeneous and particular equations of the state-costate set.

So, it is evident that the gradient technique is the easier one, while the variation of extremals algorithm needs much more programing effort.

B. REQUIREMENTS TO COMPUTER STORAGE MEMORY

In this respect the variation of extremals technique is the superior one where no trajectories are required to be stored. In the gradient, modified gradient, and modified quasilinearization n+2m, 2m, 2n trajectories, respectively, are to be stored. So, for large systems, i.e. larger n, the requirements of these last algorithms may exceed the capacity of fast access memory of any computer.

C. COMPUTATIONAL EFFORT

This is mainly to do with the required calculation per iteration and the total number of iteration after which the solution is obtained.

On comparing the 3 algorithms, the gradient method has shown to be that of less calculations per iteration (integration of 2n or 3n DEs), while 2n(1+n) DEs are to be solved in each iteration for both variation of extremals and quasilinearization algorithms. However, a real estimation of the computational effort should consider the total number of iteration required for solution. With a maximum relative error in the assumed initial conditions (w.r.t. the optimal ones) does not exceed 40%. The average number of iterations that were required to get to an optimal solution were: 14-18 for gradient, 3-5 for quasilinearization, and 4-6 for variation of extremals algorithm.

D. SENSITIVITY TO THE ASSUMED INITIAL CONDITIONS

A sensitivity index is assumed to be the ratio of the integration of the absolute values of the differences between the ICs guessed and their optimal values, to the integration of the absolute value of the optimal conditions. As shown above the initial conditions to be assumed are: the control trajectory for gradient, the state-costate trajectory for quasilinearization and the costate initial points for variation of extremals. It has been found that the value of this index (relative error in the assumed ICs) larger which the solutions diverge were about 46% in variation of extremal algorithm and 57% in quasilinearisation algorithm. For gradient algorithm ICs, a converging solution was still possible with relative error of about 300%.

This shows the high sensitivity of the variation of extremals algorithm to improper choice of ICs.

E. SENSITIVITY TO THE WEIGHTING COEFFICIENTS

It was found that the variation of extremals algorithm is the more sensitive one to improper choice of weighting coefficient, while the gradient algorithm is that one of less sensitivity in this respect.

F. MODE OF CONVERGENCE

If the problem has a solution and there are no other reasons that may cause numerical instability, the gradient algorithm yields, as usual, a very fast convergence followed by very slow one near the optimum. This may be attributed to the fact that the gradient vector is of less value and, inturn, of less effect in correcting the control vector near the optimum. It was found that while the quasilinearization algorithm converges quadratically, the variation extremal algorithm could converge, even linearily, with properly chosen ICs.

However, it should be stated that the generalization of the above mentioned results concerning the algorithms sensitivity to different factors is difficult, where the solved problem is a nonlinear one. The nonlinearity nature of problems should be of great effect when solving, using different algorithms.

To get use of the above shown advantages of the individual algorithms, the problem has been solved several times, starting with gradient algorithm and then switched to variation of extremals or quasilinearization algorithm. The automatic switching from gradient to algorithms has taken place when the solution criterion (error index) is five times its assumed value.

For the augmented gradient-variation of extremal algorithm it has been found that the average number of iterations required for solution was decreased to 3-5 iterations. For the augmented gradient - quasilinearization algorithm this number has been also decreased to 3-4 iterations. For both augmented algorithms the all other advantages of the gradient algorithms have been acquired.

Due to poor choice of control step in the expression of corrected control written for the starting gradient algorithm, the number of iterations in any augmented algorithm may be greater than the number of iterations required by any of the variation of extremals or the quasilinearization algorithms. So, it is better to calculate the value of the control step that yields the steepest descent required to get a minimum value of the given performance index.

CONCLUSIONS

According to the system size and its order of representation as well as the storage memory of the available computer, and as a result of the obtained characteristics of the three algorithms studied in this paper, it is recommended to solve the problem of optimal control of transients in electrical power systems as follows:

1. for small systems (simulated by 3-4 nonlinear DEs) any of the above mentioned algorithms may be used.

2. for not-very-large systems (simulated by 8-10 nonlinear DEs) it is recommended to use one of the augmented techniques: gradient-variation of extremals or gradient-quasilinearization. In case of limited fast-access memory in the available computer, the augmented gradient-variation of extremals algorithm could be used.

3. for relatively larger systems (up to 12-13 equations) the gradient-variation of extremals is also recommended. To overcome the problem of memory storage, the gradient algorithm can be used just for one iteration to avoid numerical instability in the variation of extremals algorithm, due to improper choice of costate initial conditions.

4. for larger systems hierarchical techniques are more suitable where multilevel solution takes place for subsystems of lower order.

REFERENCES

1. A.P.Sage and C.C.White III, "Optimum Systems Control", Prentice-Hall, Inc., 1977.

2. L.Fox, "Numerical Solution of Ordinary and Partial Differential Equations". Adison Wesley Publishing Co., Inc., 1962.

3. V.A.Venikov, V.A.Stroev and M.A.Tawfik, "An Algorithm for Solution of the Problem of Optimal Control of Transients in Electrical Power Systems", Proc. of the USSR Academy of Science on Electrical Power Systems and Transportation, 1977, No.4, pp.17-26.

4. A.R.Abul-Wafa, M.A.Tawfik, H.E.Talaat and R.Moret, "Power Systems Mathematical Model Suitable for Nonlinear Optimal Control Studies", Proceedings of the International AMSE Conference on Modelling and Simulation, Paris, July 1982, Vol.5, Group 5, pp.42-47.

5. V.A.Venikov, V.A.Stroev and Mohsen A.H. Tawfik, "Optimal Control of Electrical Power Systems Containing Controlled Reactors, part 2: Optimal Control Problem Solution". IEEE Trans., 1981, PAS-100, No.9.

LIST OF SYMBOLS

δ rotor angle

ω instantaneous angular velocity of the machine rotor

E'_q transient emf proportional the direct axis field flux

E_{q_e} forced emf proportional to the field voltage

P_t mechanical input power

X vector of n state variables

P vector of n costate variables

U vector of m comtrols

x_i^f desired steady state value of the state variable i at the end of control period

u_i^f desired steady state value of the control j at the end of control period

W_{x_i} weighting coefficient of the state variable i

W_{u_j} weighting coefficient of the control j

t symbole of transposition

T control period

THE DOUBLE-MACHINE SYSTEM PARAMETERS:

X_{d1}= 0.537 X_{d2}= 0.25 X_{q1}= 0.346 X_{q2}= 0.25

X'_{d1}= 0.191 X'_{d2}=0.032 P_{t1}= 0.4564

P_{t2}=0.3944 E_{qe1}=1.562 E_{qe2}=1.06

M_1 = 9.2sec. M_2 = 12.3sec.

T_{d1}= 4.95sec. T_{d2}= 4.27sec.

PREFAULT LOADING CONDITIONS:

Y_{10}= 0.09693 Y_{20}=0,6933 Y_{12}= 0.376

All values are in P.U.

COMMANDE ADAPTATIVE EN BOUCLE FERMEE D'UN ROBOT
DANS UN ESPACE DE TRAVAIL LOCAL DOTE D'UN SYSTEME DE VISION

M. Dufaut,   R. Husson

Centre de recherche en automatique de NANCY

ENSEM BP 850, 54011 - NANCY CEDEX, FRANCE

RESUME

Nous présentons dans cet article un système de commande avec capteur de vision permettant la surveillance, la commande en vitesse et position en boucle fermée d'un robot ainsi que l'adaptation des lois de vitesse suivant la masse de l'objet manipulé.

Le robot ne connaît sa destination dans le repère image que lors de son entrée dans le champ de la caméra. La commande en position consiste alors à annuler l'erreur entre la position actuelle et l'objectif fixé (cible) en N périodes d'échantillonnage (système à réponse finie).

1 - INTRODUCTION :

Dans le cas de postes de manutention ou d'assemblage robotisés l'utilisation de système de vision procure à l'installation des qualités essentielles telles que l'adaptabilité à l'environnement et une flexibilité véritable.

Après extraction des éléments pertinents contenus dans l'image nécessaires à l'identification et la localisation de l'objet, le processeur de vision communique au processeur de commande du robot les informations qui permettront le bon déroulement des tâches de préhension, transfert, etc...

Mais l'élaboration des commandes pour le manipulateur exige des transformations de coordonnées pour passer du repère image cartésien au repère robot pouvant être de type cartésien (cas le plus simple), cylindrique ou sphérique (cas le plus complexe). Ces changements de repères conduisent souvent à des calculs compliqués et longs.

La méthode utilisée ici permet de s'affranchir de cette transformation. Les commandes sont établies exclusivement d'après les mesures d'écarts existant entre le robot et le point cible /1/ /2/. Ceci se traduit par une perte de précision absolue (numérisation, résolution du capteur). Cependant une précision correcte et suffisante peut être obtenue en compensant les défauts de caméra (distorsion géométrique, rémanence) par correction numérique sur les pixels et en adoptant un bon compromis entre le champ et la résolution de l'image afin d'effectuer des commandes valables de vitesse et de position en temps réel /3/ /4/.

Dans le cas où on recherche l'optimisation temporelle du cycle de manutention il est également intéressant de modifier la vitesse moyenne de transfert entre postes en l'adaptant suivant la masse manipulée estimée lors de l'étape de reconnaissance de forme /5/.

Cette phase de transfert s'effectue en boucle ouverte et à la vitesse maximale possible tandis que la phase d'approche et de préhension se déroule en boucle fermée, les vitesses étant alors compatibles avec la fréquence d'échantillonnage du système de vision.

2 - DESCRIPTION DE L'INSTALLATION

L'installation (fig. 1) comporte les éléments suivants :

- un manipulateur à structure R.T.T. (1 rotation de 360°, 2 translations de course 400 mm, un préhenseur pneumatique) à coordonnées cylindriques (ρ, θ, z) (fig.2). Chaque axe est actionné par un moteur électrique pas-à-pas PATAY 16 S (400 demi-pas par tour).

Un témoin lumineux fixé à l'extrémité du bras permet un un repérage immédiat du préhenseur dans le champ de vision.

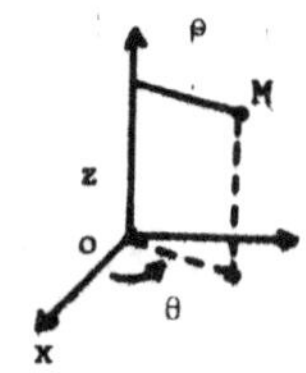

Fig. 2 Coordonnées du robot

Fig. 1 Installation

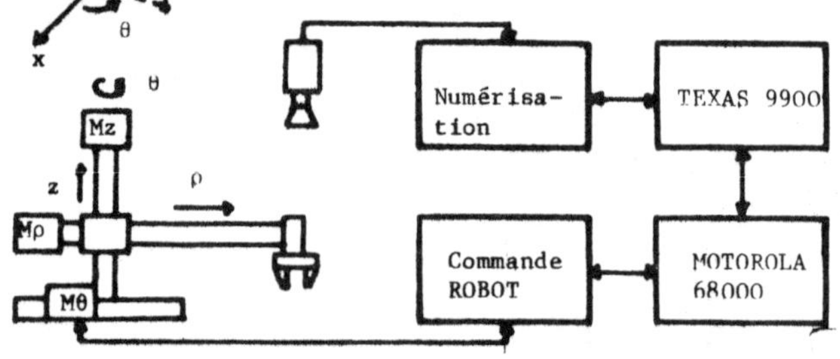

- un système de vision utilisant une caméra vidéo SONY qui fournit une image binaire en 2D de définition 256 x 256 pixels avec seurillage variable. L'image est transférée en mémoire par accès direct en 20 ms (fig. 3).

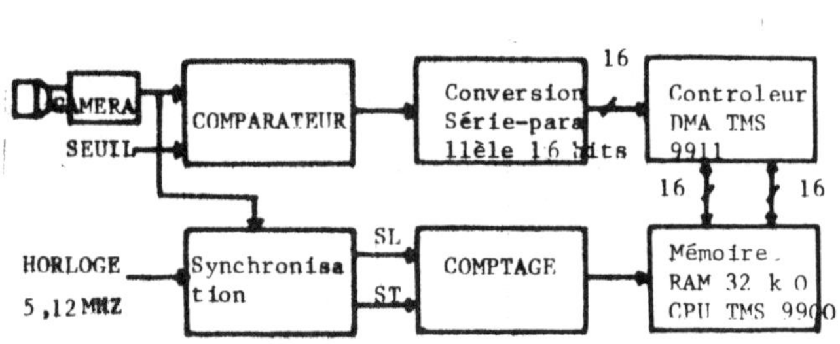

Fig. 3 Système de vision

- deux microprocesseurs à structure de 16 bits : l'un assurant l'acquisition et le traitement de l'image (TEXAS 9900), l'autre effectuant la commande en vitesse et position du manipulateur (MOTOROLA 68 000).

3 - REPERAGE ET SUIVI AUTOMATIQUE

La position finale M_F (cible) à atteindre par le robot est supposée connue lors de l'entrée du manipulateur dans le champ de la caméra. La connaissance de M_F découle d'une analyse de la scène par un algorithme de segmentation-connexité. /6/.

Le champ utile de l'espace de travail est limité à un plan de 0,5 x 0,5 m soit une résolution de 2 mm.

La vitesse linéaire maximale autorisée pour un repérage correct est de 1 m/s.

Le système de vision scrute en permanence la scène (fig.4) pour détecter l'apparition du bras dans l'espace de travail.

L'acquisition de l'image durant 20 ms et la recherche dans l'image du témoin lumineux 18 ms, la fréquence maximale de scrutation est donc de 25 HZ.

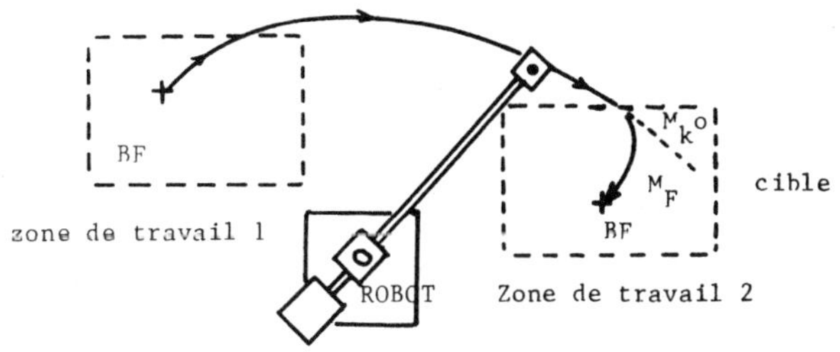

Fig. 4 Trajectoire du robot

La présence étant affirmée, la position M_k du bras dans l'image est déterminée par calcul du centre de gravité G du rectangle exinscrit de l'image du témoin lumineux.

Une fenêtre de dimension 64 x 80 pixels, centrée autour de G, est alors définie pour la recherche du témoin à la période suivante (fig.5).

L'actualisation de la fenêtre de poursuite et le calcul des coordonnées de M_k durent environ 3,5 ms.

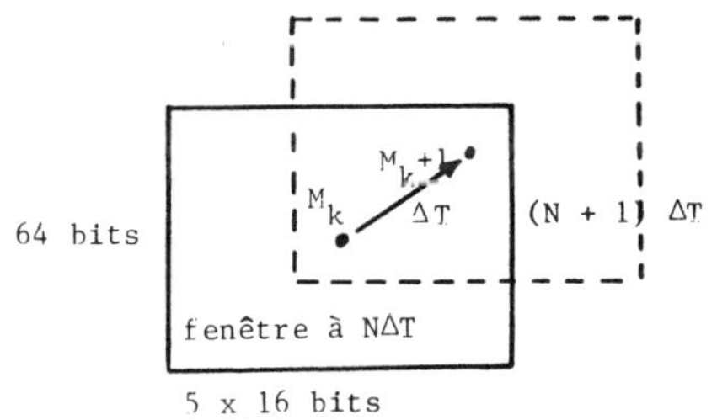

Fig. 5 Fenêtre de poursuite

4 - COMMANDE EN VITESSE ET EN POSITION

Dans le cas d'un positionnement simple sur la cible M_F sans trajectoire imposée, la description du bras par un modèle mathématique de type cinématique suffit.

En coordonnées cylindriques, le vecteur position $\vec{u}k$ s'écrit à l'instant d'échantillonnage kT :

$$\vec{u}_k \begin{cases} \rho_k = [x_k^2 + y_k^2]^{1/2} \\ \theta_k = \text{arctg} \dfrac{y_k}{x_k} \end{cases}$$

avec x_k, y_k les coordonnées cartésiennes du bras fournies par le capteur de vision.

L'écart (erreur de position) par rapport à la position finale $\vec{u}_F$ s'exprime par :

$$\varepsilon \vec{u}_k = \vec{u}_F - \vec{u}_k \quad \text{avec} \quad \vec{u}_F \begin{cases} \rho_F \\ \theta_F \end{cases} \quad \text{ou} \quad \vec{u}_F \begin{cases} x_F \\ y_F \end{cases}$$

$$\text{d'où } \varepsilon\rho_k = \rho_F - \rho_k = [x_F^2 + y_F^2]^{1/2} - [x_k^2 + y_k^2]^{1/2}$$

$$\text{et } \varepsilon\theta_k = \theta_F - \theta_k = \text{arctg} \frac{y_F}{x_F} - \text{arctg} \frac{y_k}{x_k}$$

En faisant apparaître les écarts de positions cartésiennes εx_k et εy_k, on a :

$$\varepsilon x_k = x_F - x_k \qquad \varepsilon y_k = y_F - y_k$$

$$\text{et} \quad \varepsilon\rho_k = [x_F^2 + y_F^2]^{1/2} - [(x_F - \varepsilon x_k)^2 + (y_F - \varepsilon y_k)^2]^{1/2} \quad (1)$$

$$\varepsilon\theta_k = \text{arctg} \frac{y_F}{x_F} - \text{arctg} \frac{y_F - \varepsilon y_k}{x_F - \varepsilon x_k} \quad (2)$$

Si εx_k et εy_k sont petits, c'est-à-dire M_k voisin de M_F, la relation (2) s'écrit :

$$\varepsilon\theta_k \simeq \frac{-y_F \varepsilon x_k + x_F \varepsilon y_k}{x_F^2 + y_F^2 - x_F \varepsilon x_k - y_F \varepsilon y_k} \quad (3)$$

Dans (1), (2) et (3), y_F et x_F sont définis par rapport au centre du repère qui est le centre de rotation du robot et devant être identifié au préalable.

Pour coordonner les mouvements et obtenir un positionnement en N coups, la vitesse moyenne pour chaque axe est donnée par :

$$\dot{\rho}_m = \frac{\varepsilon\rho_o}{N\Delta T} \quad , \quad \dot{\theta}_m = \frac{\varepsilon\theta_o}{N\Delta T}$$

$\varepsilon\rho_o$ et θ_o sont les écarts calculés lors de l'entrée du bras dans le champ de vision.

La mise en vitesse suit une variation trapézoïdale (fig. 6). La vitesse de palier est calculée connaissant la vitesse initiale et la vitesse moyenne sur un horizon NΔT par la relation :

$$\dot{\rho}_p = \frac{2\dot{\rho}_m N - \dot{\rho}_o N_a}{2N - N_a} \quad \text{et} \quad \dot{\theta}_p = \frac{2\dot{\theta}_m N - \dot{\theta}_o N_b}{2N - N_b}$$

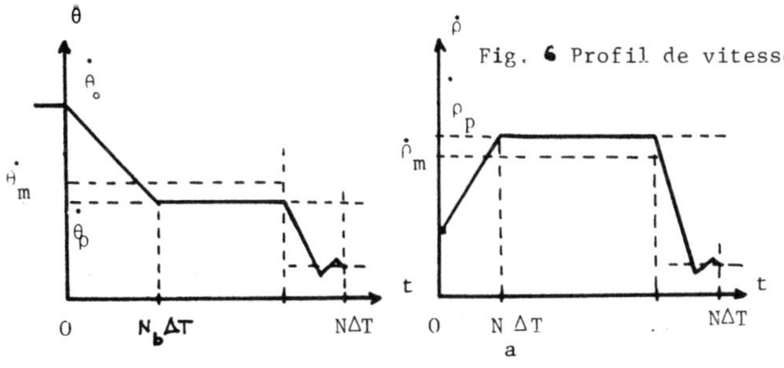

Fig. 6 Profil de vitesse

La décélération est provoquée dès la détection du bras dans un un cercle de rayon R_F centré sur la cible.

$$d_k \leqslant R_F \qquad \text{avec} \quad d_k = [\varepsilon x_k^2 + \varepsilon y_k^2]^{1/2}$$

La vitesse $\dot{\vec{u}}_k$ est alors régie par une commande de type proportionnelle telle que :

$$\dot{\vec{u}}_k = K\varepsilon\vec{u}_k \Longleftrightarrow \begin{cases} \dot{\rho}_k = K_1 \varepsilon\rho_k \\ \dot{\theta} = K_2 \varepsilon\theta_k \end{cases}$$

Dans le cas d'un positionnement avec trajectoire imposée et en temps minimum (contournement d'obstacles) une description plus complète du bras par un modèle mathématique dynamique doit être utilisée pour bénéficier de commande nettement plus performantes. /7/ /8/.

CONCLUSION

La commande en boucle fermée dans un espace limité autour des zones de préhension ou d'assemblage procure au robot le bénéfice de l'adaptabilité. En effet, elle rend possible les changements de position relatifs du robot par rapport au capteur ou du capteur par rapport au robot. A chaque nouvelle situation il suffit d'identifier le nouveau centre de rotation. Cette procédure peut être rendu automatique et s'appliquer à des systèmes mobiles et permet aussi d'éviter un calage fastidieux du capteur de vision par rapport au robot. Ce type de commande est particulièrement bien adapté aux actionneurs à moteur pas-à-pas qui par l'existence d'une boucle fermée sur l'extérieur acquièrent ainsi une sûreté de fonctionnement valable, les éventuelles pertes de pas se trouvant automatiquement compensées.

BIBLIOGRAPHIE

1. P.Y. COULON Association vision commande en robotique Thèse 3e cycle - Octobre 82. I.N.P. GRENOBLE.

2. P.Y. COULON Localisation dans un plan par caméra vidéo 2 e journées de robotique - RENNES 1981.

3. H.A. MARTINS, J.K. BIRK, R.B. KELLEY. Camera models based on data from two calibration planes. Computer Graphic and Image Processing 17, 173-180 - 1981

4. M. BRIOT, M. GHALLAB et all. Système de perception visuelle en 2 D destinée à identifier et localiser des objets partiellement observés. Journées ARA POITIERS Sept. 1982.

5. M. DUFAUT, C. GOELDEL, R. HUSSON, D. PINCHON. Génération de courbes d'accélération adaptative pour la commande d'un manipulateur équipé de moteurs pas-à-pas. 2 journées d'étude sur les moteurs pas-à-pas. LAUSANNE, Avril 1982.

6. M. DUFAUT, R. RIPP. Real time detection of defective objects in a production line by vision sensor. Congrès EUSIPCO - 80 . LAUSANNE Sept. 1980.

7. C. ROWDO, J.P. LOUIS, R. HUSSON. Etablissement d'une commande performante pour actionneur pas-à-pas appliqué à la robotique à partir d'un modèle dynamique complet.
2e congrès IASTED Robotics and Automation - LUGANO Juin 1983.

8. C. ROWDO, M. DUFAUT, J.P. LOUIS, R. HUSSON. Influence of the stepping motor torque linearisation on the dynamic control.
Congrès MECO 83 Measurement and control - ATHENES, Août 1983.

OLIGARCHY: A CONTROL SCHEME FOR FLEXIBLE MANUFACTURING SYSTEMS

Philippe R. Piche   Moshe Krieger

Dept. of Electrical Engineering
University of Ottawa
Ottawa, Ontario, Canada K1N6N5

Nicola Santoro

School of Computer Science
Carleton University
Ottawa, Ontario, Canada K1S5B6

Abstract

With the advent of advanced robot manipulators, sensory systems, and complex conveyor systems, a flexible manufacturing plant capable of producing a variety of products is now feasible. In such an automated manufacturing plant, a major problem facing the designer is the proper coordination of widely differing interacting entities (machines, sensors, robots,etc.). To achieve high reliability and quality control required in production, a scheme for systematic control and coordination is needed. This paper presents oligarchical control as the methodology for controlling and coordinating a large number of interacting entities.

1. INTRODUCTION

At present there are two major types of automated production facilities: 1) the single station facility where very complex robots are used to manufacture a single complex part; 2) assembly type facility where a number of relatively simple stations must interact to produce a finished product. The recent advances in robot manipulators, various sensors, and complex conveyor hardware have brought the state of the art to the point where both of the above types of manufacturing can be integrated into a truly Flexible Manufacturing System (FMS) (1),(2). The major limitation today in achieving this integration is the capability to control and coordinate such a very complex system.

The present day control schemes used in automated manufacturing can be divided into three main classes: fully decentralized, fully centralized , and hierarchical (3), (4), (5), (6), (7), (8), (9).

In fully decentralized systems, each system entity is assigned a controller. To allow for interaction among entities, each controller includes a coordinator which allows it to communicate with all other controllers in the system. A decision is reached by performing distributed computations, usually requiring the participation of all controllers/coordinators, and involves a large number of information transfers. Such systems have the main advantages of having relatively simple controllers/coordinators and good tolerance to local failures. The main disadvantages of these systems are: decreased system performance (slow response) as the number of entities increases; increased software overhead needed for the communication protocols; and delays in processing due to waiting time in queues at the sending and receiving ends of the communications paths. Because of these drawbacks, no major automated manufacturing system has decentralized control.

In fully centralized systems, there exists only one controller/coordinator which controls all entities. This allows fast response time for control operations at the cost of very complex software. The main disadvantages of such a system are: the system is highly vulnerable to a failure in the single controller/coordinator; controller/coordinator complexity increases rapidly with the number of entities in the system above a certain complexity; and every time a new element is added, major software redesign must be undertaken to meet the timing and interaction requirements. Furthermore, the centralized nature of the system leads to severe bottlenecks as the size of the system increases, thus effectively reducing the response time.

Hierarchical systems are in general tree structured organizations of processors where the processor capability increases with the levels of the tree. They are similar to a corporate organization structure: the lowest levels are application dependent, special purpose and dedicated to a well defined set of tasks; the top of the organization has a more general capability for controlling and coordinating the entire system. Other characteristics of hierarchical systems are: shared data bases are stored at the top rather than distributed throughout the system; repetitive real-time functions are found at the lowest levels; data rates increase as data migrate up the tree (unless data reduction techniques are used); and the information transfers are usually vertical (i.e between the different levels of the hierarchy). A major problem in hierarchical systems is the inclusion of fail-soft features, without unduly increasing system complexity; also beyond a certain complexity, hierarchical systems become very stiff with respect to any modification.

In this paper, a new general methodology for control and coordination of the interacting entities in a Flexible Manufacturing System is presented. The proposed scheme, which integrates features of decentralized, centralized, and hierarchical strategies, allows the design of control mechanisms that are highly reliable, have fast response time, and can be easily upgraded, modified and expanded. Furthermore, because of its inherent modularity, it simplifies the design process as well as the testing, verification and fine tuning of the resulting system. In the next sections, the proposed scheme is presented, its main features are described, and its application to an existing test-bed assembly line is outlined.

2. OLIGARCHICAL CONTROL SCHEME

The proposed methodology is based on the concept of an oligarchical system (10). An oligarchical manufacturing system is best described as a multilayered organization whose lowest layer is composed of controllers directing production entities, and any other layer is comprised of coordinators. The various tasks of the manufacturing entities are specified as sequences of primitives in the local controllers, and are initiated by an action from an upper level coordinator (see Figure 1).

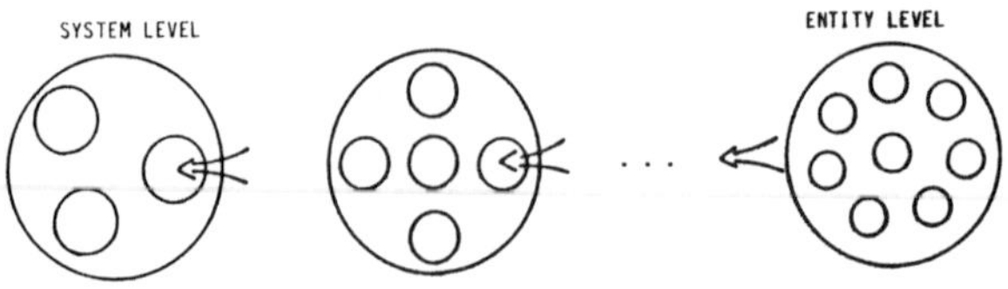

FIGURE 1 COORDINATION LEVELS.

The local controller makes all the necessary decisions for execution of a task. In the cases where interaction is required or a decision cannot be done at the local controller, a request for coordination or decision is sent to the higher coordinator. This coordination scheme enables the local controllers to work together to execute a set of complex tasks; it also allows the isolation or the take-over of a failed device, thus providing some tolerance in device or process failure. In addition, the coordination scheme allows for modular alteration of the manufacturing plant as production requirements change. This modularity is due to the fact that the coordinators are designed as simple processing devices working in an event driven manner.

In the oligarchical organization, the set of entities constitutes the base of the system. The control mechanism is organized by levels: there are $n>1$ levels of "control" and three "modes" of operation for each level.

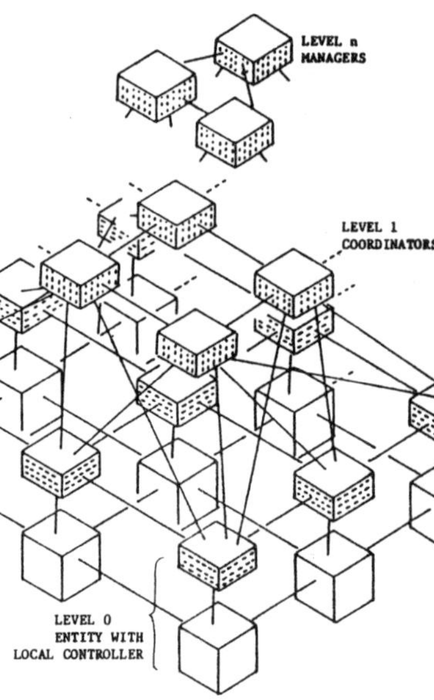

FIGURE 2 AN OLIGARCHICAL STRUCTURE

Each level i, $1 \le i \le n$, is composed of a set L(i) of coordinators; each coordinator $x \in L(i)$ is directly connected to several coordinators at level i-1 (in case i=1, the coordinators are connected to system entities), and to at least one coodinator at level i+1 (except in case i=n). An example of topology of an oligarchical system is shown in Figure 2.

Additional specifications of the topology of an oligarghical system are:

1. $|L(i)|<|L(i-1)|$; i.e. the number of coordinators is decreasing in the levels;

2. $|L(n)|\geq 1$; i.e. the top level can have more than one coordinator, called managers

3. each coordinator $x \in L(i)$ can communicate (either directly or indirectly) to all other coordinators in L(i);

4. each level is "globally connected" to the next levels; i.e. each coordinator in L(i) can communicate (either directly or via another coordinator in L(i)) to any coordinator in L(i+1), and vice versa.

The level of coordination as well as the size of the subsystem below it of which it is aware determines its control capability. Formally the view $v(x) \subseteq L(\emptyset)$ of a coordinator $x \in L(i)$ is defined as the set entities directly connected to x if i=1; otherwise, as the union of views of all coordinators $y \in L(i-1)$ directly connected to x; see Figure 3.

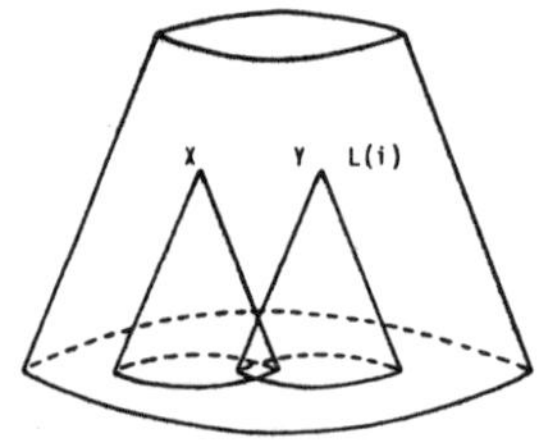

FIGURE 3 VIEWS OF COORDINATORS X AND Y AT LEVEL L(i).

There are three kinds of (control) communication primitives: requests, control signals (or commands), and transfers. Requests are sent to the higher level, control signals are sent to the lower level, and transfers are sent within the same level. Requests typically originate at the system entities level. Associated to each request there is a scope which specifies the degree of decisional capability needed to resolve. When a request is resolved at level i, appropriate commands are sent to the lower level as specified by the control function f(i). When a command is received at level i, the appropriate action is taken as specified by the processing function g(i). When an entity $E \in L(\emptyset)$ receives a command it will perform the appropriate processing activities.

Each level can operate in three different modes: hierarchical, decentralized, and failsafe. In the hierarchical mode, when a request p arrives at $x \in L(i)$, if $s(p) \subseteq v(x)$ (i.e., x has the control capability of resolving the request) it does so; otherwise, it sends the appropriate request to level i+1 for resolution. In a decentralized (or survival) mode, any request arriving from level i-1 can be resolved collectively by the coordinators in L(i) in a distributed fashion. In a failsafe mode, a coordinator triggers a shut-down mechanism. In other words, a coordinator at level i is capable of resolving immediately certain requests; for other requests, it can either defer the resolution to a higher level, or resolve it in a distributed fashion with the contribution of the other coordinators in L(i).

The decision on which mode of operation should be used is made according to the following rules:

1) coordinators at level operate in hierarchical mode unless there is a disconnection between levels i and i+1. Level disconnection can be defined to occur when there is a total disconnection (i.e., absence of any route between two levels) or just functional disconnection (eg., the average or maximum route length between the two levels exceeds a fixed "performance degree");

2) coordinators at level i enter a failsafe mode only when, in decentralized mode, a physical or functional disconnection is detected within the level.

Basic features of an oligarchical manufacturing system are reliability, fast response and low system complexity.

Reliability is achieved because, in case of failure of a coordinator x∈L(i), any coordinator y∈L(i-1) directly connected to x can still send its request to another coordinator z∈L(i) either directly or via some w∈L(i-1); recall, the coordinators in a given level can communicate with each other. Similarly, in case of direct line failure between y∈L(i-1) and x∈L(i), y can still send its requests to level i. If there is a disconnection between two levels, i and i+1, the coordinators in L(i) cannot send their requests to level i+1. In this case, they switch to a distributed mode of operation which allow them to resolve any request. Thus the system will still be operational albeit at a slower pace. Therefore, depending on the level and on the mode, fail-safe or fail-soft operation of the system can be achieved; the former at the lowest level of coordination, the latter at higher ones.

Since the number of levels in an oligarchical system can be kept small (10), the response time for a request resolution in asbence of failure is comparable to the one that would be experienced in a centralized system. In the case of serious failures, the response time is never greater (and usually much less) than in fully distributed systems; this is because the number of elements in a level (involved in the distributed resolution) never exceeds the number of entities (involved in the resolution in a fully distributed system).

Independently of their level, all coordinators are functionally equivalent in that they execute the same basic algorithm (only on different data); thus, they can be physically identical. Furthermore, by controlling the ratio |L(i)|/|L(i+1)| between successive levels, the total number of coordinators can be kept very small while still maintaining all the other features.

3. COORDINATOR IMPLEMENTATION

Functionally, the coordinators can be specified, as shown in Figure 4, by the interaction of the processes: executive process, task driver process, and intra-level communication process (11). The task driver process, upon receiving a 'routine select' command from the executive, executes the proper routine by either: 1) sending commands down to lower level coordinators, or 2) by sending control signals to the controlled entity. If proper acknowledgement is not received, a request for decision is sent to the executive process. In case of proper execution (acknowledgement), a completion message is returned to the executive process which selects the next routine. The intra-level communication process, upon receiving a routine select command from the executive, executes the proper communication routine. Routines can be handshake routines used in control interaction between coordinators of the same level. Also, routines can be either requests for distributed decision initiated by its executive, or another coordinator of the same level. As in the task driver process, it returns completion or decision requests to the executive process.

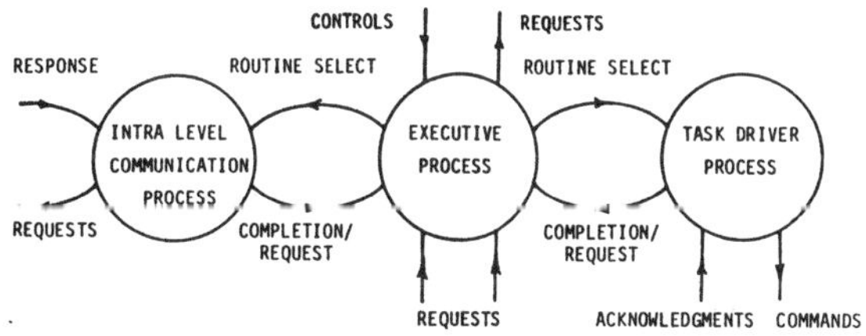

FIGURE 4 COORDINATOR PROCESSES

The executive process is responsible for the proper operation of the coordinator. Locally, it selects the proper routines to be executed by the task driver or the intra-level communication processes, and accepts completion signals and local decision requests. In the oligarchical mode, it also accepts decision requests from the lower level coordinators and either makes the decision locally, invoking the proper tasks or transmits the request upwards. Upon receiving a command (response) from and upper coordinator, it takes the proper action. If a disconnection is detected, for example by not receiving a response in the alloted time, it switches to decentralized mode. In the decentralized mode, it accepts requests from lower or intra-level coordinators, and makes the decision locally, or in a distributed fashion. In the case of a complete disconnection with no resolution, it initiates a proper shut-down sequence. Figure 5 is the flowchart representation of the operation of the coordinator.

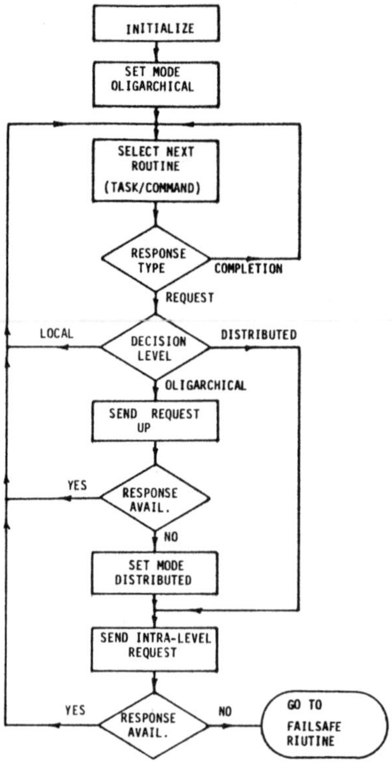

FIGURE 5 EXECUTIVE PROCESS FLOW CHART

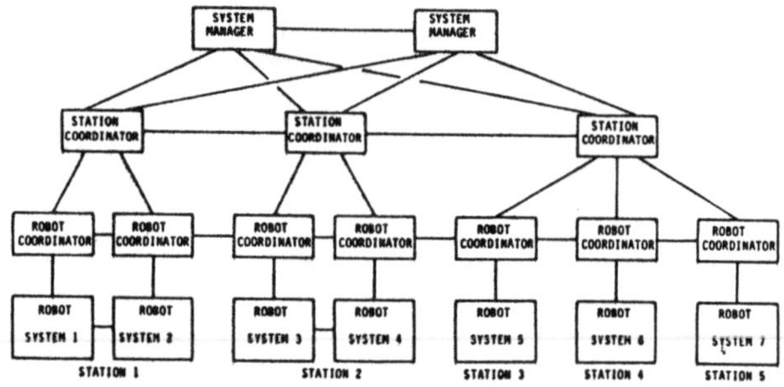

FIGURE 7 OLIGARCHICAL ORGANIZATION OF WESTINGHOUSES ADAPTABLE PROGRAMMABLE ASSEMBLY SYSTEM

As shown in Figure 7, this application has three levels of coordinators resulting in fast response, with ten coordinators, thus a simple system. Also, by including dedicated local processors for each robot allows for the use of simple processors, instead of a single but more complex processor.

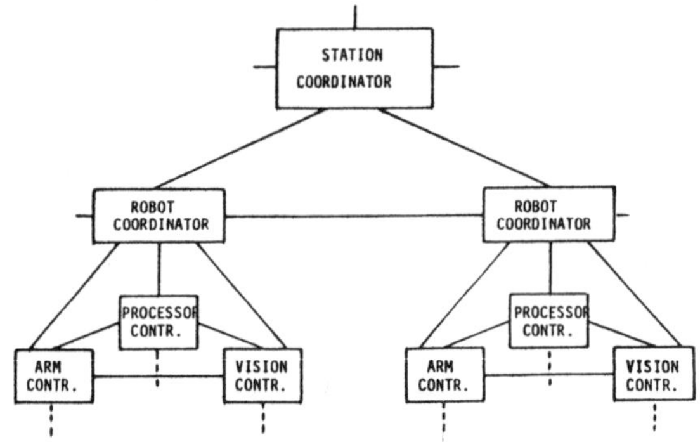

FIGURE 8 STATION ORGANIZATION

3. APPLICATION EXAMPLE

To demonstrate the applicability of the oligarchical scheme to FMS, the control of Westinghouse's Adaptable Programmable Assembly System (APAS) for small electric motor assembly using this scheme is outlined (12). The APAS consists of seven robots (manipulators with vision and sensory systems, and parts feeders) organized in five workstations on an assembly line. As shown in Figure 6, the first two stations each include two robots that execute their tasks interactively and the last three robots operate individually. Figure 7 indicates a possible mapping of the APAS for oligarchical control. As indicated in Figure 8, a processor is dedicated to every robot. This allows for local processing eliminating the need for large transfers to a central processor. The robot coodinator controls the interaction between the local processor, manipulator, vision and sensory system. At the next level are the station coordinators that control the interaction between stations. At the highest level are the assembly line coordinators that ensure proper system operation (11).

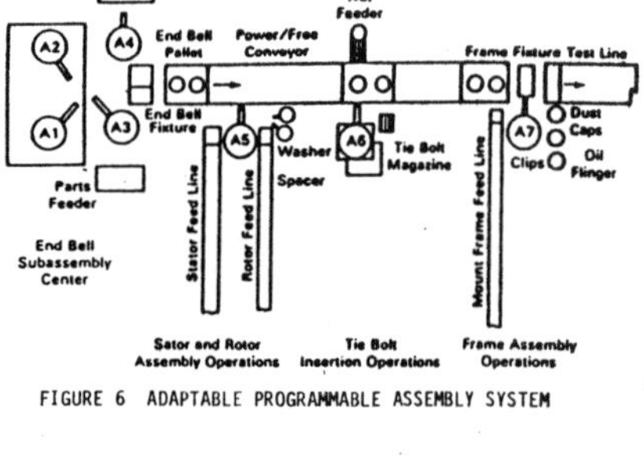

FIGURE 6 ADAPTABLE PROGRAMMABLE ASSEMBLY SYSTEM

5. CONCLUDING REMARKS

In this paper the concept of oligarchical control, a general scheme for controlling the activities of a (very) large number of interacting entities was presented. At the basis of the oligarchical concept lies the separation between the processing and coordination activities which are traditionally integrated under the ubiquitous term "control". One of the reasons for the exitsing situation has been the lack of clear distinction between entity-control amd system-control functions; while both relate to the control of entities, the former regulates the functionning of the entity as a unit, the latter manages the cooperation among entities. This lack of distinction, tollerable in small systems, becomes a major obstacle in large systems. The introduction at entity level of local controllers, which include all the necessary entity control primitives, permits separate resolution of the problems related to system control (i.e., coordination). In actual implementation, this separation, which lies at the basis of the oligarchical concept, has the immediate effect of reducing the amount of data transfers reducing, in turn, time delays.

There are several factors that must be taken into account when implementating an oligarchical control mechanism. For example, to increase the system performance the partitioning of the entities should be done according to their level of (expected) interaction; entities whose cooperation is often required (e.g., vision system and manipulator arm) should be assigned to the same coordinator. Accordingly, the partitioning of coordinators should reflect the degree of interaction of the entities within its view, Also, the overall system reliability can be greatly enhanced by connecting each system element to more than one coordinator at the next level; this obviously will increase the complexity of the coordinators and the number of connecting lines.

REFERENCES

(1) I.E.E.E. SPECTRUM; ' Special Issue on Data Driven Automation', May 1983, Vol.20, No.5.

(2) Groover, M.P.; 'Automated Production Systems and Computer-Aided Manufacturing',Prentice-Hall, 1980.

(3) Albus, J.S.; et al.;'Hierarchical Control of Robots Using Microcomputers', 9th Int. Symp. on Indus. Robots, Soc. of Manuf, Eng., 1979, pp405-422.

(4) Albus,J.S,;et al.;'A Measurement and Control Model For Adaptive Robots', 10th I.S.I.R., Soc. for Manuf. Eng., 1980, pp303-312.

(5) Albus,J.S.; et al.;'Hierarchical Control For Sensory Interactive Robots', 11th I.S.I.R., Soc. of Manuf. Eng., 1981,pp497-505.

(6) Albus,J.S.,'Brains, Behavior, and Robotics', North Quincy, Mass.: BYTE Publications Inc., 1981.

(7) Lefkowitz, I.; Buchner, M.R.; 'Distributed Computer Control for Industrial Process Systems: Characteristics, Attributes, and an Experimental Facility ' , I.E.E.E. Control Systems Magazine, Vol. 8, No. 2, March 1982, pp8-15.

(8) Mahmoud,M.S.;'Multilevel Systems Controland Applications: A Survey', I.E.E.E. Trans. on Sys., Man, and Cyber., Vol. SMC-7, No. 3, March 1977, pp125-143.

(9) Williams, T.J.; ' Hierarchical Control For Large Scale Systems - A Survey ', I.F.A.C. World Conference Proceedings, 1978.

(10) Krieger,M.; Santoro,N.;'Oligarchical Control of Distributed Processing Systems ', Proc. 20th Allerton Conf. on Communication, Control, and Computing, 1982,pp.793-802.

(11) Piche, P.R.;'Oligarchical Control Scheme for Flexible Manufacturing Systems',M.A.Sc.Elec. Eng. Thesis, University of Ottawa, Canada, (To appear fall, 1983).

(12) Abraham, R.G.;'A.P.A.S.: Adaptable Programmable Assembly System', Robots and Computer Vision, 1977, pp117-140.

A SYNTHESIS OF CONTROLLERS FOR SAMPLED-DATA SYSTEMS

J.ROZINOER

ECOLE CENTRALE DE LYON

69130 ECULLY - FRANCE

ABSTRACT

This paper deals with a procedure for synthesis in the z-plane of digital controllers. With the basic assumption of the existence of two complex dominant poles for the compensated system, it is shown that all the theoretical solutions corresponding to a given form of controller can be pointed out. The use of auxiliary curves reveals to be useful to insert constraints in the research of a convenient solution. The main feature of the procedure is to help the user to make a choice with the a priori knowledge of the consequences of the choice.

INTRODUCTION

The use of digital controllers becomes more and more widespread every day as for automation of industrial processes. In fact, the importance of such a way of control lies upon the introduction of computers in order to generate controls. For single input-single output linear systems, the synthesis of digital controllers has been studied for a long time ([1],[2],[3]), with z-or w-variable, and several methods have been pointed out. For example, a trial and error procedure can be employed; if the controlled system has to be achieved with two dominant poles, the Zdan method can furnish an acceptable solution; in the case of special temporal specifications, a procedure can be developed which implies a deadbeat response controller and finally, there are frequency methods. Except for the deadbeat response method which implies very accurate specifications, the other ways of determination of controllers present a certain lack of efficiency since it looks evident that the result depends on the quantity and quality of information about the system.

A different approach has been developed in a recent past ([4],[5],[6],[7],[8]) for analog controllers, based on the choice of a geometric parameter. The purpose of that method is to make easier, more rational and more direct the determination of the parameters of a controller by an evaluation a priori of the consequences of the choice over the system performances.

This paper deals with the application of that method above mentioned to the synthesis of digital controllers.

PART 1. STATEMENT OF THE PROBLEM.

Of consider here is a linear, invariant, single input-single output system described by its transfer function $G_p(s)$. In order to improve its characteristics in closed loop configuration, a digital controller with $D(z)$ as transfer function is introduced in the forward path (Fig.1)

An extrapolator (usually a Zero Order Hold) appears to transform the input of the plant into a continuous signal. The Z-transform of $G(s)$ is noted $G(z)$. The fundamental hypothesis is the use of the dominant poles method, i.e. we assume that transient response, only considered at the sampling times, approximates those of a second order system with two complex conjugate poles P_d and P_d' which must belong to the root locus of the closed loop system.

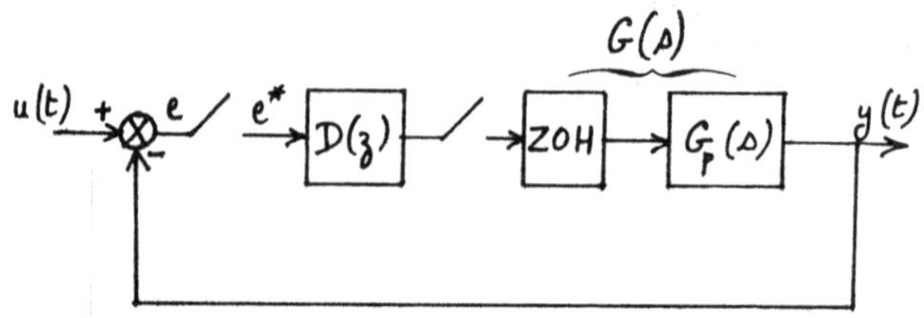

Figure 1.

These two complex poles can be chosen either in the left half plane, either in the right half plane but inside the unit circle (Fig.2).

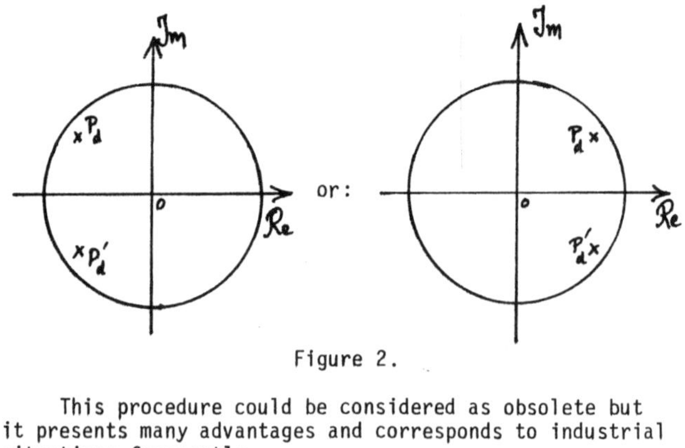

Figure 2.

This procedure could be considered as obsolete but it presents many advantages and corresponds to industrial situations frequently.

PART 2 - DESCRIPTION OF THE METHOD

From the knowledge of $G(z)$ and the dominant poles P_d and P_d' ($z \equiv z_d$), $|G(z_d)|$ and $\angle G(z_d)$ can easily be calculated, from which $|D(z_d)|$ and $\angle D(z_d)$ are deduced with the conditions for P_d to belong to the root locus of the controlled system. The sign of $\angle D(z_d)$ added to the steady-state constraint (nullity or limitation) induces the choice of a controller. Then, the problem of determination of the controller coefficients remains, which is equivalent to a pole and (or) zero assignment. For common controllers, the choice of one or two geometric parameters (in the z-plane) enables to find all the solutions by the means of plotting few auxiliary curves which explicit the controller coefficients vs. the geometric parameters.

Moreover, in the practical case of constraints for realisability, these constraints can be superimposed on the curves. Hence, the useful variation domain of the parameters is obtained, even when it is empty. The following organigram resumes the procedure (Fig.3):

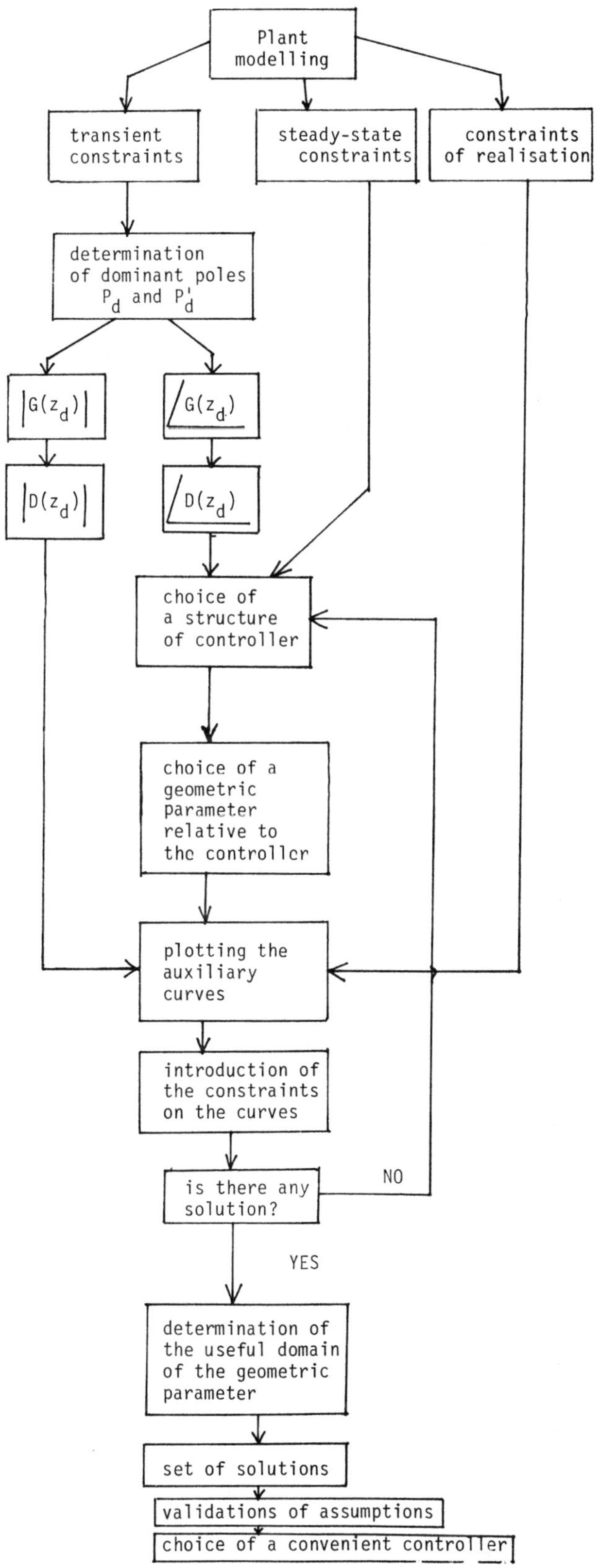

Figure 3.

PART 3 - EXAMPLE OF APPLICATION

From the defined position of the two comples dominant poles, $|G(z_d)|$ and $\angle G(z_d)$ are calculated. Let us assume that:

$$\begin{cases} |G(z_d)| \neq 1 \\ \angle G(z_d) \neq -\pi \end{cases} \qquad (1)$$

which means that P_d and P'_d do not belong to the root locus of the controlled system.

If there is no condition of nullity for the steady-state error, the transfer function of the controller can be taken such as:

$$D(z) = A\,\frac{z+a}{z+b} \qquad (2)$$

The relative assignment of zero and pole of D(z) depends on the value of $\angle G(z_d)$:

-if $\angle G(z_d) < -\pi$, the controller is of the lead type

-if $\angle G(z_d) > -\pi$, the controller is of the lag type

Let us assume to be in the first case. Then:

$$D(z) = A\,\frac{z+\frac{1}{\tau}}{z+\frac{\alpha}{\tau}} \qquad (3)$$

$$A > 0 \qquad \tau > 0 \qquad \alpha > 1$$

For P_d to be a complex pole of the compensated system, the parameters A, α, τ have to take the values:

$$A = A_d \qquad \alpha = \alpha_d \qquad \tau = \tau_d$$

such as:

$$|G(z_d)| \cdot |D(z_d)| = 1 \qquad (4)$$

$$\angle G(z_d) + \angle D(z_d) = -\pi \qquad (5)$$

or:

$$|G(z_d)|\; A_d \left| \frac{z_d+\frac{1}{\tau_d}}{z_d+\frac{\alpha_d}{\tau_d}} \right| = 1 \qquad (4')$$

$$\angle D(z_d) = -\pi - \angle G(z_d) = \phi_c \qquad (5')$$

The indetermination points out in these two equations with three unknown parameters.

A possible geometric description of the problem in the z-plane is the following (Fig.4). Since the dominant poles P_d and P'_d can be positioned either in the left half plane, either in the right half plane, but inside the unit circle, θ can vary from 0 to π.

Hence, it can be demonstrated that:

$$\alpha_d = \frac{\sin\beta}{\sin(\beta-\phi_c)} \cdot \frac{\sin(\beta-\phi_c+\theta)}{\sin(\beta+\theta)} \qquad (6)$$

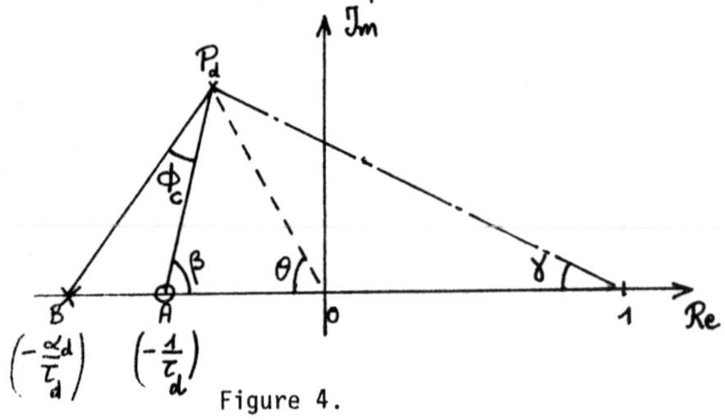

Figure 4.

Thus,the parameter α_d is considered as a function of β .All the α_d functions,defined over:

$$\beta \in \left]\phi_c, \pi-\theta\right[$$

have a minimum for:

$$\beta = \beta_0 = \frac{\pi}{2} + \frac{\phi_c}{2} - \frac{\theta}{2}$$

and are symmetrical with respect to β_0 (Fig.5).The corresponding curve is one of the auxiliary curves plotted to determine a convenient controller.

Another one is drawn from:

$$A_d \left|G(z_d)\right| = \frac{\sin\beta}{\sin(\beta-\phi_c)} \qquad (7)$$

and is also represented in Figure 5.

Finally,let us define the error constant of the uncompensated system by:

$$K_x = \lim_{z\to 1} (z-1)^x \, G(z)$$

(x:type of the system)

If the objective of the controller is to limit the steady-state error and not to annul it,the compensator does not modify the system type.The steady-state error constant of the compensated system is:

$$K'_x = A_d \frac{1+\frac{1}{\tau_d}}{1+\frac{\alpha_d}{\tau_d}} K_x \qquad (8)$$

Thus,it seems quite natural to deal with a relation between

$$A_d \frac{1+\frac{1}{\tau_d}}{1+\frac{\alpha_d}{\tau_d}}$$

and β.

$$A_d \frac{1+\frac{1}{\tau_d}}{1+\frac{\alpha_d}{\tau_d}} \left|G(z_d)\right| = \frac{\sin(\beta+\gamma)}{\sin(\beta+\gamma-\phi_c)} \qquad (9)$$

An example of representative curve appears in Fig.5.

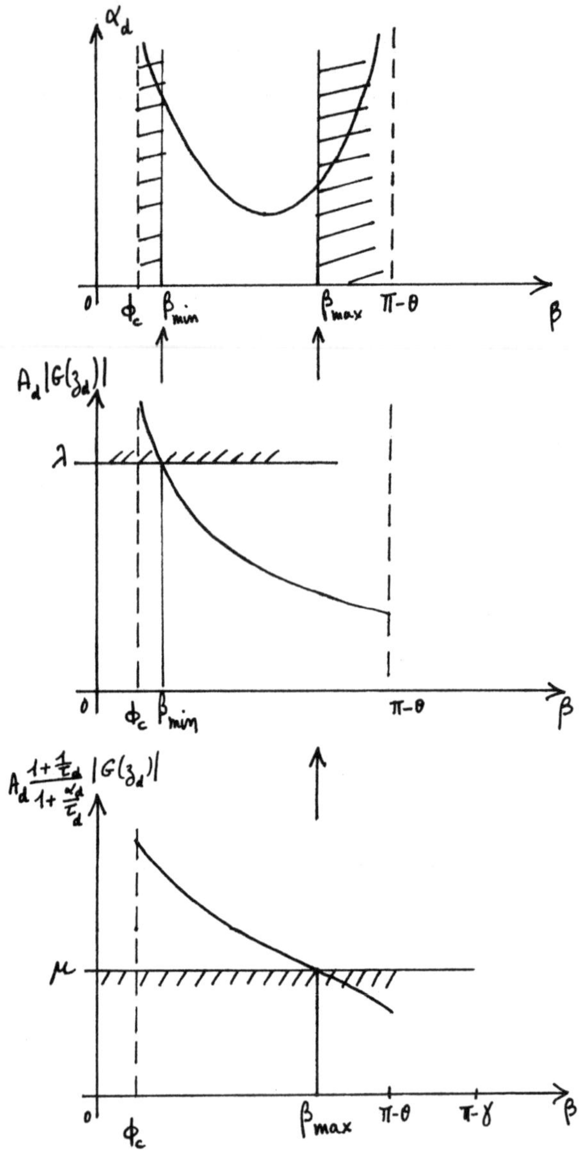

Figure 5.

Practically,if there is an upper bound to A_d,the useful part of the corresponding curve is restrained by:

$$\lambda = A_{dmax} \left|G(z_d)\right|$$

If the steady-state error must be reduced by a factor $\frac{1}{\mu}$,the steady-state error constant has to be multiplied by a coefficient greater than μ which reduces the usable part of the curve.

Then,the design of the compensator appears as a selection of a value of β satisfying the constraints:

$$A_d \left|G(z_d)\right| < \lambda \qquad (10)$$

$$A_d \frac{1+\frac{1}{\tau_d}}{1+\frac{\alpha_d}{\tau_d}} \left|G(z_d)\right| > \mu \qquad (11)$$

So,we are enabled to conclude that the set of possible controllers is characterized by the following set of values of β :

$$\beta_{min} < \beta < \beta_{max}$$

CONCLUSION.

A methodology has been presented for the synthesis of digital controllers.The objective is to make the user free of choosing a controller which achieves all the constraints by the knowledge of all the solutions.With the use of auxiliary curves in relation with the parameters of the compensator,it is possible to introduce directly the constraints.An example illustrated the general organigram in the case of the determination of a first order controller.

BIBLIOGRAPHY.

[1] D.P.LINDORFF "Theory of sampled-data control systems" Wiley.1965

[2] Y.SEVELY "Systèmes et asservissements linéaires échantillonnés" Dunod 1973

[3] H.BUHLER "Réglages échantillonnés" Presses polytechniques romandes 1982

[4] J.ROZINOER;G.THOMAS "A global approach to the design of lead controllers by the root locus method" Proc. of the 3rd Brazilian Congress of automatics 1980 pp.25-30

[5] J.ROZINOER;G.THOMAS "Une méthode de détermination paramétrique des correcteurs à avance et retard de phase" Intern. Symposium IASTED AMS'82 .Paris 1982

[6] J.ROZINOER;G.THOMAS "Une approche paramétrique de la correction des systèmes asservis linéaires" Le Point en Automatique Chap.15,pp.287-304,Ed.Lavoisier 1981

[7] J.ROZINOER;G.THOMAS "Ageometric design of PID controller " Control and computers Vol.10,n°3,1982, pp.67-70

[8] J.ROZINOER;G.THOMAS "Une nouvelle méthodologie pour la détermination des correcteurs PID" Proc.of the Intern.Symposium AMSE ,Gif-sur-Yvette, 1982,pp.28-31,Vol.S1

A Prototype robot workstation for the motif application process within the garment manufacturing industry.

D.R.Kemp, P.M.Taylor, G.E.Taylor
Department of Electronic Engineering
University of Hull
HU6 7RX
England

Abstract.

The garment manufacturing industry is potentially a fertile area for robot applications. During the manufacture of a garment about half of an operative's time is spent moving material from one place to another. Before further automation can take place the basic problem of fabric handling must be tackled, and it can be expected that sensory feedback to the handling device will be required for reliable operation. The motif application process was chosen for study because it contains both handling and sensory problems.

A prototype workstation has been constructed which comprises two parts. A vision system is used to correctly orientate motifs whilst simultaneously a robot equipped with a novel gripper is used to separate a single ply of fabric from a stack of cut pieces. The motif is placed in the correct position on the fabric for presentation to the bonding machine. After bonding the garment piece with a motif fused onto it is removed and stacked. This paper will describe the prototype workstation and discuss further developments.

Section 1. Introduction.

Fabrics are inherently awkward to handle being non homogeneous and behaving in a nonlinear and live manner. The garment manufacturing industry therefore provides a difficult but challenging area for research into robot applications (Reference 1). and so a collaborative project, funded by the United Kingdom Science and Engineering Research Council robots iniative, is being undertaken by the University of Hull Robotics Research Group, Marks and Spencer plc (a large British retail company) and Corah plc (one of their suppliers).

The motif application task was chosen, not for its importance within the industry, but because it contains the need for fabric handling and visual feedback which are seen as essential ingredients of further automation within the industry.

The task is to pick up an individual piece of fabric from a stack using a specialized sensory gripper (Reference 2). The gripper uses an air jet to vibrate the uppermost piece on a stack of cloth and this vibrating piece is detected by the gripper sensors and then separated from the stack. The material is then laid out and on to it is placed a motif in the correct position and orientation. A simple low resolution vision system (Reference 3) and a controlled rotation table are used to orientate the motif. The motif is a separately manufactured item with special backing material. The two items, motif and cloth, are then to be joined together under high temperature and presure in a special purpose bonding machine. The material must then be removed and stacked.

A positional accuracy of about +/-1mm and a rotational accuracy of +/-1 degree is required together with a cycle time for the complete operation of the order of one minute.

This paper describes the prototype workstation in its layout and operation. In section 2 the layout of the system components and the interconnections between them are detailed and in section 3 the system operation under normal circumstances is then described and further to this how the system deals with errors. The later sections assess the performance of the system and discuss improvements that could be made.

Section 2. Layout of the system components.

It should be stressed at this point that this is a prototype laboratory system and the layout and interconnection of the system components has been dictated by the need to have fast and simple interchange of some of the system components which are shared with other projects within the research group.

The robot system (Figure 1) consists of a Unimate PUMA 560 robot equiped with a special sensory gripper for handling single pieces of material and for picking motifs. In the area in front of the robot there is space for a stack of material and for the robot to separate and lift a single piece of material and lay it down.

A vision system, camera, rotation table and motif bin form a module located to one side of the robot. The camera is mounted above the rotation table to monitor the orientation of the motif placed on the rotation table which , under the control of the vision system, can be steped in increments of 0.375 degrees.

The workstation.

The workstation consists of two discrete subsystems one being the robot system the other the vision system and task controller (Figure 2).

The robot subsystem consists of the robot, robot controller, air jet and vacuum control valves, gripper sensors and sensor interface. The robot controller is connected to the control valves and sensor interface via the robot input-output signal lines which can pass information to and from VAL programs.

The Vision system and task controller.

This subsystem is controlled by a DEC LSI 11/2 microcomputer which perfoms the vision processing and task control operations. The LSI 11/2 is connected to an image digitizer and robot bus interface , Robus (Reference 4), via a DRV11 16 bit parallel interface card. The microcomputer can address either the image digitizer or the Robus interface and the Robus interface can in turn address many other devices. There are currently two devices connected to the Robus, the rotation table controller and a comunications port connected to the signal lines of the robot controller.

Section 3. Operation of the workstation.

The workstation functions using a master slave protocol. The system master, which is in all cases the vision system, passes job numbers to the robot slave. The robot interprets the job number as a "task" for

example : " Get a motif from the motif bin and place it on the rotation table ". When the task has been completed the robot signals this by sending the system master an acknowledge signal. Similarly the system master can command the rotation table to move to a certain angle at a specified speed rotating either clockwise or aniclockwise.
If no errors are detected the normal operation of the workstation would be as follows. The vision system sends the robot a job number so that the robot will get a motif from the motif bin and place it on to the rotation table. When the robot has completed this task the vision system checks the motif and then instructs the robot to separate a piece of material from the stack and lay it out, while simultaneously the vision system orientates the motif by commands to the rotation table. When the robot has completed its task it sets an acknowledge signal. When the vision system has orientated the motif it waits for this acknowledge signal and then instructs the robot to acept the coordinates of the motif on the rotation table and use them to pick the motif and place it onto the piece of material. The vision system then waits for the robot to finish. This is the point at which the end of the cycle is currently defined.
The system has to deal with various errors these are either generated within the robot system or within the vision system.
The robot system deals with errors internally. Recovery from errors can be accomplished by retrying the particular part of the task in which the error was found. Errors which cannot be corrected by the robot system cause the current task to be terminated and human operator intervention is required. In this case the robot does not send an acknowledge signal to the vision system and thus the vision system after completing its task waits for the robot to be restarted.
The hardware and software organisation of the system allows the robot to inform the vision system of an error if necessary.

The vision system can detect and recover from known errors by a robot task specific to the particular error. On detection of an error the error number is sent as a job number to the robot, along with any other relevant information, so that the robot can interpret the job number as a VAL subroutine task. A list of the vision system errors and their appropriate recovery tasks is given below. The first five errors are found by a simple check routine within the vision system. This is carried out before orientation of the motif is attempted and before the robot system is given any other task. The last error is more serious because of the time taken to recover from it.

1. error number 11.
Motif out of field of view.
This error is detected when the motif is found at the extreme edges of the field of view and is thus is in danger of moving out of the field of view. This is caused by the motif falling from the motif picker in such a way that the motif is immediately found to be out of the field of view or such that the motif moves out of the field view when the rotation table is operated. The robot job 11 takes the coordinates of the centre of area (COA) of the remaining part of the motif in the field of view from the vision system and converts them to robot coordinates of the motif on the rotation table. The robot can then pick up the motif and move it to a point at the centre of the rotation table and release the motif.

2. error number 12.
Motif area too small.
This can be caused by a damaged, deformed or incorrect motif being placed under the camera.
Job number 12 causes the robot to take the COA coordinates from the vision system convert them to robot coordinates, as above, and then pick up the motif and move it to a pile of rejects. It then fetches another motif and places it on the rotation table.

3. error number 13.
Too many motifs in field of view.
This error is caused by the motif picker collecting two or more motifs from the motif bin.
Job 13 instructs the robot to clear the rotation table by using the COA coordinates from the vision system. The robot may clear one motif or two motifs in one clear operation and thus the table is checked after each clearing operation to see what, if anything, remains.

4. error number 14.
Motif backing side uppermost.
This error is caused by the motif picker collecting the motif from the motif bin backing side uppermost.
Job 14 consists of the robot using the COA coordinates to clear the backing side up motif from the rotation table to place the motif on a predefined reject stack.

5. error number 15.
No motif in view.
The motif picker is non sensory thus it cannot detect a motif on the vacuum picker and may not deliver a motif to the rotation table. This error has also been generated when two motifs have been cleared simultaneously from the rotation table. This error causes the robot to fetch another motif from the motif bin.

6. error number 13.
Failure to orientate the motif.
This case arises when the vision system has commanded the rotation table to rotate the motif by a total of 360 degrees and not found the correct orientation. This can arise when two motifs are placed on to the rotation table by the robot, one of the motifs being the correct side uppermost and the other motif partially covering the first and backing side uppermost. The resultant image area of this combination is within the limits defined for the vision sytem and thus the vision system attempts to orientate an object which cannot be orientated.
Note that the job error number/job number is the same as that in case 3 and the same job is performed to recover from this error.

Section 4. System performance.

The cycle time of the truncated process as defined above is approximately 1 minute. However the essential elements of the process are within the desired specifications. The vision system software is capable of achieving orientation within 12.5 seconds. The robot system can achieve fabric separation, that is detecting a piece of material between the gripper jaws, in 4.5 seconds. The large cycle time is caused by the slow speed of the robot movement when used with the fabric gripper and by the vision system being slow at orientating the motifs.
It should be noted that the gripper is 450mm long with a significant proportion of the gripper mass being located at the end furthest away from the gripper actuator mechanism and bearing surfaces. Thus speeding up the robot movement could damage the gripper actuator in its present form. A new design of gripper and actuator is needed to overcome the robot speed problem. The speed of the rotation table limits the speed of the vision system in orientating the motifs a new or modified type of rotation table would be needed in an industrial system.
The vision system sometimes fails to detect two motifs on the rotation table because of one motif being the correct side uppermost and the other being backing side uppermost and covering the first. It is proposed to include more visual recognition parameters within the system so that the error can be detected before the system attempts to orientate the motif.
At a later date the system may be equiped with a simple device to assist the robot in turning a backing side up

motif over to its embroidery side.
The motif picker is non sensory thus it cannot detect a motif on the vacuum picker and may not deliver a motif to the rotation table. This problem is simply solved by detecting the change in vacuum presure when a motif is attached to the motif picker.
Further experiments are in hand to obtain sensory information about the material whilst it is vibrating under the air jet the aim being to speed up the separation process.

Section 5. Conclusion.

The prototype system can separate a single piece of material from a stack and lay it out while simultaneously orientating a motif so that it is ready to be applied to the material. The cycle time of the process up to and including the placement of the motif on the material but not the bonding process is 55 seconds. The system can currently separate many types of materials from stacks when the correct parameters have been programed in and it may be possible by adding more sensory input to make the system adapt to different types of material. The system, excluding the robot, is not expensive to implement since wherever possible simple techniques, such as low resolution vision, have been used.

References

1. Kurt Salmon Associates, Altringham Cheshire England. " The 1980's : The decade for technology ? " Prepared for The commission of the European Economic Community. 1979.

2. D.R.Kemp, G.E.Taylor, A.Pugh. "A sensory gripper for handling textiles" 13 th Int Sym on industrial robots, Chicago USA, April 1982.

3. G.E.Taylor, D.R.Kemp, P.M.Taylor, A.Pugh "Vision applied to the orientation of embroidered motifs in the textile industry" 2 nd Int Conf on robot vision and sensory controls Stuttgart Germany. 1982.

4. P.M.Taylor, C.A.Stubbings "Software and hardware aspects of a flexible workstation for assembly tasks using sensory robots" 2 nd IASTED International symposium on industrial robotics and automation. Lugano, Switzerland, June 1983.

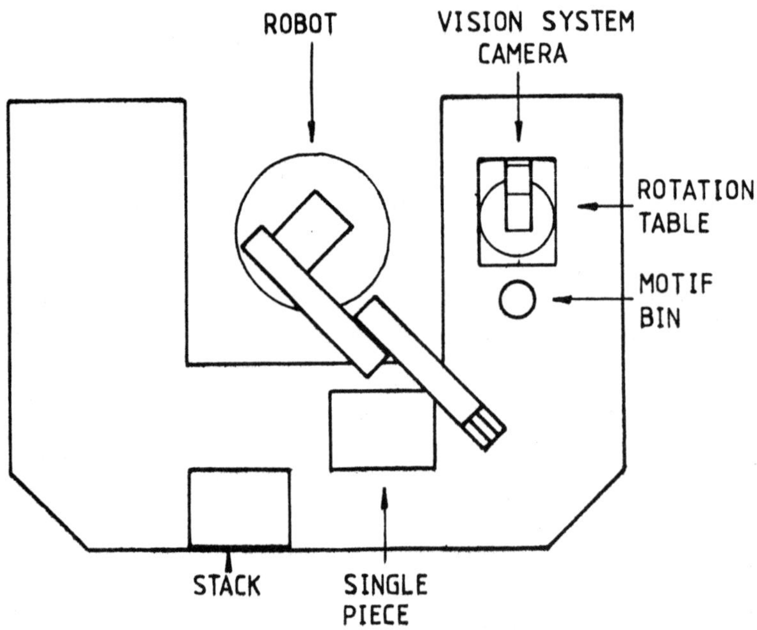

FIGURE 1
LAYOUT OF THE WORKSTATION

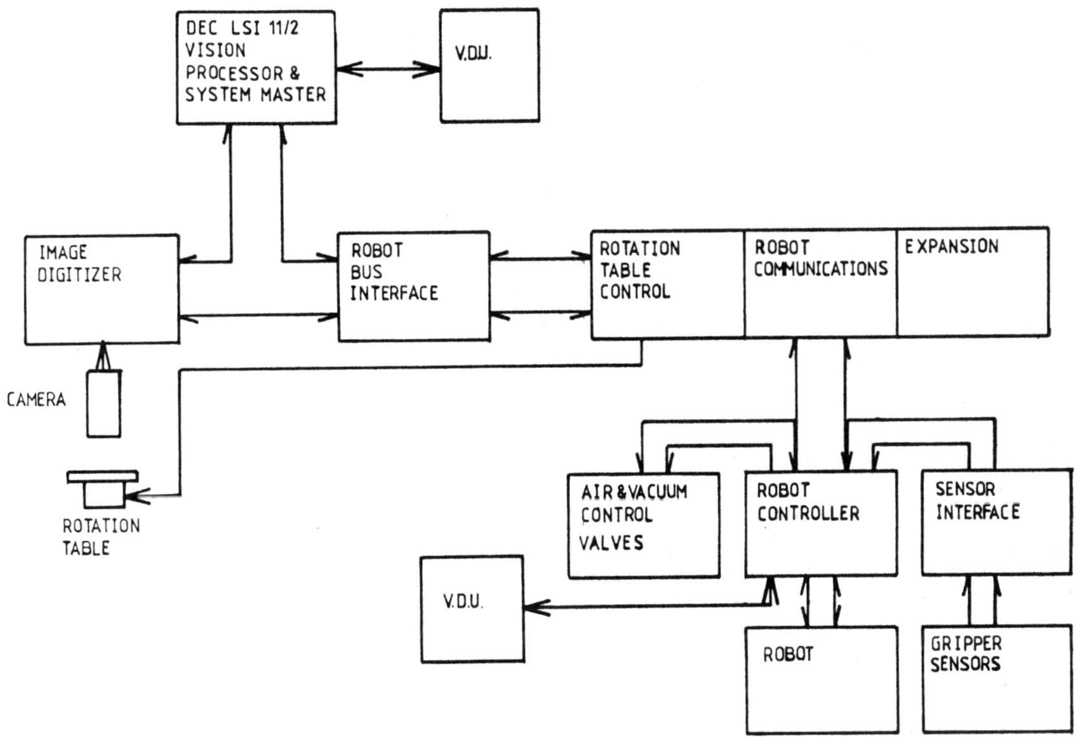

FIGURE 2 WORKSTATION SCHEMATIC

Software and Hardware Aspects of a Flexible Workstation for Assembly Tasks Using Sensory Robots

P. M. Taylor, C. A. Stubbings,
Department of Electronic Engineering,
University of Hull,
Cottingham Road,
HULL. HU6 7RX.
United Kingdom.

Abstract

If robots are to be used extensivly in industry for 'small' batch assembly tasks it is desirable that they form part of a flexible workstation which can be reconfigured and reprogrammed as simply and as quickly as possible when the assembly task is changed or the workstation hardware is modified.

Such a system is currently being implemented within the robotics group at Hull. The assembly task is defined in a language implemented in a supervisory processor which controls the whole assembly process, initiating requests for robot movements and sensory processing and defining communication paths between devices. Closed loop sensory feedback and multi-tasking are allowed. The robot and sensory devices are connected together using a bus system and the real time interdevice communication is controlled from a dedicated bus controller acting under commands from the bus supervisor. This bus, its controller and the interface modules which plug into it form a well defined communications system, and the use of a different robot or sensor requires at worst either the use of a new interface module or the reprogramming of an existing module.

This paper will describe the hardware and software aspects of the system together with a simple example of its use.

1: Introduction

It is predicted [1] that there will be a trend in industry towards the use of robots to solve assembly problems that require sensory feedback. This aim is difficult to achieve mainly because there is very little compatability between different robots and sensors which makes interconnection and programming both difficult and time consuming.

Various solutions or partial solutions have been proposed and these can be divided up into three main sections

1. Commercially available systems - The main contender here is the PUMA and VAL [2] with the promise of VAL2. VAL makes simple programming easy but has little structure. Sophisticated sensory feedback using a variety of sensors is not possible except in cases such as the SRI vision package [3] which has been grafted into VAL.

2. In house development systems - A number of different systems are being used and developed for in house research and development work. These are not generally available. (eg. Philips [4], GEC [5], Macdonnell Douglas [18], IBM [19] etc.)

3. Research systems - Various high level systems are under development. Generally availability is limited or they are tied to particular hardware or both. Also, in many cases there is a requirement for powerful and expensive computer resources (over £100K). [6-19]

At the University of Hull there are a number of research projects on the use of robots for sensory assembly. [20-24] These projects have identified a need to be able to program very different tasks using a wide variety of sensory and robot hardware. None of the above systems was both available and met the project requirements so a flexible workstation was designed based on the following requirements:

1. The hardware should be low cost, modular, easily modified and reconfigured.

2. There should be simple yet powerful communications between workstation modules.

3. The assembly task should be defined by a language that is easy to read and understand.

4. The system should ultimately be able to communicate with a higher level computer system used for design.

2: System hardware

In a sensory robot system data flow varies considerably from one part of the system to another. Obvious examples of this are vision systems and robot controllers. In both cases any high speed processing would be done in a dedicated processor which would return relatively simple results to the commanding processor. Hence the mechanics of the vision processing and robot dynamics are hidden to the commanding processor which only needs to send a simple command to the interface and wait for any reply.

This leads to a heirarchical structure (fig 1) The heirarchical structure gives the paralellism required to deal with the sensory data in real time and also makes closed loop and multi-tasking at the robot level possible because each processor is delegating tasks to those below and hence is sending and receiving the minimum amount of data necessary. This also helps in forcing a common and concise data format for communication between the processors.

As an example of data flow in a simple operation consider using closed loop feedback with a gripper mounted camera trying to position the gripper over a hole. (fig 2) The camera controller accepts data from the camera at up to video rate, the camera controller uses this data to calculate a centre of area and hence the required displacement of the robot. This is then passed to the robot at between 10 and 100 samples per second. The robot controller then expands this data and tries to servo the robot to the required point doing considerable calculation to translate the world coordinates to joint coordinates. At completion the task control processor is informed that the move is complete, perhaps after 2 or more seconds. The data rate has varied from about one event in 2 seconds at the top of the heirarchy to 1M samples per second in the video processor. The inter-device communication at 100 samples each of perhaps 10 bytes requires a data rate of about 1K bytes per second. In the future it is likely that greater than 100 samples per second will be required and it will also be necessary to have more than one closed loop operation active at a time.

To implement this system requires some form of interconnection structure. Three main options exist.

1. A commercially available bus
2. A network
3. An in house bus

The design chosen was an in house structure on the following grounds.

1. Commercial buses were not considered suitable since the requirement is for a communications bus with each device having its own independent resources. Using a commercial bus such as Multibus with commercial circuit cards would place considerable restrictions on those cards that could be used. Unless special cards were to be designed this would restrict the facilities available.
2. When this design was started network tecknology was not readily available and is still going to be quite an expensive approach for a few years.
3. An in house design gives freedom to specify the data and control signals as best suit the requirements giving flexibility at low cost with mimimal hardware.

The University of Hull design, called Robus, is a simple robot-sensor communication bus. In normal use data is transfered with relatively long gaps between bytes since the controller and interfaces restrict the data rate. Normal transfers occur at about 100K bytes per second usually in bursts. This easilly meets the requirements for data transfer given above. However the bus is capable of transfers up to 1M bytes per second if a faster controller is used.

An eight bit data highway is used, capable of talking to a maximum of 255 devices, each device having up to 256 common memory locations accessible to the master processor. [26]

The task supervisor unit (TSU) is a 16 bit standalone DEC SBC 11/21 and is used as the main language processor. It deals with interpretation of the language and the control of the rest of the processors in the heirachy. At the moment the Task supervisor is being supported from a DEC LSI 11/23 multi-user system but any convenient form of backing store could be used if there was no host available. The Task supervisor, having decided on the action required, sends commands to the Task Supervisor Interface. This processor is an INTEL 8085 and serves primarily as a smart buffer but it is intended to also use it as a switching node to distribute commands and control between communications control units (CCU) in a multi CCU system. The Communications Control processor is the main inter device communications processor. It deals with all the real time requirements of the connected robots and sensors and controls all local inter-device communication.

Since robotic devices are so varied it was felt that the easiest way to interface them was to have a simple standard interface that would make all devices look the same to the controlling processor. The method chosen was to fit an interface processor to each device which would then be completely responsible for all comunication to and from that device. [25]

3: System Software

The software can be divided into three main sections. Firstly the main languagee interpreter, secondly the Task supervisor interface and communications control processor software and finaly the software for the individual device interfaces.

1. Main language interpreter

   The main language interpreter is written to make programming as easy and flexible as possible. (an example is given in section 4) For this reason commands have been divided into two groups, firstly system commands that are always known to the system (ie. Print, If, Then etc.) and secondly the device specific commands (ie. Open, Close etc). This allows the device commands to be defined and added to at run time.

   Each device command is valid for a particular device or group of devices. It is possible to have the same command meaning totally different things to diferent devices. All device commands are stored in a table together with the parameters required to be sent to the device and data on those parameters that will be returned. This makes it possible for the system to to be largely self configuring. However for the system to self configure it must be able to get the data about the device from somewhere. There are two options. Either the data is available from the device itself or else it is given in the program.

   At startup the system first looks at all devices and tests their status to see how much data they can supply about themselves. Part of the protocol specification for Robus includes the facility for this data.

2. Task Supervisor Interface and Communications Control Processor

   The Task supervisor interface and Communications Controller software are both written in assembler. Of the two the Task Supervisor Interface is rather simpler. The main requirements are to accept command packets from the TSU and send them to the correct CCU and also to take data packets from one CCU and send them to another.

   The CCU is more complex in that it is required to deal with real time ROBUS communication. Data packets arrive at the CCU from either the TSU or another CCU via the Task Supervisor Interface. Each packet contains a number of commands for the CCU to execute and details of parameters to expect returned. The CCU then deals with those commands which can include closed loop moves and co-ordinate transforms and returns any requested data values to the TSU or to another CCU via the Task Supervisor Interface.

3. Device interfaces

The devices are very varied but usually the software is relativly simple and much is common between one unit and the next. it is possible to make 'smart' interfaces which could perform some simple sensory processing or coordinate transforms.

4: Example of Use

To use the system to implement a sensory assembly task the following steps are required

1. Define task
2. Choose hardware
3. Plug hardware into system. If new hardware it will also be necessary to configure an interface card first.
4. Write definition files for any new devices
5. Write assembly program (offline if required)
6. Run program
7. Online debug

Simple example

This simple example uses a single device to illustrate the general structure of the language. Note that some of the functions used are still at a crude state of development.

This example uses a simple sensor (in this case an LED array) to control the movement of an indexing (X-Y) table. The LED array comprises four LED's mounted in a square to with simple electronics to give UP/DOWN/LEFT/RIGHT/FOUND signals depending on its position over the hole in a block on the indexing table. These five simple commands return a TRUE/FALSE status to the interpreter and are a simple way of solving the problem. The code below shows how these commands can be used to centre the hole in a block on the indexing table. The TASK shown would be a part of a program which would also need to define the devices used and the five device commands above.

```
o
o Definition of the XYTABLE
o    device commands - SEND
o
o definition of the LED_ARRAY
o    device commands - FOUND
o                    - UP, DOWN
o                    - LEFT, RIGHT
o

TASK FIND_HOLE {

   WHILE NOT FOUND HOLE WITH LED_ARRAY {

      IF LEFT LED_ARRAY THEN SEND LEFT TO XYTABLE;
         ELSE IF RIGHT LED_ARRAY THEN
                  SEND RIGHT TO XYTABLE;
      IF UP LED_ARRAY THEN SEND UP TO XYTABLE;
         ELSE IF DOWN LED_ARRAY THEN
                  SEND DOWN TO XYTABLE

   }
}
```

This is not a particularly elegant solution to the problem as closed loop control is acheived at a relatively low rate via the Task Supervisor. It is assumed that the hole is on the XYTABLE and that the frames of reference of the XYTABLE agree with those of the sensor.

This problem could be solved in a different way by implementing closed loop control directly between the devices on Robus. The above program could then become:

```
o
o Device definitions
o Device commands
o
  CENTRE LED_ARRAY ON HOLE
```

This requires the system to know what the hole is, and where it is, how to generate movement between the hole and the sensor and the relative frames of reference of the hole and sensor. It is also necessary to know when to stop trying to centre the hole.

5: Conclusions

The system described is still in an early stage of development. Most facilities work but need considerable refinement. It is intended to have a fully operational system including data typing, structured commands, co-ordinate transforms to deal with device alignment, and multi communications controller software in the next 12 months. Additional enhancements beyond that include a compiling option, and a full online screen editor. An editor is essential if the whole system is not being run from a suitable host computer and it is felt that a good editor will improve the user friendliness of the whole system. It also makes online debuging practical.

References and Bibliography

1. 'Industrial Robots', A Report Presented by Ingersoll Engineers, Vol 1: Industrial Appreciation, DoI, National Engineering Laboratory.
2. 'User's Guide to VAL - A Robot Programming and Control System', Unimation Inc., Version 12, June 1980.
3. 'Unimation - SRI Vision System User's Manual', Unimation Inc, Shelter Rock La., Danbury, CT, USA, 10th March 1982.
4. Saraga P., Burnett D.J., Jones B.M., 'ROBOS - Towards a Robot Operating System', IEE Colloquium on Robot Operating Systems, London, 24th May 1983.
5. Taylor P.W., 'A Procedural Operating System', Abstract, IEE Coloquium on Robot Operating Systems, London, 24th May 1983.
6. Salmon M., 'SIGLA - THe Olivetti Sigma Robot Programming Language', Proc. Eigth Int. Symp. on Industrial Robots, Stuttgart, May 1978, p 358-363
7. Finkel R., Taylor R., Bolles R., Paul R. and Feldman J., 'An Overview of AL, a Programming System for Automation', Proc. of 4th IJCAI, Tbilisi, USSR, pp 733-735, 1975.
8. Goldman R., 'Recent Work with the AL System', Proc of 5th IJCAI, Boston ,Mass, 1977.
9. Mujitaba S. and Goldman R., 'AL User's Manual', Stanford Artificial Intelligence Laboratory, Memo AIM-323, Stanford University, January 1979.

10. M Mujitaba S., 'Current Status of the AL Manipulation Programming Language', Proc of 10th ISIR, Milan, pp 118-127, March 1980.

11. Lieberman L.I., Wesley M.A., 'Autopass: An Automatic Programmigng System for Computer Controlled Mechanical Assembly', IBM Journal of Research and Development, Vol. 21, No 4, pp 321-333, July 1977.

12. Lozano-Perez T., Winston P.H., 'LAMA: A Language for Automatic Mechanical Assembly', Proc of 5th IJCAI, Boston, Mass, pp 710-716, 1977.

13. Latombe J.C., Mazer E., 'LM: A High-level Programming Language for Controlling Assembly Robots', Paper presented at Tripartite Robotics Seminar, Stuttgart, Feburary 1981.

14. Popplestone R.J., Ambler A.P., Bellos I.M., 'RAPT: A Language for Describing Assemblies', The Industrial Robot, Vol 5, No 3, September 1978.

15. Popplestone R.J., Ambler A.P., Bellos I.M., 'An Efficient and Portable Implementation of RAPT', Proc of first Conference on Assembly Automation, IFS Ltd, 1981.

16. 'Short Reference of the ROBEX Language, a Language for Programming NC-Robots', Laboratorium fuer Werkzuegmaschinen un Betriebslehre, Aachen, West Germany.

17. Weck M., Eversheim W., Zuhlke D., 'ROBEX - An Off-line Programming System for Industrial Robots', Laboratorium fuer Werkzuegmaschinen un Betriebslehre, Aachen, West Germany.

18. Oldroyd A., 'MCL: An APT Approach to Robotic Manufacture', SHARE 56, Mar, 1981.

19. Darringer J.A., Blasgen M.W., 'MAPLE: A high level language for research into mechanical assembly', IBM Research report RC 5606, IBM T.J. Watson Research Center, Yorktown Heights, N. Y. Sept 1975.

20. Taylor P.M., Taylor G.E., Kemp D.R., Soim J., Pugh A., 'Software and Hardware Aspects of a Sensory Gripping System', 11th Int. Conf. on Industrial Robots, Tokyo, Japan, 1981.

21. Kemp D.R, Taylor G.E, Pugh A,"A Sensory Gripper for Handling Textiles", 13th Int. Symp. on Industrial Robots, Chicago, USA, April 1983.

22. Selke K.K.W., 'MEDL Thyristor Assembly Using Low Resolution Camera for Sensory Feedback', Internal Report, Dept Electronic Engineering, University of Hull.

23. Mitchell I., 'Progress Report on the Hotpoint Research Project', Internal Report, Dept. Elec. Eng., University of Hull, Jan 1983.

24. Burgess D., Hill J.J., Pugh A.,'Vision processing for Robotic Inspection and Assembly', SPIE Proc. on Robot and Industrial Inspection, Vol 360, 1982.

25. Mitchell I, Whitehead D.G, Pugh A, 'A Multiprocessor System for Sensory Robotic Assembly', Sensor Review, April 1983.

26. Stubbings C.A., 'Robus - A Cheap Multiprocessor Robot and Sensor Control Bus', Internal Report, Dept. Elec. Eng. University of Hull, May 1983.

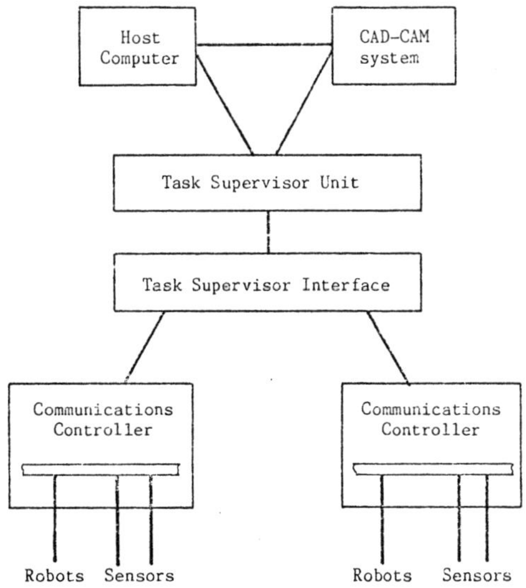

Figure 1

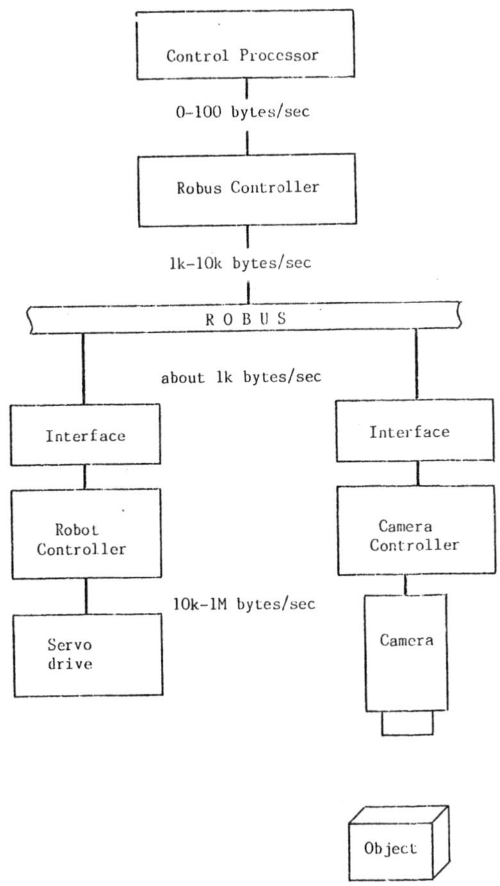

Figure 2

ETABLISSEMENT D'UNE COMMANDE PERFORMANTE POUR ACTIONNEUR PAS A PAS APPLIQUEE

A LA ROBOTIQUE A PARTIR D'UN MODELE DYNAMIQUE COMPLET

C. ROWDO, J.P. LOUIS, R. HUSSON

Centre de Recherche en Automatique de Nancy de l'ENSEM

B.P. 850, 54011-NANCY CEDEX FRANCE

RESUME

On présente la modélisation d'un manipulateur à trois degrés de liberté en vue de sa commande dynamique. A la modélisation classique (Lagrange) on associe des techniques d'identification (moindres carrés) pour intégrer au modèle divers phénomènes : frottements visqueux, frottements secs, jeux, couples non linéaires des actionneurs (moteurs pas à pas). Le modèle est ensuite utilisé pour l'étude d'une commande performante des actionneurs que l'on compare aux méthodes traditionnelles.

INTRODUCTION

Lorsque dans un mécanisme déformable, les frottements ne sont pas négligeables, le modèle classique de Lagrange ne suffit plus pour décrire les mouvements.
Un modèle dynamique complet étant établi, une commande optimale des actionneurs est alors possible.

Le modèle dynamique de la chaîne articulaire étudiée (voir figure 1), est déterminé à partir de la représentation de Lagrange [1] . Ce modèle est donné par trois équations (nombre de degrés de liberté de notre manipulateur), sous la forme matricielle classique [2] .

$$\underset{3x3}{[A]} \frac{\ddot{\theta}}{3x1} + \underset{3x3}{[B]} \frac{\dot{\theta}^2}{3x1} + \underset{3x3}{[C]} \frac{\dot{\theta}\,\dot{\theta}}{3x1} = \frac{Q(\theta)}{3x1} + \frac{\Gamma(\theta)}{3x1} \qquad (1)$$

où [A] , [B] , [C] sont les coefficients dynamiques du système et Q (θ) et Γ (θ) les couples ou les forces * appliqués au système dynamique. Notre objectif étant de poursuivre une trajectoire optimale en temps et en énergie, il nous faut tenir compte des couples de frottements secs <u>fs</u> (θ) et visqueux <u>fv</u> (θ) s'opposant aux couples <u>Γθ</u> des actionneurs et dont vont dépendre les vitesses des déplacements θ et les temps d'accélération.
Les actionneurs utilisés sont des moteurs pas à pas hybrides alimentés par des ponts à transistors en H. Les conditions de fonctionnements des moteurs (couple maxi pour les translations 1,5 Nm et 2,5 Nm pour la rotation) et de nos alimentations continues (V_{cc} = 35 V, 50 A) nous imposent de tenir compte dans le modèle du couple des moteurs, des modèles des courants dans les enroulements

$$\Gamma(\theta_i) = K\,[I_1 \sin(Z_r x) + I_2 \cos(Z_r x)] - \Gamma_o \sin(4 Z_r x) \qquad (2.1)$$

$$U_1 = K \dot{x} \sin\left(Z_r \dot{x} + \text{arctg} \frac{L\dot{x} Z_r}{R}\right) + L \frac{dI_1}{dt} + RI_1 \qquad (2.2)$$

$$U_2 = K \dot{x} \cos\left(Z_R \dot{x} + \text{arctg} \frac{L\dot{x} Z_r}{R}\right) + L \frac{dI_2}{dt} + RI_2 \qquad (2.3)$$

N.B. Les différents paramètres sont définis en annexe.
Le modèle dynamique complet de notre manipulateur est donc un modèle classique auxquels viennent s'ajouter les termes de frottements et des équations électriques nécessaires pour simuler des déplacements rapides. Ce modèle est donc le suivant

$$\underset{3x3}{[A]} \frac{\ddot{\theta}}{3x1} + \underset{3x3}{[B]} \frac{\dot{\theta}^2}{3x1} + \underset{3x3}{[C]} \frac{\dot{\theta}\,\dot{\theta}}{3x1} + \frac{F_s(\theta)\,\text{sign}\,\dot{\theta}}{3x1} = \frac{Q(\theta)}{3x1} + \frac{\Gamma(\theta)}{3x1} \qquad (3)$$

Certains coefficients des équations (2.1) et (3) sont mesurables (cas de Γ_o, K, I, x) ; d'autres sont calculables comme par exemple Q (θ) et plus particulièrement [B] et [C] . Ces derniers se déduisent des équations de Lagrange à partir du calcul de l'énergie cinétique T_i et des tenseurs d'inerties II (i) dont seul le moment I_{33} (i) intervient.

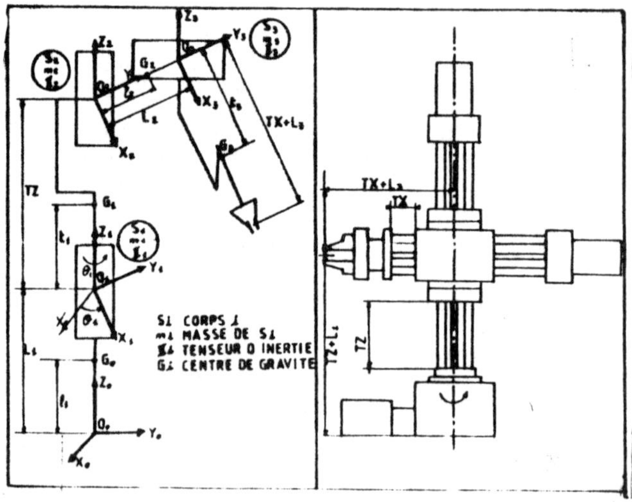

Figure N° 1 : Représentation de notre manipulateur.

Les différents calculs des énergies cinétiques des corps S_i (figure 1) nous donnent les formules

$$2\,T_1 = m_1 \dot{t}_1^2 + I_{33\,1}\,\dot{\theta}_s^2 \qquad (4.1)$$

$$2\,T_2 = m_2 \dot{T}Z^2 + (I_{33\,2} + m_2 l_2^2)\,\dot{\theta}^2 \qquad (4.2)$$

$$2\,T_3 = m_3(t_3^3 + TZ^2) + [I_{33\,3} + m_3(t_3^2 + L_2^2)]\,\dot{\theta}_1^2 - 2\,m_3 L_2 t_3 \dot{\theta}_1 \qquad (4.3)$$

En supposant rigides les segments de notre chaîne articulaire, le Lagrangien représentant le mouvement de cette chaîne se réduit à la somme des énergies T_λ des corps S_λ interconnectés. Les différents paramètres de Lagrange sont alors donnés par

$$\left\{ \sum_{\lambda=1}^{3} \frac{d}{dt}\left(\frac{\partial T_\lambda}{\partial \dot{\theta}_i}\right) - \frac{\partial T_\lambda}{\partial \theta_i} = Q_i + \Gamma_i \right. \quad i = 1,3 \qquad (5)$$

Les équations de Lagrange résultant de (5) sont alors

$$a_{11}\ddot{\theta} + a_{13}\ddot{TX} + C_{12}\dot{\theta}\,\dot{TX} = \Gamma_1(\theta) \qquad (6.1)$$

$$a_{22}\ddot{TZ} = g\,(m_2 + m_3)\,\partial Z_{G2,3}/\partial TZ + \Gamma_2(T_Z) \qquad (6.2)$$

$$a_{31}\ddot{\theta} = a_{33}\ddot{TX} + b_{31}\dot{\theta}^e = \Gamma_3(TX) \qquad (6.3)$$

* force si θ est une translation
couple si θ est une rotation

Dans lesquelles les différents coefficients peuvent être facilement calculés. Cependant "vus" du côté actionneur, les coefficients de Lagrange sont incomplets car il nous faut tenir compte des rapports de réduction ou de conversion rotation-translation $P\theta_1$ ainsi que de l'inertie propre du moteur et des divers frottements.

Nous identifions des macanismes régis par des équations différentielles du second degré non linéaires s'appliquant à des problèmes de déplacement mécaniques. Ces équations sont de la forme

$$(J_m + \frac{a_{ii}}{p_i^2})\ \ddot{X}_1 + F_V \dot{X}_1 + F_s \text{ sign } \dot{X}_1 = KI \sin Z_R X_1 - {}_o \sin 4 Z_R X_1 \quad (7)$$

La résolution de l'équation (7) se fait par intégration numérique selon la méthode de Runge-Kutta d'ordre 4. L'essai nous permettant de mesurer la position du rotor est la reponse à un échelon (1 pas de rotation). Pour des raisons de rattrapage de jeux, l'essai pris en compte est le $i^{\text{ème}}$ essai (figure 2).

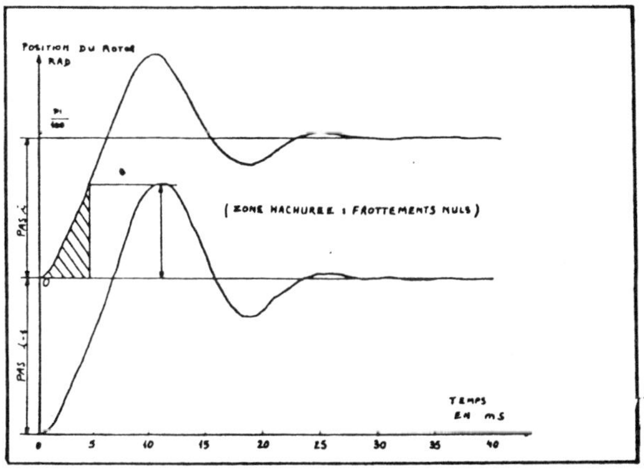

Figure N° 2 : Essais pour l'identification pour le bras vertical en montée.

La mesure de la position du rotor étant X_1 nous posons $\theta_1 = X_1 - \theta_o$. La position du rotor prise à l'arrêt est fonction du couple sec statique. Celui-ci ayant pour signe celui donné par le signe de la vitesse au cours de la dernière oscillation de l'essai précédent, ceci nous permet d'approcher la valeur initiale de la position prise par le rotor. Mis sous forme de vecteurs d'état l'équation (7) s'écrit

$$\dot{\theta}_1 = \theta_2 \quad (8.1)$$

$$\dot{\theta}_2 = - \frac{F_V}{J_T}\ \theta_2 - \frac{F_s}{J_T} \text{ sign}\theta_2 + \frac{KI}{J_T}[\sin Z_R(\theta_1 + \theta_o) - \frac{\Gamma_o}{KI} \sin 4 Z_r (\theta_1 + \theta_o)] \quad (8.2)$$

Le problème est alors d'identifier les quatre paramètres Q_i

$$Q_1 = - F\,V/J_T, \quad Q_2 = - F\,s/J_T, \quad Q_3 = KI/J_T \text{ et } Q_4 = \theta_o$$

Par rapport aux différents paramètres Q_i les équations (8.1, et 8.2) sont deux fois différentiables et l'existence de la solution recherchée pour l'identification est résolue par des théorèmes de convergence dans le cas des méthodes de Newton.
J (q) étant une fonctionnelle définie par un critère quadratique.

$$J\ (q) = \frac{1}{2} \sum_{i=1}^{N} (X_i - \theta_1\ i)^2 = \frac{1}{2} \sum_{i=1}^{N} \varepsilon_i^2 \quad (9)$$

Le vecteur variation des paramètres ΔQ pour minimiser la forme quadratique J (q) est donnée par

$$\underset{4x1}{\Delta Q} = - \ \underset{4x4}{|J_{qq}|}^{-1} \ \underset{4x1}{J_q} \quad (10)$$

$|J_{qq}|$ la matrice des dérivées secondes dont tous les termes doivent être positifs pour assurer la convergence, se met sous la forme

$$\begin{vmatrix} \sum_{i=o}^{N} [(\frac{\partial \varepsilon_i}{\partial Q_1})^2 + \varepsilon_i \frac{\partial^2 \varepsilon_i}{\partial Q_1\,\partial Q_1}] & \text{- - - - - - - -} \\ \vdots & \vdots \\ \sum_{i=o}^{N} [\frac{\partial \varepsilon_o}{\partial Q_1} \cdot \frac{\partial \varepsilon_i}{\partial Q_4} + \varepsilon_i \frac{\partial^2 \varepsilon_i}{\partial Q_1\,\partial Q_4}] & \text{- - - - - - -} \end{vmatrix}$$

$:|J_{qq}|$ matrice symétrique (4x4)

Notre méthode pour l'identification des paramètres est alors classique. Pour des raisons que nous indiquerons par la suite, nous n'identifions pas toutes les courbes mesurées de la même manière. C'est pourquoi les programmes d'identifications de celles-ci diffèrent quelque peu. Nous donnons l'organigramme type et nous indiquons les différences cas par cas. Pour certains mouvements la présence de jeux nous oblige à représenter le mouvement d'un axe par deux équa. différentielles : Une quand il y a contact, l'autre quand il n'y a pas contact.

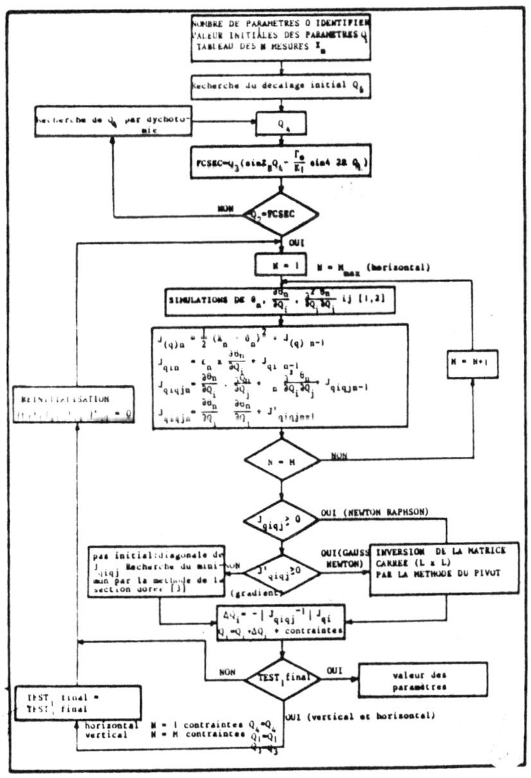

Examinons les courbes de la figure 3, nous constatons que les amplitudes des oscillations et leur amortissement diffèrent beaucoup.

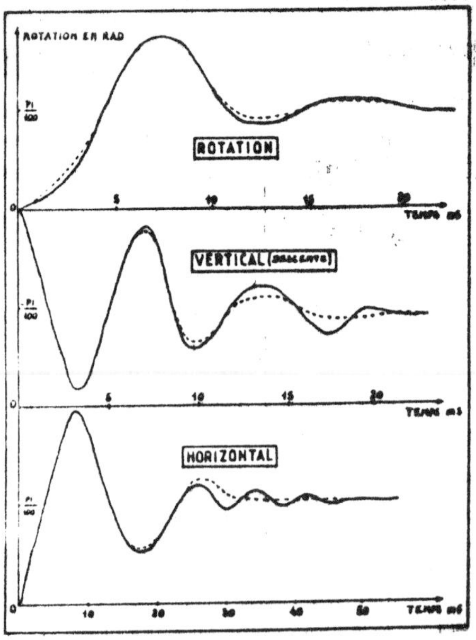

Figure 3 - Courbes d'identifications des 3 axes du robot (en pointillé les courbes obtenues par simulation)

Par expérience nous pouvons déduire les informations regroupées dans le tableau ci-dessous.

	ROTATION (1)	VERTICAL (2)	HORIZONTAL (3)
JEUX MECANIQUES	NON	PEU	OUI
COURBES	3 pour des TX : différents	1	1
NOMBRE DE PARAMETRES	4	5	7
PARTIE DE COURBES A IDENTIFIER ET DEFINITION	1 : toute la courbe	2 : {1ère montée, 1ère descente	2 : {1ère montée, reste de la courbe
DECOUPAGE EN ZONE DES PARTIES DE LA COURBE ET NOMBRE	non	non	oui 2 zones dans la 1ère partie de courbe
LISTE DE PARAMETRES IDENTIFIES	Frottement visqueux Frottement sec Décalage intitial Inertie et ses coefficients en fonction de TX Frottement sec statique	Frottement visqueux Frottement sec Force de gravité Inertie Décalage initial	Frottement visqueux Z_1 et 2 Frottement sec Z_1 et 2 Inertie zone Z_1 et 2 Décalage initial Frottement sec statique

De ce tableau il ressort que pour les axes verticaux et horizontaux, les courbes de mesure doivent être identifiées en deux temps, mais la manière de procéder diffère.

	VERTICAL	HORIZONTAL
CONDITIONS INITIALES	$\theta_i = 0$ $\dot{\theta}_i = 0$ $N = 1$	$\theta_i = X_{max}$ $\dot{\theta}_i = 0$ $N = M_{xmax}$
TEST 1 FINAL	$\theta_f = X_{max}$ $\dot{\theta}_f = 0$ $N^f = M_{xmax}$	$\theta_f = X$ $\dot{\theta}_f = 0$ $N^f = M$
CONTRAINTES POUR L'IDENTIFICATION DE LA PARTIE SUIVANTE	$Q_2 = Q_1$ $Q_3 = Q_3$	en ZONE 1 pas d'identification SEUIL (ZONE 1/ZONE 2) défini par l'amplitude du dépassement et de la courbe précédente EN ZONE 2 $Q_4 = Q_4$
TEST 2 FINAL OU CONDITIONS FINALES	Précision < 10^6 de la 2ème partie	$\theta_f = X_{max}$ $\dot{\theta}_f = 0$ $N^f = M_{xmax}$

Les résultats obtenus par simulation numérique nous permettent alors de calculer les coefficients dus aux forces centrifuges et de Coriolis à partir des formules données par (6) et compte tenu des rapports de réduction ou de conversion P (θ_i). Les résultats obtenus par identification mesure et calcul pour les trois axes sont alors les suivants.

ROTATION : θ_R

$$(279,316-222,96\ TX + 455,50\ TX^2)10^{-6}\ddot{\theta}_R - 0,466\ 10^{-6}\ \ddot{\theta}_X + (1,782\ TX - 0,251)\ 10^{-6}\ \dot{\theta}_R\ \dot{\theta}_X + 9,8\ 10^3\dot{\theta}_R + 308\ 10^{-3}\ \text{sign}\ \dot{\theta}_R = \Gamma \quad (12)$$

VERTICAL : θ_Z

$$274,80\ 10^{-6}\ \ddot{\theta}_Z + 10,42\ 10^{-3}\ \dot{\theta}_Z + 122,10^{-3}\text{sign}\dot{\theta}_Z = 189,38\ 10^{-3} + \Gamma \quad (13)$$

HORIZONTAL : θ_x

$$-33,599\ 10^6\ \ddot{\theta}_R + 288,548\ 10^{-6}\ddot{\theta}_X + 161,15TX-9,058)10^{-6}\ \dot{\theta}_R^2 + 19,165\ 10^{-3}\dot{\theta}_x + 57,61\ 10^{-3}\ \text{sign}\ \dot{\theta}_X = \Gamma \quad (14)$$

COUPLE : Γ

$$\Gamma = 0,172\ 645[I_1 \sin(50.\theta) + I_2 \cos(50.\theta) - 0,28382\sin(200.\theta)] \quad (15)$$

pour le calcul de I_1 et I_2 $R = 1,629\ \Omega$, $L = 0,0013075\ H$

CONCLUSION

La méthode présentée ici est avantageuse car il est difficile d'établir des essais typiques lorsque les bras se déplacent à grande vitesse pendant des temps très courts. Les équations obtenues sont suffisamment précises pour nous permettre de simuler le fonctionnement du robot sur plusieurs pas.

Une application est réalisée en simulant le déplacement du bras horizontal sur 125 pas, en tenant compte des jeux pour trois types de commandes(voir figure 4):

- Commande en boucle ouverte consistant à appliquer au moteur une suite d'impulsions de durées prédéterminées, dont l'algorithme est du type BANG-BANG [4] la loi en fonction du temps étant la suivante

$$X = 2.\ D\ \left(\frac{t}{T}\right)^2 \quad o \leq t \leq 0,5\ T$$
$$X = D\ [-1 + 4\frac{t}{T} - 2\ \frac{t^r}{t^r}]\ 0,5\ T \leq t \leq T \quad (16)$$

De cette loi nous tirons une suite d'impulsions pour nos moteurs dont les fréquences N sont données par la relation (17)

$$N(i) = \frac{2.D}{T^2 \times PAS} \sum_{k=1}^{i-1} \frac{1}{2.N(k)} \quad o \leq t \leq 0,5\ T$$
$$N(i) = \frac{2.D}{T.PAS} - \frac{2.D}{T^2.PAS} \sum_{k=1}^{i-1} \frac{1}{2.N(k)}\ 0,5T \leq t \leq T \quad (17)$$
$$\text{et}\ T = \frac{1}{2.N(o)}\sqrt{\frac{2.D}{PAS}}$$

Cet algorithme est théoriquement le plus rapide en boucle ouverte. Sur la courbe de simulation, nous constatons de nombreuses oscillations et en particulier à basse vitesse où on observe des phénomènes de résonance.

- Commande en boucle classique [5] . Les instants de commutations sont définis par la position de l'angle électrique du rotor. On suppose que la forme du couple des moteurs pas à pas, en fonction de l'angle électrique est une sinusoïde parfaite et que les courants dans les enroulements s'établissent instantanément. Il en découle que le couple maximum est obtenu pour un angle de commutation égal à $\pi/4$. La courbe de montée en vitesse est une exponentielle qui montre un gain de 52 % sur la vitesse max en B.O.

- Commande en boucle fermée performante : Le modèle du couple et des équations électriques associés nous permet de calculer le couple maximal en fonction de la vitesse du rotor. La formulation est simple ; il suffit d'écrire que la pointe du courant (celui-ci étant haché) doit se produire lorsque le sinus est maximal. Connaissant le temps de montée T_m du courant et la vitesse du rotor θ, l'angle de commutation α_c sera défini de la manière suivante

$$\alpha_c = \pi/4 \text{ si } \dot{\theta}T_m \geq \pi/4$$
$$\alpha_c = \dot{\theta}T_m \text{ si } \dot{\theta}T_m \leq \pi/4$$

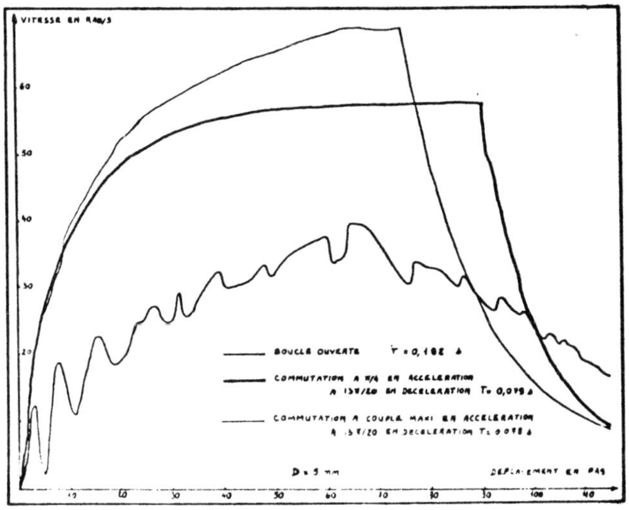

Figure 4 - Vitesses obtenues par simulation pour divers types de commandes.

Cette technique s'apparente aux méthodes dites "d'avance à l'allumage". La vitesse maximale atteinte est alors près de 20 % plus élevée que pour une commande en boucle fermée classique.

ANNEXE :

Z_R : Nombre de dents au rotor du moteur pas à pas.

K : Constante de proportionnalité couple courant en $N.m\,A^{-1}$

Coefficients de Lagrange

$$a_{11} = I_{331} + I_{332} + I_{333} + m_2 l_2^2 + m_s (t_3^2 + l_2^2)$$

$$a_{13} = - m_3 L_2$$

$$a_{22} = m_1 + m_2 + m_3$$

$$a_{31} = -2 m_3 L_2$$

$$a_{33} = m_3$$

$$C_{12} = 2 m_3 t_3 + \partial I_{333}/\partial (TX+L_3)$$

$$b_{31} = - C_{12}$$

BIBLIOGRAPHIE.

[1] - COIFFET
Les robots tome (1) Modélisation et commande

[2] - KHALIL
Contribution à la commande automatique des manipulateurs avec l'aide d'un modèle mathématique des mécanismes. Thèse d'Etat.

[3] - BERNHARD
Commande optimale, décentralisation et jeux dynamiques. Page 124.

[4] - ZABALA ITURRALDE
Commande des robots manipulateurs à partir de la modélisation de leur dynamique. Thèse de 3° cycle.

[5] - GRANDIDIER
Conception et commande de robots manipulateurs équipés de moteurs pas à pas. Thèse de Docteur-Ingénieur

CONTOUR AND REGION FILLING METHODS

C.Gambaro - P.Sinigaglia
Istituto Matematico
Universita' di Genova
ITALIA

ABSTRACT - One of the most common problems in Image Processing and Pattern Recognition is determining the interior of a region when the border is known. This problem can be solved in many different ways, which can be subdivided in two major classes according to the contour input type : polygon based, if the contour is given as an ordered sequence of vertices, pixel based if the pixels which belong to the border are directly supplied. Furthermore tecniques can be distinguished for the two criteria, parity check or connectivity filling, which are used to decide if a point is in a region. In this paper we will examine and evaluate some algorithms belonging to the two previously mentioned categories. We will illustrate the best algorithm of the polygon based class which uses a parity check criterion to fill both multiply connected and disjoint but not overlapping polygons. We will also illustrate and compare the most representative algorithms of the pixel based class which use connectivity filling tecniques, and we will describe in detail three methods: Smith, Pavlidis and Shani.

INTRODUCTION

Images generated on a video raster can be made up not only of lines and alphanumeric texts but also of coloured areas. These areas can represent lines of different widths coloured geometric figures or the faces of three - dimensional objects. The generation of these images on the video screen requires the establishement of three properties for each area: the mask, the shadow and the priority. The mask signifies the representation which defines which pixels are in the area. Usually it is made up of a matrix of binary values,where 0 indicates the external pixels and 1 stands for those pixels internal to the area. The shadow specifies how to calculate the intensity of each pixel inside the area. The priority is used to determine if and to what extent two or more areas overlap, in the case of an image made up of more than one area.

Determination of the mask of a region, i.e. finding the interior when the border is known, is one of the most common problems in Computer Graphics. For example, any algorithm for defining the shadow presumes the solution of this problem, in Pattern Recognition many algorithms calculate integrals on the area of a region and require knowing its interior, in fototypesetting the characters are often described in terms of contour which are then filled to determine the final copy.

This problem has been solved in different ways which can be divided into two large classes. The methods which belong to the first class start from the description of the contour as a polygon and decide what part of the plane lies within the interior, based on the equations of the lines. These tecniques have been labelled polygon or edge filling methods. Methods from the second class project the contour on the discrete plane and then determine the interior by examining the values of the pixels. These techniques are called pixel based or region filling.

There is a notable difference between the filling of polygons and the filling of the regions. In the first case the region is defined by the vertices of the polygon and do not make hypotheses on the initial value of the buffer. In the second case the region is defined by the values of the pixels in the refresh buffer.

In addition, the techniques differ in the principle which they utilize to decide if a point is internal to an area. Some use the parity check while others use the criteria of connectivity. The first method is based on the fact that a straight line intersects a closed curve (such as the contour of a region) an even number of times. If we know that the first point of the line is outside the region we can then establish exactly which segments are in the interior by counting the number of intersections. If this number is odd the segment is in the interior, otherwise it is outside. All the algorithms of the first class (edge or polygon filling) use techniques based on this principle. The second method (connectivity filling) presumes that an internal point (seed) is given in addition to the contour and all the points are found that can be reached without intersecting it. This technique is suitable only for pixel based algorithms because it requires random access to the internal pixels. The major advantage is that it is not affected by the complexity and the irregularity of the contours which are to be filled. Its principal disadvantage is the necessity of knowing a priori an internal point.

We shall now examine some algorithms which belong to the two classes indicated above.

1. POLYGON BASED ALGORITHM

We shall first illustrate the most general algorithm of the polygon based class /1,5,2/ which uses a parity check criterion to fill both multiply connected and disjointed but not overlapping polygons. To more clearly illustrate the procedure we shall refer to a region defined by a single polygonal contour.

The algorithm works on the list of the vertices which determine the contour of the area to be filled and considers a horizontal line (scan-line) at a time repeating the procedure for all lines that intersect the polygon (scan-polygon).

The filling of the part of the line interior to the polygon is achieved by exploiting the coherence existing between the pixels belonging to the same line, in other

words the property by which if a pixel is interior (exterior) to the polygon it is extremely probable that the immediately adjacent pixels are also interior (exterior). There will be a change between two adjacent pixels only if the contour of the polygon passes through them.
This coherence can therefore be utilized by looking for only those pixels at which changes occur.

Determining these pixels means:

1) finding the intersections between the scan-line and all edges of the polygon;
2) putting them in order according to increasing x-ordinates;
3) assigning the required value to all pixels included between the intersections.

Filling the entire polygon is based on the property of coherence of the edges of the polygon, i.e. if an edge intersects the horizontal line y = i, it most probably also intersects the line y = i+1.

The algorithm operates, for each scan-line, only on the edges that intersect it, constructing an active edge table which is updated time to time by the addition or elimination of some edges.

In passing from one scan-line to the next, new intersected edges are introduced while those which are not intersected are eliminated.

The procedure ends when all the edges of the structure have been analyzed and when all the edges of the active edge table have been eliminated.

The algorithm described has the advantage, when compared to others of the same class, of optimizing both A) the computation of the intersections, thereby drastically reducing the comparisons between the scan-line and edges of the region with the management of the active edge table and B) the treatment of the intersections which fall in the vertices, counting one or two intersections depending on whether the edges adjacent to the vertex are monotonous or not, through the shortening of one of the two incident edges in the first case.(fig. 1 and fig. 2).

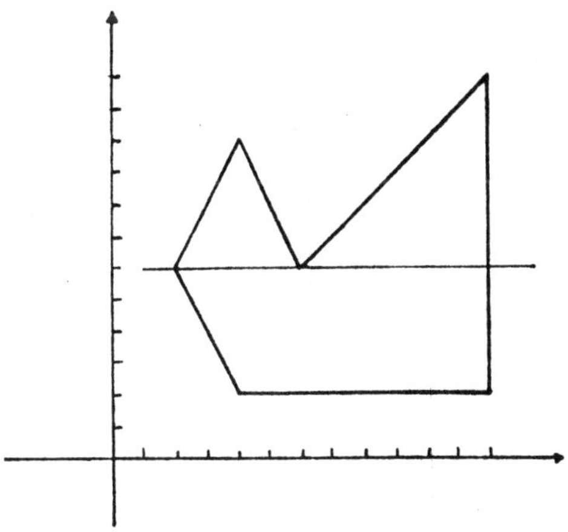

fig. 1

The complexity of the entire algorithm is O(N $\log_2$N), where N is the number of edges of the entire structure, since the construction of the active edge table and the computation of the intersections have a complexity of order N while for the sorting of the edge table and the sorting of the intersections algorithms with complexity of order O(N $\log_2$N) can be used.

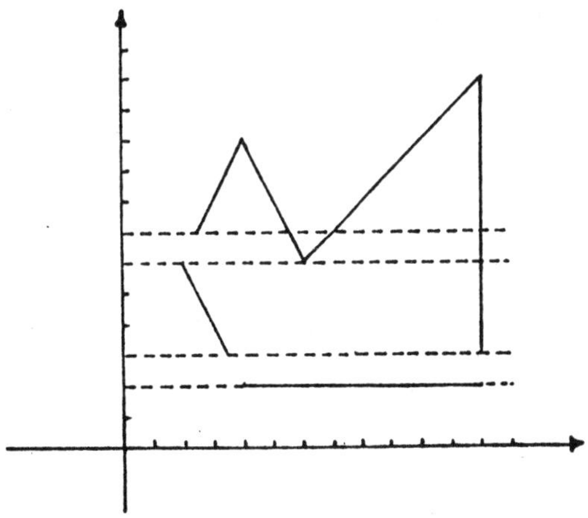

fig. 2

2. THREE PIXEL BASED CONNECTIVITY ALGORITHMS

We shall now move on to illustrating and subsequently comparing the basic concepts of three of the most representative algorithms of the pixel based class which use techniques of connectivity filling /2,6/, omitting treatment of the parity check algorithms /6/ of this class since they are generally very slow even for small regions and not robust enough respect to the irregularity of the contour.

2.1. SMITH'S ALGORITHM

Smith's algorithm /5/ effects the filling of regions with colour assigned to the interior (i-defined) but it also has been implemented for c-defined regions (colour assigned to the border) which allows comparison with the other algorithms, all of which have been developed for c-defined regions. It also permits us to demonstrate the problems involved in adapting the same algorithm to different ways of representing the regions.

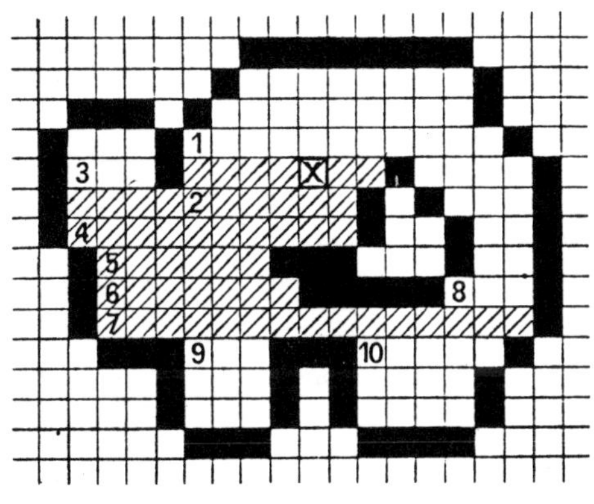

fig. 3

The basic idea is to fill the horizontal segment containing the current pixel and to examine the adjacent lines looking for new segments to colour. One looks for the pixels connected with the just filled segment which belong to the region and have not yet been filled. Once they are found, they are put on the stack to await correct completion of the filling.

An example of filling by this algorithm is shown in fig. 3, where the numbers indicate the order in which the pixels are pushed on the stack.

2.2. PAVLIDIS'S ALGORITHM

Pavlidis's algorithm /6,7/ is based on the treatment of a graph (C-LAG Contour Line Adjacency Graph) constructed on the contour of the region (fig.4) and on some relations between the structure of the graph and the region to be filled. The degree of the node (number of entering arcs, number of leaving arcs) of the C - Lag to the left of the current segment (containing the pivot pixel) is examined. If it does not present a maximum or minimum a pixel is saved from the next line (in the current direction) and the contour on the right is then examined. If this does not present any singularities the procedure is repeated with the saved pixel from the previous step taking the place of the new pivot.

This algorithm also uses a stack on which the pixels are memorized with a direction such that filling of the zones left suspended is guaranteed. A pixel and relative direction are also taken from this stack when filling in the current direction is finished.

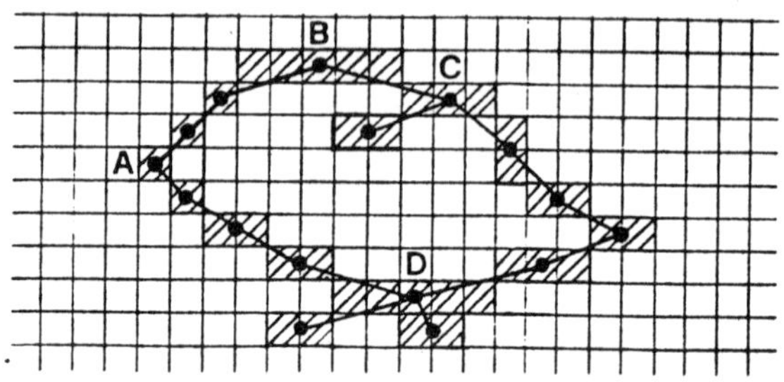

fig. 4

2.3. SHANI'S ALGORITHM

Shani's algorithm /4/ raises the problem of filling as a variant of traversing a graph constructed on the region (fig. 5).

The region is subdivided in regular regions (the nodes of the graph) and each boundary between two regular regions corresponds to an arc of the graph.

The procedure consists of traversing the graph in successive phases characterized by one direction, colouring the nodes and pushing the nodes left suspended in the current direction on the stack.

To avoid eventually returning on previously coloured nodes, the incident arcs on the node in the direction opposite to the current direction, excluding the so-called ' entrance arc', are blocked (this corresponds to a tracing of temporary contours) and memorized in a queue from which they will be withdrawn only when the stack is empty.

The stack and the queue are combined in a single structure called Queue-Stack from which it is possible to remove an element regardless of the position occupied, in addition to performing the operations of management of the stack and the queue.

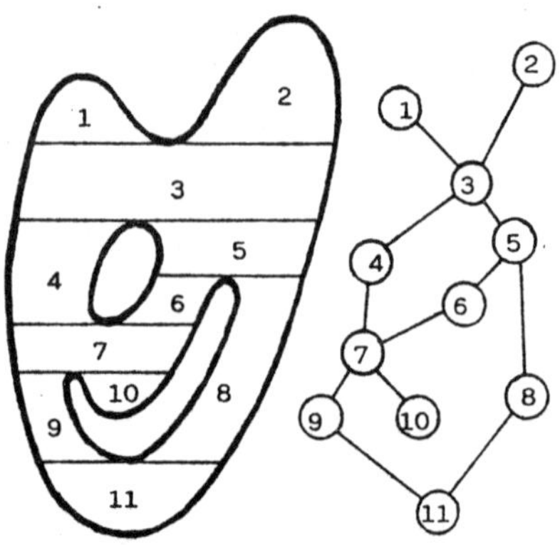

fig. 5

2.4. COMPARISION

We shall now compare the pixel based algorithms described above according to various criteria.

The first involves the range of application, in other words the class of regions which each algorithm is capable of correctly filling.

The second criterion deals with the complexity of the algorithms, which is not always based on comparable parameters, and the third considers the filling times obtained on some significant regions.

2.4.1. COMPARISON BASED ON RANGE OF APPLICATION

The factor which can cause problems for some of the algorithms are: A) the presence of pixels of the filling colour inside the region, B) the possibility of filling with the boundary colour, C) the possibility of filling open regions (Tab. 1).

Some algorithms, to avoid wandering infinitely around any holes present in the region, recognize the already filled zone solely by the colour of the pixels. Therefore if the filling colour is already present inside the region the filling may stop in some directions if zones with pixels of the new colour are interpreted as already coloured zones.

This limitation is present in the algorithms of Smith and Pavlidis while it is bypassed in Shani's algorithm through the blocking operation, which also permits fillings with non homogenous colours (e.g. textures).

Pavlidis's algorithm, based on the analysis of C-Lag, is the only one which does not permit filling with the boundary colour or filling region with open contours. In the first case the C-Lag graph is changed during the filling, thereby incorrectly altering the result; in the second case the C-Lag graph is not connected, producing an interruption in the running of the algorithm.

	Smith	Pavlidis	Shani
Internal points of filling colour.	N	N	Y
Filling with boundary colour.	Y	N	Y
Region with open contour.	Y	N	Y

tab. 1

2.4.2 COMPARISON BASED ON COMPLEXITY

The greatest difficulty encountered in trying to make a correct comparison in this field is due to the lack of uniformity of the parameters used for evaluation of each algorithm, derived from the different abstract structures used and the lack of input which can be directly utilized as parameters of comparison (e.g. the number of vertices for polygon based algorithms) . One could consider the area of the region in pixel as a uniform parameter of comparison, but the difference in treatment of two regions with equal areas, even using the same procedure, demonstrates how the complexity of these algorithms depends not only on the area but also on the form and the orientation of the region to be filled.

The major problems arise when one wants to compare the algorithm of Shani with the others. The complexity of this algorithm is evaluated solely on the abstract structure of the graph and is O(2 M + N K) where N is the number of nodes of the graph, M is the number of the arcs of the graph and K is the average queue-stack size. Transformation in terms of the number of pixels examined, a uniform parameter for the other two algorithms, in this case introduces assumptions that give too approximate results, which have not been considered since they are unreliable.

Smith's algorithm examines from 2(I+N) to 3(I+N) pixels, where I is the area in pixels of the region and N is the number of interior segments. To make the comparison easier lets assume that on an average 3 I pixels are examined, while Pavlidis's algorithm examines 3 C + I pixels where C means the area in pixels of the contour. Of the two algorithms, Pavlidis's is the more suitable if 3C is less than 2 I , i.e. if the ratio C/I between the area of the contour and the area of the region is less than 2/3. Generally, even in presence of fairly tortuous contours, the C/I ratio remains significantly below the threshold value 2/3. For example, for a square region of N pixels side the ratio equals 4/N and therefore Pavlidis's algorithm is more suitable than Smith's if N is greater than 6, as is usually in the case.

The only case in which Smith's algorithm may be preferable is when the contour has dimensions which are comparable to those of the region. This can occur, other than with extremely small regions, in the presence of regions drawn on a homogeneous background which acts as the contour. In this case the ratio C/I can be significantly greater than 2/3, thereby making Pavlidis's algorithm less efficient than Smith's.

2.4.3 COMPARISON BASED ON FILLING TIME

Since an overall comparison in terms ofcomplexity was not possible we decided to consider the filling times obtained for some significant regions (the algorithms are implemented in FORTRAN IV on a PDP 11/40 with a graphic display RAMTEK 9351 at the Institute for Applied Mathematics of the National Research Council of Genoa).

The fastest turned out to be Shani's algorithm followed by Pavlidis's with a time that is increased by 5-10%. Smith's algorithm has filling times that are 60-70% greater than Shani's. Similar results are obtained for regions characterized by contours of areas significantly greater than that of the interior. In fact the times obtained with Pavlidis's algorithm are almost double those obtained on regions with opposite characteristics.

CONCLUSIONS

The main advantage of the polygon based algorithms is their speed, since the number of I/O operations between computer and frame buffer is reduced to the minimum.
On the other hand their use is limited since sometimes it is impossible to dispose of a contour in terms of its geometrical coordinates.

In those applications where the only information about contour comes from the displayed image, it is necessary to adopt pixel based techniques.
These methods are notably slower since the number of I/O operations necessary for the processing of the image is much higher than for polygon based methods.

Of the pixels based methods studied Shani's algorithm was superior to the others for speed and flexibility.
It allows both the filling of regions with open contours (derived for example from clipping operations) and the filling with the colour of the contour or with a colour already present in the interior of the region.

As we have already seen, these limitations regard the other algorithms to varying degrees (tab. 1).
Since it is an algorithm of connectivity it is also correct for regions which present irregularities in the contour (for example due to scale factors) and therefore never goes out of the region which is to be filled.

ACKNOWLEDGEMENTS

The authors wish to thank B. Falcidieno and C. Pienovi for their helpful comments.
This work was supported by the "Progetto Finalizzato Informatica, P1-SOFMAT" of the Italian National Research Council.

REFERENCES

/1/ Falcidieno, B., Gambaro, C., Pienovi, C.
Riempimento di regioni poligonali su terminali raster.
Il progettista industriale - 1-2/1983, pp. 105-111

/2/ Foley, J.D., Van Dam, A.
Fundamental of interactive computer graphics.
Addison Wesley - 1982, pp. 446-460

/3/ Shani, U.
Filling regions in binary raster images: a graph-theoretic approach. Siggraph proc. - 1980, pp. 321-327

/4/ Smith, A. R.
Tint fill.
Siggraph proceedings - 1979, pp. 276-283

/5/ Newmann, W.N., Sproull, R.F.
Principles of interactive computer graphics.
II ed. Mc Graw-Hill 1979

/6/ Pavlidis, T.
Contour filling in raster graphics.
Computer graphics 15 (3) - Aug 1981

/7/ Pavlidis, T.
Algorithms for graphics and image processing
Springer Verlag 1982

AUTOMATED DEVELOPMENT OF PROGRAM MODULES IN ROBOT CONTROL OPERATING SYSTEMS

M. K. AVRAMOVA, B. MANAHILOV, D. KARASTOJANOV

INSTITUTE OF INDUSTRIAL CYBERNETICS AND ROBOTICS,IICR,BAS, 1113 SOFIA, BULGARIA

ABSTACT

SOME MAJOR PROBLEMS TO RESOLVE IN THE ROBOT SOFTWARE DEVELOPMENT ARE EXAMINED.ROBOTS FOR TECHNOLOGICAL PROCESSES AUTOMATION ARE SELECTED. THE LOGICAL RELATIONSHIP BETWEEN SEPARATE TASKS ARE DISCUSSED AS WELL AS THE SEQUENCE OF THEIR RUNNING IN TIME.

THE DECOMPOSITION OF EVERY TASK TO A CONSTANT AND A VARIABLE PARTS IS GROUNDED.THE CONSTANT PART IS INVARIANT TO THE WIDE RANGE OF PROBLEMS AND THE VARIABLE PART DESCRIBES THE PARTICULAR REALISATION CONDITIONS.

HIGH- LEVEL PROBLEM - ORIENTED LANGUAGES FOR THE DIALOGUE WITH AN OPERATOR ARE PROPOSED.THE WORK CONDITIONS AND NECESSARY PROCEDURES FOR A PARTICULAR PROCESS CONTROL ARE DESCRIBED BY MEANS OF THE LANGUAGES.SYSTEM RESOURSES ARE DESIGNED,WHICH INCLUDE AN EDITOR,A TRANSLATOR AND AN INTERPRETER FOR CALCULATING PROCEDURES DESCRIBING,AS WELL AS AN TRANSLATOR FOR DESCRIBING THE PARTICULAR PROCESS MODEL.

A PARTICULAR REALISATION OF AN INDUSTRIAL ROBOT SOFTWARE IS DESCRIBED.

INTRODUCTION

THE USE OF ROBOTS OR ROBOTIZED COMPLEXES FOR TECHNOLOGICAL PROCESSES AUTOMATION HAS PROVED ITS EXPEDIENCE AND EFFECTIVITY.THE AUTOMATION OF SUCH A PROCESS BY MEANS OF A ROBOT COMES TO PERFORMING OF POSITIONING,TRANSPORT MOVEMENTS AND OPERATIONES ON PREDEFINED TRAJECTORIES. THE MAIN PROBLEM IS STICKING TO THE FÜLL CYCLE OF THE TECHNOLOGY.THE TASK DESCRIPTION IS PERFORMED WITH THE PROBLEM - ORIENTED HIGH - LEVEL LANGUAGE.THE FUNCTIONS,WHICH REFER TO MOVEMENT CONTROL, SYNCHRONISATION OF PROGRAMS AND HARDWARE ARE PERFORMED BY MEANS OF AN OPERATION SYSTEM SPECIALLY DESIGNED FOR THIS PURPOSE.

THE SEPARATION OF THESE FUNCTIONS AMONG THE SOFTWARE MODULES CAN BE DONE ON THE BASIS OF:
-FUNCTIONALLY INDEPENDENT ROBOT'S WORKING MODES
- WORK OF CREATING AND FULLFILLING THE USER TASK
A CONCRETE CLASSIFICATION IS GIVEN ON FIG. 1 . THE CONNECTIONS IN THIS HIERARCHICAL SYSTEM ARE REALIZED BY THE OPERATIONAL SYSTEM RESOURCES AND/OR BY DATA FIELDS. THE DISTRIBUTION OF THE MEMORY, TIME AND PRIORITY IS CONFINED TO THE CONTROL PROGRAM MONITOR, WHICH IS THE OPERATIONAL SYSTEM NUCLEUS. THIS PROGRAM CONSIST OF TWO PARTS: SCHEDULER, WHICH IS WRITTEN IN ASSEMBLER AND IS INDEPENDENT OF CONCRETE SYSTEM AND TABLE DESCRIPTION, WHICH IS CONNECTED WITH THE PARTICULAR REALIZATION.

MODULES AND MODES

FOR A GREAT NUMBER OF TASKS CONNECTED WITH THE TECHNOLOGICAL PROCESSES AUTOMATION FUNCTIONS CAN BE DESCRIBED IN A STANDARD WAY:

1. CREATING USER PROGRAM

THE MAN-MACHINE INTERFACE IS HELD BY A SPECIALIZED DIALOGUE HANDLER. INDEPENDENT OF THE PARTICULAR TECHNOLOGICAL PROCESS, ONE AND THE SAME INFORMATIONAL PROCESSING IS DONE, OF THE TYPE:

- SYMBOL BY SYMBOL PROCESSING
- TRANSLATION TO AN INNER PRESENTATION FORM.

LANGUAGE CONSTRUCTION HAVE A SIMPLE STRUCTURE OF THE TYPE:/1/
<COMMAND>::=<OPERATIONAL CODE><ARGUMENT> .
THEN THE PROGRAM GENERATIONAL MODULE HAS THE FUNCTIONS OF A SYSTEM EDITOR. STANDARD FUNCTIONS ARE READING,STORING, INSERTING, SCRATCHING, SEARCHING, COMMAND INPUT, TERMINATION OF THE INPUT WHEN A LEXICO-SYNTACTICAL ERROR IS FOUND, MESSAGE DISPLAY.

- MESSAGE DISPLAY AS A REACTION TO THE INFORMATION RECEIVED.

After information input an analysis for recognizing a right language phrase begins. In case of analysis failure a syntactical error message is displayed if necessary. In case of success a display is done or not, the symbol received is stored or not in a user buffer, specific informational processing is done or not. For this purpose it -s convenient to create an operator-structure with control parameters./6/. This is a multiplace IF...THEN operator with a control parameters dynamic adjustment.

To simplify the lexico-syntactical analysis of the information introduced, it is necessary to built a particular language model. All kinds of informational processing connected with the particular realisation must be included in this model, too.

2. Program control

Having in mind the operator's low qualification and the fact that incorrect programming can make the robot dangerous to man and his surroundings an analysis of the processes described in the user program should be maid. These functions realized by comparing the generated and permissible by the process technology states.If it is necessary the process status is modified during commad execution. For this purpose the connections among the separate commands and their technological consequence must be known. This information called process model can be given in a declarative or procedural way. In the memory it is a reference table. /2/.

3. High-level language interpretation

The interpretation module compares to each language command a combination of functions leading to technological process transition from one state to another, following the program execution. Standard interpreter functions are:

- interpreter status control
- user program branches
- subroutine calls in the user program
- servo regulator and generator control
- auxiliary functions output
- calculation movement parameters
- preliminary logical processing of asynchronous events and signals and synchronization

Dependent interpreter functions are:

- user program decoding and data distribution
- particular command description and user program cycles
- mode control
- error check and display

These functions must be given in a concrete realization model. If this functional set is full enough, this is a model of the interpreter's structure./3/.

4. Adaptation to environment inaccuracies

Two approaches are possible:

- creating adaptive functions as a higher hierarchical level over a non-adaptive controller
- paralel work of an adaptive contour and heterhierarchical connection of the results from its action in time and place, considering the characteristics of the realisation. The most general way is to create environment model in a teaching mode on a standard environment and configuration of parts. Then in an executive mode when one and the same signals are input to the model and to the real object /4/, the differences in their reactions can be used to generate corrective control. The necessary corrections are made in correspondence with a chosen algorhithm for processing and an optimal criterion.

5. A module for calculating procedures

As the analytical dependences description in Assembler is non-effective, there exist several ways of coping with this problem:

- setting of suitable hardware
- creating a high-level language
- writing a macro-language for the computing processes where the high-speed execution is of greatest importance.

The calculation procedures are programmed on a high-level language and translated into a pseudocode. The source program execution is done by an interpreting module in an interactive mode. The program is created and edited by an editing module with its own interactive command set. The result is an object code which can be stored on disk or PROM. The software processor can operate standalone - managing other modules or as a part of large system./7/. When the source program is debugged, data are input and results are output in decimal system.

6. PROGRAM MODULES STRUCTURE

A WIDE VARIETY OF PRACTICAL PROBLEMS CAN BE UNITED IN ONE CLASS WITH RESPECT TO THE SEPARATE MODULES FUNCTIONS. IT IS POSSIBLE TO ACHIVE A SEPARATION OF EACH MODULE IN THE FOLLOWING WAY:

- OPERATIONAL PART - PERFORMS STANDARD FUNCTIONS AND INVARIANT TO THE PARTICULAR REALIZATION
- A MODEL - CLOSELY CONNECTED WITH THE PARTICULAR REALIZATION

IN THIS WAY THE SAME OPERATIONAL PART, ONCE CREATED, WHEN WORKING ON DIFFERENT MODELS OF PARTICULAR SYSTEMS WILL PERFORM THE SAME FUNCTIONS IN A DIFFERENT WAY. THIS CORRESPONDS TO USING DIFFERENT HIGH-LEVEL SPECIAL-PURPOSE LANGUAGES.

7. REAL TIME CONTROL

THE REAL-TIME SYSTEM PERFORMS INFORMATIONAL PROCESSING OF ONLY ONE PROCESS AT A TIME INTERVAL. FOR DOING THIS THE PROCESS RECEIVES INFORMATION AND PERMITION TO USE THE PROCESSOR AND AFTER FINISHED PROCESSING LEAVES PROCESSOR TO ANOTHER PROCESS. THUS, HERE ARE DONE: /5/

- TRANSFER INFORMATION (SYSTEM DATA)
- ACTIVATE PROGRAM MODULES FOLLOWING THE PROCESSING

SOME OF THE COMMON CONTROL FUNCTIONS ARE:

- SET A SYSTEM IN A STATUS ACCORDING TO THE PREVIOUSLY DECLARED SYSTEM STATUSES
- INPUT INFORMATION HAVING IN MIND BEHAVIOUR OF THE SYSTEM IN THE CASE WHEN DATA ARE BEING WAITED FOR OR NOT
- SIMULTANEOUSLY SET STATUS AND INPUT SYSTEM DATA
- TRANSFER DATA TO A CHOSEN PROGRAM HAVING IN MIND ITS STATUS
- TRANSFER DATA AND ACTIVATE THE PROGRAM
- DATA REQUEST AND IF DATA EXIST THE PROGRAM IS ACTIVATED
- FORBID PROCESSOR TIME TO THE PROGRAM
- ENABLE PROCESSOR TIME TO THE PROGRAM HAVING THE HIGHEST PRIORITY AND WAITING FOR PERMITION TO USE THESE SYSTEM RESOURCES
- INACTIVATE PROGRAM
- ENABLE A PROGRAM TO COMPETE FOR PROCESSOR TIME, DEPENDING ON ITS STATUS AND PRIORITY

EACH PROGRAM MODULE STATUS CAN BE AS FOLLOW:

- READY OR ACTIVE
- SUSPENDED - WAITING FOR DATA DURING INFORMATIONAL REVISION OF THE MODULE
- INACTIVE

ONLY A PROGRAM IN STATUS READY CAN USE PROCESSOR.

REX OPERATION SYSTEM

A GENERATING AND DEVELOPING REAL-TIME CONTROL SYSTEM OF THIS TYPE IS THE REAL-TIME EXECUTIVE (REX) SYSTEM WORKED-OUT IN THE ISOMATIC LABORATORIES AT THE INSTITUTE FOR CYBERNETICS & ROBOTICS OF BULGARIAN ACADEMY OF SCIENCES. IT PROVIDES AN OPPORTUNITY FOR EFFECTIVE USSAGE OF READY PROGRAM PRODUCTS AS WELL AS FOR CREATING NEW MODULES. TO SIMPLIFY THIS, A MONITOR HAS BEEN INCLUDED IN THE SYSTEM, WHICH ALLOWS STANDARTIZED REQUESTS OF THE DIFFERENT PROGRAMS ONE TO THE OTHER.

UP TO NOW REX INCLUDES THREE MAIN PARTS:

1. REX-JL (GENERATING LANGUAGE) - GIVES THE OPPORTUNITY TO DESCRIBE THE INTERACTIONS BETWEEN CONCURRENT PROCESSES IN A USER PROGRAM.
2. REX-SL (SYSTEM LIBRARY) - THIS IS A SYSTEM MODULES COMPLEX, AIMING THE ENGINEER IN DESIGNING A PARTICULAR PROGRAM COFIGURATION.
3. REX-RD (PROCESS DESCRIPTION) - HELPS THE SYSTEM DESIGNER IN DESCRIBING THE PROCESSES IN THE SYSTEM.

THE REX-GL TRANSLATOR YIELDS A PROCESSOR - INDEPENDENT PRODUCT, INTERPRETED BY THE SYSTEM SCHEDULER, A REX-GL RESOURSE.

THE REX-PD TRANSLATOR YIELDS A PROCESSOR - INDEPENDENT PRODUCT, INTERPRETED BY THE REX-SL'S RESOURCES. REX-PD CONTAINS ALL THE USUALL ARITHMETIC AND TRIGONOMETRIC PROCEDURES WITH A SINGLE AND DOUBLE PRECISION, LOGICAL PROCESSING PROCEDURES AND REAL TIME PROCEDURES. THE AVERAGE TIME FOR PROCESSING A SINGLE OPERATION IS BELOW 1MS. THE REX-SL LIBRARY IS AVAILABLE FOR USE IN THE 6800 FAMILY AND IS PRESENTLY UPDATED FOR USE IN THE INTEL 8086 FAMILY.

REX'S MONITOR CAN CONTROL UP TO 68 PARALLEL PROCESSES HAVING IN MIND THEIR STATICAL AND/OR DYNAMICAL PRIORITY. THE SYSTEM ALLOWS UP TO 128 DIFFERENT INPUT DATA STRUCTURES. SYSTEM MODES NUMBER CAN BE UP TO 32. THE AVERAGE TIME FOR SWITCHING FROM ONE PROGRAM MODULE TO ANOTHER IS UNDER 1MS FOR MC 6800.

BESIDES THE ABOVE-MENTIONED PROGRAMMES IT CONTENTS INPUT/OUTPUT DRIVERS, A FLOPPY DISK

CONTROLLER, A DIALOG HANDLER, TRANSLATOR AND INTERPRETER OF A USER ORIENTED LANGUAGE FOR TECHNOLOGICAL PROCESSES DESCRIPTION.

CONCLUSIONS

THE ROBOT DEVELOPMENT SETS A SERIOUS PROBLEMS WITH RESPECT TO SOFTWARE DESIGN. IT IS NECESSARY TO STANDARTIZE THE DIFFERENT EXISTING SYSTEMS AND CONSEPTIONS. THE INPUT AND OUTPUT FORMATS AND REALIZED FUNCTIONS SHOULD BE EQUALIZED. THERE IS ALSO A TENDENCY TO MAKE EASY FOR THE PROGRAMMER DESIGNING THE NEW PROBLEM, WHICH TURNS TO ROUTINE AFTER SEVERAL REALISATIONS. THE PRESENTED PAPER GIVES AN APPROACH TO THIS ASPECT, WHICH HAS BEEN EXPERIMENTED PRACTICALY IN ISOMATIC LABS AT IICR, WHEN DESIGNING AN ARC WELDING ROBOT IZATRON RB-251. NOW, THE WORK ON THE REX-SYSTEM IS CONTINUED FOR USE IN CNC MACHINES, ARTICULATED ASSEMBLY ROBOT AND ANTHROPOMORPHIC WELDING ROBOT.

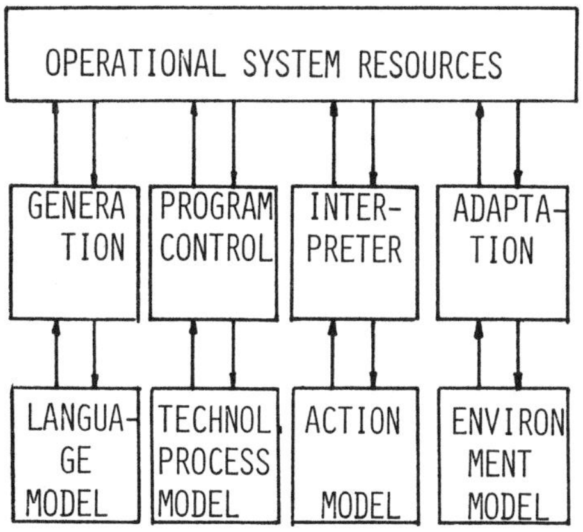

FIG. 1

REFERENCES

1.BLAGOEV L., MINI-COMPILER IN INTERACTIVE MODE FOR THE ROBOT CONTROL SYSTEM, PROCEEDINGS OF THE INT. SYMPOSIUM, BRATISLAVA, 1980

2.KARASTOJANOV D., L. BLAGOEV, SEMANTIC ANALYSIS IN THE INDUSTRIAL ROBOTS LANGUAGES, ANNUAL REPORT, IICR-BAS, VOL. 14, 1981

3.KARASTOJANOV D., V. CHRISTOV, A. ANDREEV, PROBLEM-ORIENTED HIGH-LEVEL REAL-TIME LANGUAGES INTERPRETATION, ANNUAL REPORT, IICR-BAS,V. 13, 1981

4.NATCEV G., D. KARASTOJANOV, GEOMETRIC ADAPTATION IN THE CONTROL OF ARC-WELDING ROBOT, 2ND NATIONAL SESSION, INDUSTRIAL ROBOTS, S. ZAGORA, 1980

5.CHRISTOV V.,L. BLAGOEV, E. NICOLOVA, D. KARASTOJANOV, B. MANAHILOV, M. AVRAMOVA, AN APPROACH TO REAL-TIME CONTROL SYSTEM DEVELOPMENT, INTERN. SYMP. " INDUSTRIAL ROBOTS CONTROL PROBLEMS",VARNA 1981, OCT.

6.PETROV E., L. BLAGOEV, AN APPROACH TO GENERATION A HIGH-LEVEL LANGUAGE FOR A DIALOGUE MONITOR DESIGH, 2ND INT. SYMP. ASI, VARNA, 1983

7.MANAHILOV B., M. AVRAMOVA, HIGH-LEVEL LANGUAGE FOR AUTOMATEG DESIGH OF MATHEMATICAL PROCESSING WITH MICROPROCESSOR, 2ND INT. SYMP. ASI, VARNA, 1983.

A ROBUST MRAC FOR INDUSTRIAL ROBOTS

Professor Fernando Nicolò
Department of Systems and Computer Science
University of Rome
Via Eudossiana 18
00184 ROME, ITALY

James Katende
Department of Elect. Engineering
University of Lagos
Lagos, NIGERIA

ABSTRACT

Investigations have shown that the main problem in continuous path control of Industrial Robots, is taking into account the unpredictable parametric variations in the robot dynamics especially with respect to friction and varying payloads.

The structure of the dynamic equations of industrial robots allows the application of a new robust Model Reference Adaptive Control (MRAC) technique.

The proposed MRAC algorithm, based on hyperstability theory, may be combined with the use of constant feedback on the tracking error. Here a case study is presented and the results of the simulation runs are discussed highlighting the robust performance with respect to parameter variations.

INTRODUCTION

As industrial robots gain acceptance in terms of applicability and reliability, higher performance requirements are increasing demanded. Robot control theory is currently at a satisfactory level of advancement, however, some work remains to be done in the domain of designing efficient methods for real-time dynamic control. In general the control problem is complicated by the highly coupled nonlinear dynamics of industrial robots and the unpredictable variation of the mechanical characteristics. Several available robot control techniques have a draw-back of not only requiring the detailed description of the model but also an apriori knowledge of the payload characteristics [1-5].

As an alternative, in recent years, adaptive control techniques have been used to maintain good performance over a wide range of robot motions and payloads. Among various adaptive schemes, the MRAC approach has enjoyed considerable attention [6-10].

The salient feature of this approach is that one can specify the desired performance in terms of the characteristics of a given linear system [Reference Model]. The controller is then designed in such a way that the closed loop control system has the same characteristics as those of the reference model.

In order to design a robust MRAC system for linear time-invariant plants, it is necessary to fulfill the perfect model-matching conditions and a suitable condition on the input matrix [11]. For the case of a nonlinear plant it is enough to introduce one additional condition [12].

In this paper an application of this MRAC technique to the dynamic control of industrial robots is presented. Simulation results for a three joint robot show its feasibility in real-time control.

PROBLEM FORMULATION

We consider the dynamic control of an industrial robot, Fig.1, whose end-effector is required to move along a continuous path such that velocity, acceleration and deceleration along the path are controlled.

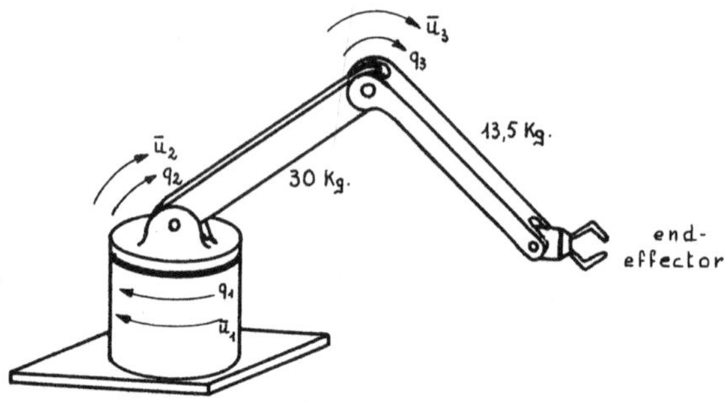

Fig.1

The dynamic equation of the robot is based on the Lagrangian formulation:

$$\bar{u} = \frac{d}{dt}\left(\frac{\partial L}{\partial \dot{q}}\right) - \frac{\partial L}{\partial q} \tag{1}$$

where

$\bar{u}$ are generalized forces to be supplied by the joint actuators

q and $\dot{q}$ are generalized positions and velocities.

The Lagrange function, L, is expressed in terms of the system kinetic energy, $T(q,\dot{q})$ and potential energy $P(q)$:

$$L = T(q,\dot{q}) - P(q) \tag{2}$$

The kinetic energy is given by

$$T(q,\dot{q}) = \frac{1}{2}\dot{q}'M(q)\dot{q} \tag{3}$$

where the inertia matrix $M(q)$, is symmetric and positive definite (the prime indicates transpose).

Thus, including a friction force vector, $H\dot{q}$, we have for the manipulator motion

$$\ddot{q} = M^{-1}(q)\left[-f_q(q,\dot{q}) + G(q) + \bar{u}\right] \tag{4}$$

where the generalized gravity force is

$$G(q) = \frac{-\partial P(q)}{\partial q} \tag{5}$$

and the components of $f_q(q,\dot{q})$ represent friction, centrifugal and Coriolis forces given by

$$f_{qi}(q,\dot{q}) = \dot{q}'F_i(q)\dot{q} + H_i\dot{q}_i \tag{6}$$

The relationship between matrices $M(q)$ and $F_i(q)$ is

$$F_i = \frac{1}{2}\left[\frac{\partial m_i}{\partial q} - \frac{\partial M(q)}{\partial q_i} + \left(\frac{\partial m_i}{\partial q}\right)'\right] \tag{7}$$

where

m_i is the i-th column of $M(q)$.

The functions in eqns.(4)-(7) depend on some physical parameters which are arranged in vector α.

Choosing state vector x such that $x_i=q_i$ and $x_{i+j}=\dot{q}_i$ for $i=1,2,\ldots j$, where j is the number of joints, eqn.(4) can be put in the general from

$$\dot{x} = f(\alpha,x) + B(\alpha,u)u \qquad x(o) = x_o \tag{8}$$

where

$x \in \mathbb{R}^n$, $\alpha \in \mathcal{P}$, $f \in \mathbb{R}^n$ $\{f: (\mathcal{P}\times\mathbb{R}^n) \to \mathbb{R}^n\}$ is a nonlinear vector function, $B(\alpha,x) \in \mathbb{R}^{n\times r}$ is a matrix function and $u \in \mathbb{R}^r$ is a new control vector such that $\bar{u}=M_o u$, ($M_o \in \mathbb{R}^{r\times r}$ being a constant nonsingular matrix).

If the path of the end-effector is prescribed then one can know exactly the values of desired position, y, desired linear velocity, $\dot{y}$ and the desired linear acceleration $\ddot{y}$ of the end-effector on the entire path. Suitable kinematic transformations exist such that the desired generalized position q_d, angular velocity $\dot{q}_d$ and acceleration $\ddot{q}_d$ can be computed, thus defining the desired state trajectory.

We incorporate the desired trajectory in a linear time-invariant reference model represented by

$$\dot{x}_m = A_m x_m + B_m g \tag{9}$$

where

$x_m \in \mathbb{R}^n$ is the model state vector

$g \in \mathbb{R}^r$ generates the desired motion

A_m and B_m are constant matrices of proper dimensions.

In particular, generator g and the model system are to be selected such that $x_{mi}=q_{di}$ and $x_{m_{i+j}}=\dot{q}_{di}$ $(i=1,2,\ldots j)$.

The end-effector trajectory tracking error is then expressed as the error between the plant and the model states;

$$e = x - x_m \tag{10a}$$

and

$$\dot{e} = A_m e+B(\alpha,x)u+f(\alpha,x)-A_m x-B_m g \tag{10b}$$

with

$$e(o) = e_o$$

Thus the control task is to ensure that $\dot{e}$ and e approach zero asymptotically.

It is known that a MRAC system which is robust with respect to variations in α can be designed for linear time-invariant plants which fulfill perfect model-matching conditions and suitable conditions on the input matrix [11]. These conditions can be extended to the nonlinear case through the following new condition [12].

Separability Condition:

Matrix and vector functions $L(\alpha)$ and $z(x,g)$ exist of proper dimensions such that

$$B^{\#}(\alpha,x)\left[f(\alpha,x)-A_m x-B_m g\right]=L(\alpha)z(x,g)+u_o(x,g) \tag{11}$$

for $\alpha \in \mathcal{P}$, $x \in \mathbb{R}^n$ and $g \in \mathbb{R}^r$

where $B^{\#}$ is the Penrose pseudoinverse of $B(\alpha,x)$.

The extensions are then given as

Perfect model-matching condition:

The considered plant and reference model are such that

$$\left[I-B(\alpha,x)B^{\#}(\alpha,x)\right]\left[f(\alpha,x)-A_m x-B_m g\right] = 0 \tag{12}$$

for all $\alpha \in \mathcal{P}$, $x \in \mathbb{R}^n$ and $g \in \mathbb{R}^r$

Input matrix condition:

A constant matrix B_o and a positive definite symmetric matrix $R(\alpha,x)$ exist such that

$$B(\alpha,x) = B_o R(\alpha,x) \tag{13a}$$

and

$$2R(\alpha,x) - I > 0 \tag{13b}$$

where I = identity matrix of proper dimensions.

If the conditions are satisfied eqn.(10b) may be rewritten as

$$\dot{e}=A_m e+B_o R(\alpha,x)\left[u+L(\alpha)z(x,g)+u_o(x,g)\right] \tag{14}$$

which represents a linear time-varying system whose dynamics are perturbed by the signal $L(\alpha)z(x,g)+u_o(x,g)$.

The strategy is thus to synthesize the control signal u which compensates the disturbing effect of the signal $Lz + u_o$.

CONTROLLER DESIGN

We consider an adaptive controller for the end-effector movement of an anthropomorphic robot having three revolute joints.

The same reference model is chosen for each joint in the form,

$$\ddot{x}_{mi}+2\xi\omega_n\dot{x}_{mi}+\omega_n^2 x_{mi}=\omega_n^2 g_i \qquad i=1,2,3 \tag{15a}$$

Equivalently putting $a_o=\omega_n^2$ and $a_1=2\xi\omega_n$ we can write,

$$g = (\ddot{q}_d + a_1\dot{q}_d + a_o q_d)/a_o \tag{15b}$$

From the robot model, parameter matrix $L(\alpha)$ and vector $z(x,g)$ are chosen so that

$$L(\alpha)z(x,g)=\begin{bmatrix} \alpha_2 & \alpha_3 & \alpha_4 & 0 & 0 & 0 & 0 & 0 & 0 & \alpha_1 & \alpha_2 & \alpha_3 & \alpha_4 & 0 & 0 & 0 & 0 & \alpha_{11} & 0 & 0 \\ 0 & 0 & 0 & \alpha_2 & \alpha_3 & \alpha_4 & 0 & \alpha_9 & \alpha_{10} & 0 & 0 & 0 & 0 & \alpha_4 & \alpha_5 & \alpha_6 & \alpha_7 & 0 & \alpha_{12} & 0 \\ 0 & 0 & 0 & 0 & \alpha_3 & 0 & \alpha_4 & 0 & \alpha_{10} & 0 & 0 & 0 & 0 & \alpha_7 & \alpha_6 & \alpha_8 & 0 & 0 & 0 & \alpha_{13} \end{bmatrix} \cdot \begin{bmatrix} \dot{x}_1\dot{x}_2\sin(2x_2) \\ (\dot{x}_1\dot{x}_2+\dot{x}_1\dot{x}_3)\sin(2(x_2+x_3)) \\ \dot{x}_1\dot{x}_2\sin(2x_2+x_3)+\dot{x}_1\dot{x}_3\sin(x_2+x_3) \\ -\dot{x}_1\dot{x}_1\sin(2x_2)/2 \\ -\dot{x}_1\dot{x}_1\sin(2(x_2+x_3))/2 \\ -\dot{x}_1\dot{x}_1\sin(2x_2+x_3)/2+(x_2+x_3/2)x_3\sin x_3 \\ -\dot{x}_1\dot{x}_1\cos x_2\sin(x_2+x_3)/2-\dot{x}_2\dot{x}_2\sin x_2 \\ \cos x_2 \\ \cos(x_2+x_3) \\ -G_1 \\ -G_1\cos^2 x_2 \\ -G_1\cos^2(x_2+x_3) \\ -G_1\cos x_2\cos(x_2+x_3) \\ -G_2\cos x_3 \\ -G_2 \\ -G_3 \\ -G_3\cos x_3 \\ -\dot{x}_1 \\ -\dot{x}_2 \\ -\dot{x}_3 \end{bmatrix} \tag{16}$$

where $G_i = \ddot{q}_{di} + a_o e_i + a_o \dot{e}_i$

$$i = 1,2,3$$

$\dot{x}_i$ and $\dot{x}_i$ are components of x

In this case $u_0(x,g)=0$. Putting g as a function of $e=x-x_m$ simplifies the computation of some components of $z(x,g)$ and we note that with the chosen control structure, the inputs to the system are q_d, $\dot{q}_d$ and $\ddot{q}_d$ and only these need be changed whenever a new trajectory is prescribed.

With reference to eqn.(15) it is clear that we may arbitrarily choose a_0 and a_1. Assume that with a suitable choice of A_m it is possible to consider $R(\alpha,x_m)$ to be constant during the variation of e. Thus if e is negligible compared with x_m, we can assume that $R(\alpha,x)$ is also constant, and the usual adaptation loop can be designed.

Obviously, A_m would be stable and hence we can find a symmetric positive definite matrix P such that

$$A_m' P + P A_m < 0 \tag{17}$$

The condition on the input matrix, given above is fulfilled by

$$R(\alpha,x) \stackrel{\Delta}{=} M_o M^{-1}(\alpha,x) M_o \tag{18}$$

$$B_o = \begin{bmatrix} 0 \\ \hline M_o^{-1} \end{bmatrix}$$

with a suitable M_o.

An adaptive signal, u_a, is synthensized, for compensating the disturbance due to parameter variations, manifested in signal Lz (cfr. eqn.(14).

$$u_a = \hat{L}(t)\ z(x,g) \tag{19}$$

$\hat{L}$ is the estimate of L given by

$$\hat{L} = \int_0^t \phi(v_1(\tau),t,\tau)d\tau + L(\alpha_o) \tag{20}$$

where

$$v_1 = B_o' Pe \tag{21}$$

and ϕ is a matrix of proper dimensions.

Let L_{ij} be the entries of $L(\alpha)$ and define constant matrices S_j with rows $\sigma_{ij}=0$ for each pair of indices i,j such that $L_{ij}=0$. The columns of ϕ_j are given by

$$\phi_j(\tau) = S_j\ v_1(\tau)\ z_j(x(\tau),g(\tau)) \tag{22}$$

The non-zero elements of matrix S_j are the adaptation gain factors and may be chosen all equal.

By arranging that

$$u = -u_a \tag{23}$$

we have the adaptation loop depicted in Fig.2, where the adaptation mechanism is described by (20) and (22) and the variable w_1, is given by

$$w_1 \stackrel{\Delta}{=} R(\alpha,x)\left[-u_a + L(\alpha)\ z(x,g)\right] \tag{24}$$

The adaptation loop is stable because the linear time variant system with input w_1 and output v_1 is strictly positive real, as a consequence of (17) and (21), and the nonlinear time varying system with input v_1 and output w_1 satisfies the Popov integral inequality:

$$\int_0^{t_1} v_1'(t)w_1(t)dt = \int_0^{t_1} v_1'(t)R(\alpha,x)\left[-u_a(t) + L(t)z(x(t),g(t))\right]dt < 0 \tag{25}$$

Since $R(\alpha,x)$ is positive definite, this inequality is obviously satisfied.

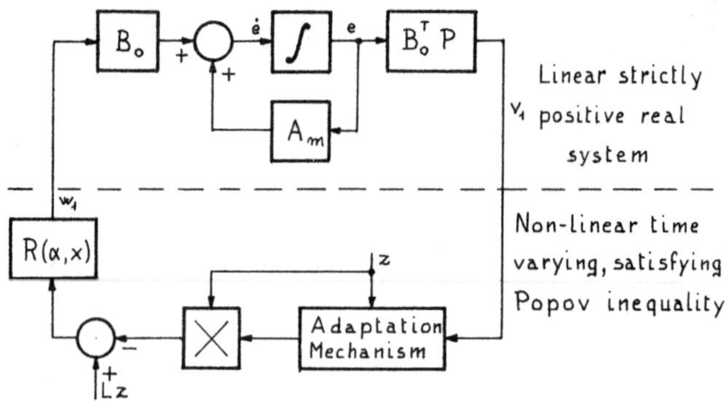

Fig.2

The actual control system is represented by Fig.3.

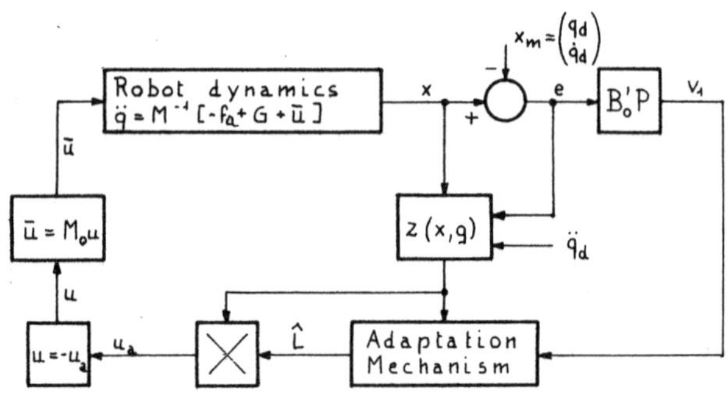

Fig.3

As an alternative to choosing of A_m such that $R(\alpha,x_m)$ can be assumed to be constant, we propose the use of an additional linear feedback given by

$$v = K\ B_o'\ Ne \tag{26}$$

where $N' = N > 0$ is a solution of the steady state algebraic Riccati equation

$$(A_m+\gamma I)' N + N(A_m+\gamma I) - KNB_oB_o'N + Q = 0 \tag{27}$$

Constant $K > 0$ and matrix $Q' = Q > 0$ are arbitrary.

The control signal is then given by $u=-v-u_a$ with u_a obtained from (18) and (19) by using v instead of v_1.

The resulting error system is shown in Fig.4.

Parameter $\gamma \geq 0$ is chosen to prescribe the degree of stability of the linear closed loop system in order to validate the hypothesis that $R(\alpha,x)$ is constant during the adaptation process. Actually the closed-loop dynamic matrix is given by

$$A_c(\alpha,x) = A_m - K\ B_o\ R(\alpha,x)\ B_o'\ N \tag{28}$$

and its eigenvalues $\lambda_i(\alpha,x)$ satisfy the inequality

$$\mathrm{Re}\left[\lambda_i(\alpha,x)\right] < -\gamma \tag{29}$$

It can be shown that the linear time-invariant system and the non-linear time-varying system satisfy the same conditions as before [12] and thus the adaptation loop is stable as a consequence of the input matrix condition (13).

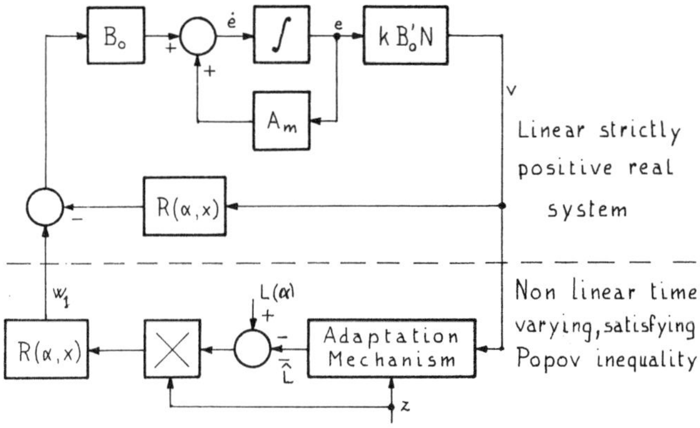

Fig.4

The control structure is shown in Fig.5.

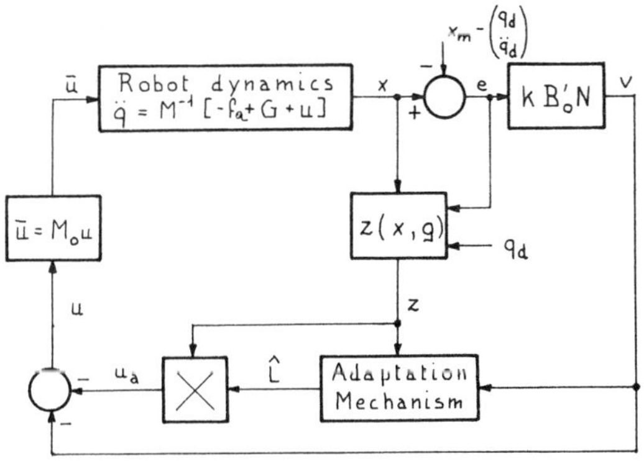

Fig.5

SIMULATION RESULTS AND CONCLUSION

Digital simulation was performed to investigate the effectiveness of the control system. The controller design was based on a dynamic model of the robot which was assumed to have an electrical-torque motor actuating each degree of freedom through a reduction gear without backlash. The physical parameters of the robot are relative to a structure designed to carry a nominal payload of 5 kg with a maximum linear speed of 0.75 m/s.

The upper arm of length 0.5 m weighs 30 kg while the lower arm of same length weighs 13.5 kg (without payload).

The hypothetical simulated problem was that the end-effector has to travel along a straight continuous path of 1.5 m in three seconds. The end-effector has to acquire maximum velocity with constant acceleration of 0.75 m/s^2 and is brought to a stop with the same amount of deceleration.

The first set of simulation runs were based on the control system of Fig.3, for which the reference model parameters were $a_0=100$, and $a_1=14$. ($\omega_n=10$ rad/s and $\xi=0.7$). The second set of simulation runs was for the control system of Fig.5 in which a reference model with smaller parameters was used.

In both sets of runs we studied the effect of parameter variations also for the case in which the continuous control is approximated with a discrete control up to a maximum sampling rate of 10 ms. The adaptation gain factors are all equal to a value σ.

Simulation results are summarized in Table I in which the maximum errors during the actual trajectory are reported.

With continuous controls the results are good with respect to parameter variations. In the case of discretization sampling periods lower than 10 ms give satisfactory results, but like the table shows, for the case of T=10 ms, the situation worsens mainly because σ has to be lowered, since large values in this case result in instability.

In general the best choice of σ is much lower with linear feedback, but the closed-loop system is more oscillatory, for this reason $\xi=1$ has been used in the model.

Finally we note that the results appear to be better than those obtained in the case of nonlinearity compensation and decoupling [4].

Table I

MAXIMUM TRACKING ERROR IN mm

Nominal payload: 5 kg Maximum linear speed: 0.75 m/s Maximum linear acceleration: 0.75 m/s^2	CONTR. SYSTEM FIG. 3 (NO LINEAR FEEDBACK)		CONTR. SYST. FIG. 5 LINEAR FEEDB. $\gamma = 0.5$	
	Continuous Control σ=1.0E+6	Discrete Control T=10ms σ=50.000	Continuous Control σ=6500	Discrete Control T=10ms σ=3500
Nominal condition	.29	.29	.23	.72
Load variation -5 kg	.29	2.9	.23	2.8
Load variation +5 kg	.32	3.1	.27	2.9
Friction(10% max torque at max speed)	.29	.57	.39	.92
Friction and load variation (-5 kg)	.35	2.3	.44	3.8
Friction and load variation (+5 kg)	.32	2.3	.43	4.3
Reference Model Parameters	a_o=14	a_1=100	a_o=1.5	a_1=2.5

REFERENCES

[1] D.E. Whitney: *Resolved Motion Rate Control of Manipulators and Human Prostheses*. IEEE Trans. MMS-10, No.2, June 1969.

[2] M.H. Raibert and B.K. Horn: *Manipulator Control Using the Configuration Space Method*. Industrial Robot, Vol.5, June 1978.

[3] M. Renaud and J. Zabala: *Robot Manipulator Arm Control*. Proc. 9th ISIR Washington USA 1979.

[4] S. Nicosia, F. Nicolò and D. Lentini: *Dynamical Control of Industrial Robots with Elastic and Dissipative Joints*. 8th IFAC Triennal World Conf. Kyoto, Japan, August 1981.

[5] J. Luh, W. Fisher: *Joint Torque Control by a Direct Feedback for Industrial Robots*. IEEE Trans. AC-28, No.2, February 1983.

[6] S. Dubowsky and D.T. Des Forges: *The Application of Model Referenced Adaptive Control to Robotic Manipulators*. Trans. ASME, J. Dynam. Syst. Meas. and Contr., Vol.101, September 1979.

[7] M. Tomisuka, R. Horowitz and I.D. Laudau: *On the Use of the Model Reference Adaptive Control Techniques for Mechanical Manipulators*. 2nd IASTED Symp. on Modelling, Identification Control and Robotics, March 1982, Davos.

[8] M. Takegaki and S. Arimoto: *An Adaptive Trajectory Control of Manipulators*. Int. J. Contr., Vol.34, No.2, 1981.

[9] A. Balestrino, G. De Maria and L. Sciavicco: *Feedback and Feed Forward in Adaptive Model Following Control of a Class of Nonlinear Time-Varying Plants*. Workshop on Adaptive Control, Florence-Italy, 377-389, October 1982.

[10] G.S.G. Lee: *Robot Arm Kinematics, Dynamic and Control*. Computer, Vol.15, No.12, December 1982.

[11] I.D. Laudau: *Adaptive Control: The Model Reference Approach*. Marcel Dekker, New York, 1979.

[12] I.V. Miroshnik, S. Nicosia, F. Nicolò: *Robust Model Reference Adaptive Control*. Workshop on Adaptive Control, Florence-Italy, October 1982.

Supported with a grant of M.P.I. Italy

AN ADAPTIVE SEAM-TRACKING SUBSYSTEM FOR ARC-WELDING ROBOTS

D. KARASTOJANOV G. NACHEV B. MANAHILOV

INSTITUTE OF INDUSTRIAL CYBERNETICS AND ROBOTICS, IICR, BAS, 1113 SOFIA, BULGARIA

ABSTRACT

THIS ARTICLE DEALS WITH THE PROBLEM OF SUPPLYING AN ARC-WELDING INDUSTRIAL ROBOT WITH ADAPTIVE FUNCTIONS BY ADDING AN ADAPTIVE SUBSYSTEM, WHICH WORKS PARALLEL TO THE BASIC CONTROLLER OF THE ROBOT AND GENERATES CONTINUOUSLY CORRECTING SIGNALS TO THE TAUGHT-IN SEQUENCE OF MOVEMENTS DURING THE WELDING PROCESS. A SOLUTION TO THE COMPOSITE SUBPROBLEMS OF OBSERVATION, BUILDING THE ENVIRONMENT'S MODEL, DECISION-MAKING AND TAKING THE CORRECTIVE ACTION IS PROPOSED. A REALISATION, MADE IN IICR, SOFIA IS ALSO PRESENTED, AND THE EXPERIMENTS WITH IT ARE DESCRIBED.

INTRODUCTION

THE APPLICATION OF MODERN INDUSTRIAL ROBOTS FOR PROCESSING, A GIVEN LOT OF PARTS, IS CHARACTERISED BY TWO STAGES-TEACHING AND EXECUTION. THE FIRST STAGE IS NECESSARY BECAUSE THE ROBOT IS NOT ABLE TO SET GOALS TO ITSELF AND BECAUSE ITS CONTROL SYSTEM HAS A LIMITED CAPACITY TO ACCEPT AND PROCESS DIFFERENT TYPES OF INFORMATION. THEREFORE EVERY ROBOT IS AN AUTONOMUS EXECUTOR WITH VARYING DEGREE OF CAPACITY TO PERFORM TASKS WITH CORRESPONDING COMPLEXITY. HERE THE PROBLEM FOR THE IDENTITY OF CONDITIONS DURING THE EXECUTION AND TEACHING PHASE ARISES. THESE CONDITIONS CONCERN THE STATES OF THE ROBOT AND ITS WORKING ENVIRONMENT. MOST EXISTING ROBOTS WORK IN A STATIC AND STRICTLY DEFINED ENVIRONMENT, IDENTICAL TO THAT AT THE TEACHING PHASE. THE CHANGES IN THE STATE OF THE ROBOT ARE KNOWN AND ARE PART OF ITS CONTROL ALGORITHM, WHICH CAN BE CONSIDERED AS OPTIMAL ACCORDING TO THE RESPECTIVE CRITERION,(2).

TASKS OF THE ADAPTIVE CONTROL

THE PRESENT MANUFACTURING TECHNOLOGY AND THE LIMITS OF PRODUCTION COSTS IT IMPOSES LEAD TO NON-COMPLIANCE WITH THE ABOVE MENTIONED FORMULATION. IT IS WELL KNOWN THAT IN ARC WELDING CONSIDERABLE DEVIATIONS FROM THE TAUGHT-IN MODEL MAY OCCUR DUE TO MANUFACTURING TOLERANCES, INADEQUATE FIXTURING, HEAT DEFORMATIONS AND PROGRAMMING. THE SYSTEM IN THIS CASE IS QUASIOPTIMAL AND PASSIVELY ADAPTING WITHIN THE RANGE OF SMALL DEVIATIONS IF THEY BECOME LARGE, TO ACHIEVE HIGH QUALITY WELDS INTRODUCTION OF ADAPTIVE PROPERTIES IS NECESSARY,(3).

WE ASSUME FOR THE FOLLOWING DISCUSSION, THAT THE ROBOT CONTROL PROGRAM CONTAINS THE POINT COORDINATES AND THE TRAJECTORY TYPE (STRAIGHT LINE, CIRCLE ARC ETC.). THE MOVEMENT ALONG A PATH IS REALISED BY SIMULTANEOUS CONTROL OF ALL JOINTS WITH PREDETERMINED SPEED AND DIRECTION. ANY CHANGE IN THE SPEED AND DIRECTION OF JOINT MOVEMENT LEADS TO DEVIATION FROM THE PROGRAMMED TRAJECTORY. THUS THE ADAPTATION TO THE GEOMETRICAL DEVIATIONS IS REDUCED TO A SEQUENCE OF DECISIONS ABOUT THE TYPE, SIZE AND DIRECTION OF THE JOINT-MOVEMENT CORRECTIONS, MADE ON THE BASIS OF INFORMATION FROM THE ENVIRONMENT AND AIMED AT COMPENSATING INACCURACIES OF A KNOWN CHARACTER,(1).

ON THE BASIS OF SEQUENTIAL AND LOGICAL CONNECTIONS, THE GENERAL TASK FOR ADAPTIVE BEHAVIOR OF AN INDUSTRIAL ROBOT CAN BE DECOMPOSED INTO THE FOLLOWING SUCCESSIVELY SOLVED SUBTASKS,(4):

-INFORMATION GATHERING FROM THE ENVIRONMENT
-INFORMATION PROCESSING AND DECISION-MAKING
-REALISATION OF THE CORRECTIVE ACTION

THE FOLLOWING PARAGRAPHS WILL ILLUSTRATE SOME SOLUTIONS TO THESE PROBLEMS, EXPERIMENTED ON THE ARC-WELDING ROBOT ISOTRON RB 250 IN THE IICR, SOFIA.

Observation of the object

Information gathering from the environment is based on measurement of an observable physical phenomenon, (cf Table 1). The arc itself can serve directly as a signal source (D), or some indirect (I) sources can be used. For convenience the result of the measurement is categorized as:

-ratio between signals
-distances less than 10 mm
-distances more than 10 mm

It is obvious that not all combinations from these three characteristics can be practically realised or applied in arc welding. Some possible means for information gathering being applied now are as follows:

-light sensor - group 1aD
-TV camera - group 2aD or 1cI
-inductive sensor - group 4bI
-mechanical tactile sensor - group 9bD or 9bI

IICR, Sofia developed a sensor for information gathering during welding based on a laser range finder, scanning the welded part's surface near the welding torch. The device is compact and comparatively cheap. It can be mounted remotely on the torch and the emitting range is chosen not to be influenced by the arc disturbances. Thus using a low power laser emitter distances within 50-300 mm can be measured accurately enough. Further discussion will be based on this sensor.

Deviation measurement

The processing of information received during the program execution depends on the characteristics of the enviroment and the sensor. We assume that the characteristic parameters describing the sensor realisation are previously known. If the sensor signal is an extremal function of environment parameters it is possible to build up an adaptive subsystem for searching a drifting extremum on a known type of signal characteristics. Otherwise it will be necessary to build an environmental model during teach-in, in particular by using the same observation mechanism as during programm execution. The task for determining the model parameters represents an object identification procedure based on a known object class type.

Fig. 1a and 1b show a particular configuration of parts for welding and the necessary characteristic parameters of the environment and the sensor. The type of the measured signal for one observation cycle is show on Fig. 2a. Based on the model, it is possible to generate the expected value of the signal for each measurement point in execution mode. Due to changes in the environmental parameters for the specific part, the actual value of the signal may vary within a sertain range. Fig. 2b shows the differences between the expected and the real signal value for one measuring cycle. Depending on the varying parameter khe following components of deviation for each part can be distinguished:

-translation - modification of parameter L
-rotation - modification of parameter K
-deformation - modification of parameters I or J respectively

Fig. 3a, 3b and 3c show the typical difference between those signals for each of those three cases during the execution phase in one observation cycle.

Corrective action

With the help of said information an adaptive subsystem has been synthesed which allows to compensate every component of the deviation for each joint, based on the minimum value of the quadratic mean error. The adaptive functions are realised by an additional contour, working in parallel to the basic robot control program and generating corrections for it. It is also possible to build an extra level which modifies the task for the highest hierarchical level of the control system.

IICR, Sofia has experimented an adaptive subsystem, working in parallel to an unadaptive controller. It processes the error between the actually measured signal and the signal generated on the basis of a previously built environmental model during teach-in. This error is further decomposed into translation, rotation and deformation components for each of the two parts being welded together. The results are transformed into the basic coordinate system of the robot and are added to the program control. The gain of the servoloops for the corrections are chosen in such a way that the error is compensated in the time of the next observation cycle.

Fig. 4 shows a flow-chart of an adaptive subsystem for seam-tracking in welding robots. The decision-making about the type of the part deviation from its taught-in position is realised according to the combination of deviation components for each of the two parts in the basic coordinate system of the robot.

The first robot to make use of the unique features of the new sensor is ISOTRON RB 250 - a cartesian coordinates robot, having a 5 axis on the torch manipulator and up to 4 servo-controlled axis for part orientation (fig. 5).

Conclusion

The adaptation to inaccuratelly produced or positioned parts in arc welding robots increases considerably their capabilities and the quality of welds. The noise generated by the arc and the uneven surfaces of the parts are a barrier to the successful realisation of adaptive functions. The advanced methods, developed in IICR however permitted to solve this problem and give good and reproducible results for straight seams between flat parts regardless of their configuration for deviations within the range of 10 mm. Work is being carried on for developing appropriate real-time models for dealing with non-flat parts.

References

1. Karastojanov D. and N. Bonev., Geometric adaptation in the control of arc welding robot., Automatic and computer design, v 4 1981, Sofia
2. Kuropatkin P., Optimale and adaptive systems., Moskow, Viysha scola, 1980
3. Nachev G., D. Karastojanov and B. Petkov., On the following of the weld and the control of robots for arc welding., Zavaryavane, 1982, No 1
4. Saridis G., Self-organizing control of stochastic systems., New York, Marsel Dekker, 1977

	A Ratio	B Distance than 10mm	C Distance than 10mm
1 Light-Normal, UV	D + / I	D / I	D / + I
2 Light - IR	D + / I	D / I	D / + I
3 Microwaves	D / I	D / I	D / + I
4 Radiofrequenses	D / I	D / + I	D / I
5 R, L, C of the Arc	D + / I	D / I	D / I
6 R, L, C of the part	D + / I	D / + I	D / I
7 Ultrasonic	D / I	D / + I	D / I
8 Pneumo-effects	D / I	D / + I	D / I
9 Tactile elements	D / I	D + / + I	D / I
10 Radiation	D / + I	D / I	D / I

Table 1

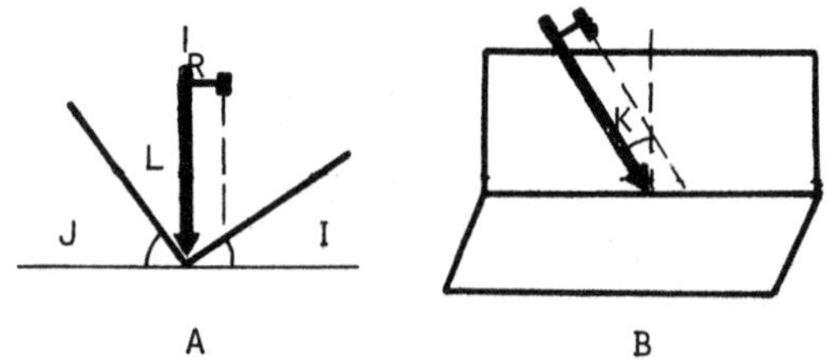

Fig. 1

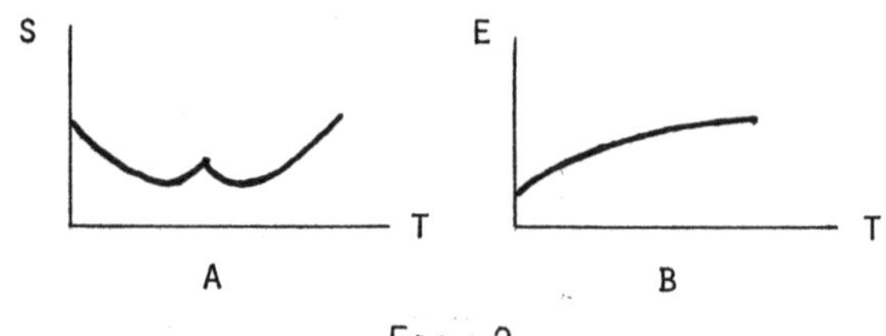

Fig. 2

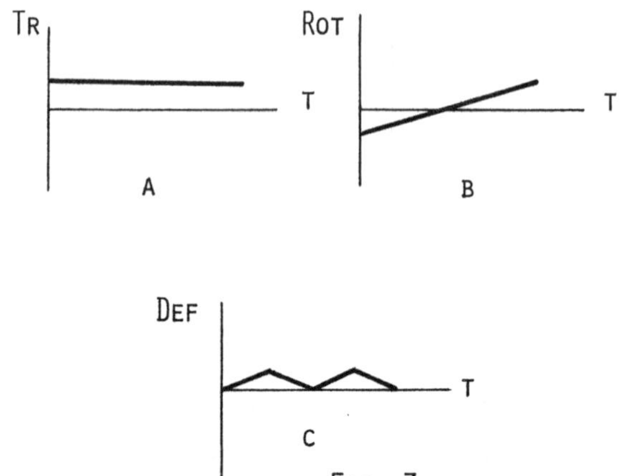

Fig. 3

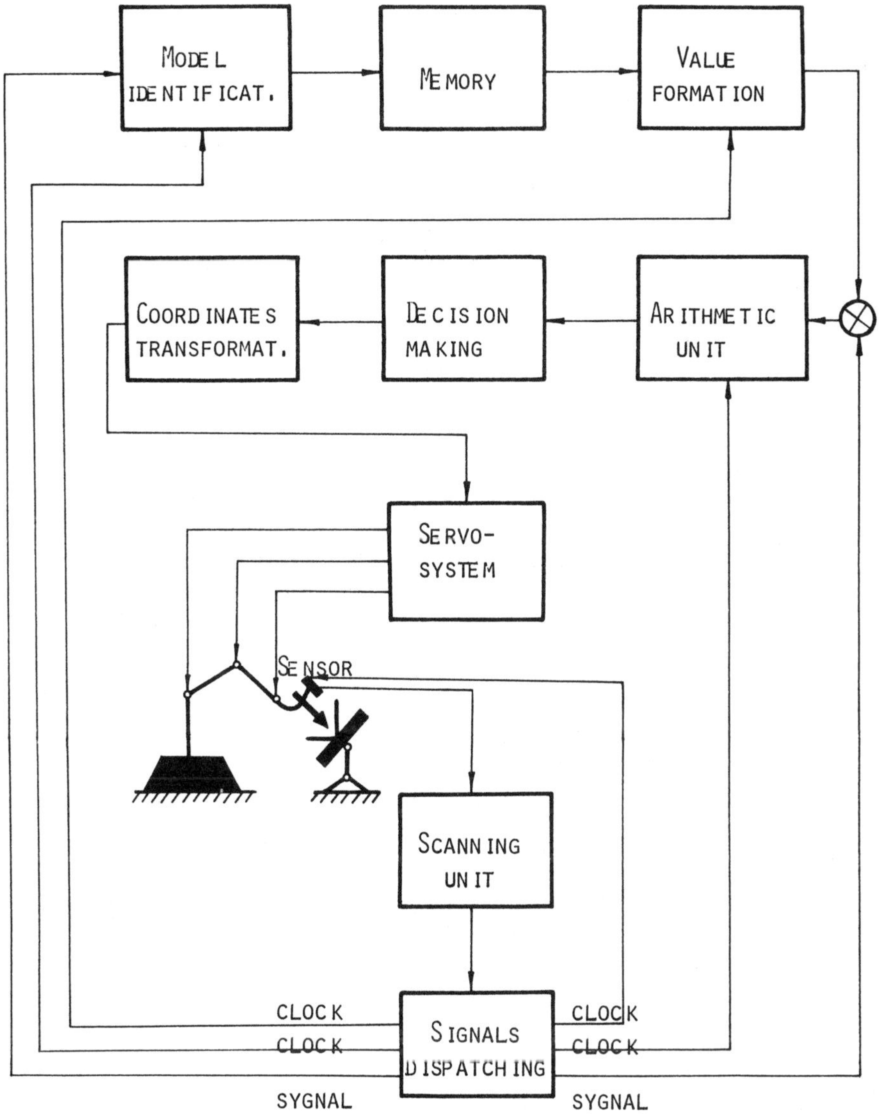
MODEL IDENTIFICAT.
MEMORY
VALUE FORMATION
COORDINATES TRANSFORMAT.
DECISION MAKING
ARITHMETIC UNIT
SERVO-SYSTEM
SENSOR
SCANNING UNIT
CLOCK
CLOCK
SIGNALS DISPATCHING
CLOCK
CLOCK
SYGNAL
SYGNAL

FIG. 4

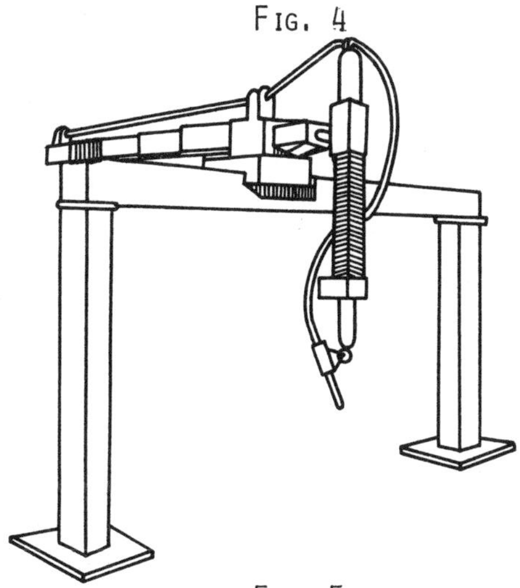

FIG. 5

STABILITY OF NON-LINEAR CONTINUOUS SYSTEMS

by

F. ROTELLA[1], J.P. RICHARD[2], I. ZAMBETTAKIS[1]

(1) Laboratoire de Systématique
UER d'IEEA
Université des Sciences et Techniques
59655 VILLENEUVE D'ASCQ - Cedex
FRANCE

(2) Laboratoire d'Automatique et d'Informatique Industrielle
Institut Industriel du Nord
B.P. 48
59651 VILLENEUVE D'ASCQ - Cedex
FRANCE

ABSTRACT. A special decomposition of the state space representation is proposed for continuous systems. This leads to the systematic definition of a candidate Liapounov's function.
After stating general conditions of stability, several important particular cases are pointed out. Lastly, the results are applied to Lure' Postnikov type systems.

INTRODUCTION.

The Liapunov's direct method allows to study the stability properties of a system equilibrium by introducing a comparison function.

For some system classes, such as Lure ' Postnikov type ones, particular constructions of Liapunov candidate functions have been proposed. Unfortunately, the choice of such a function is still a problem for more complex systems.

Considering the free motion state space representation of a general continuous differential system, we propose a special decomposition of evolution matrix coefficients. This particular configuration then induces a candidate function. Using appropriated majoration technics on its eulerian derivative, we determine sufficient conditions of stability.

The worked out conditions are very easy to check when reducing the study to a special conditioning of evolution matrices. In a first time we define "pseudo-antisymmetrical" matries and in a second time, "pseudo-symmetrical" ones. These two cases induce several important simplifications both on the considered model and on the stability methodology.

The presented work appears as a generalisation of classical criteria and of canonical representations.

Lastly, the particular case of a linear system with a monovariable non-linear feedback is considered as an example of application .

I - SYSTEM DESCRIPTION.

This study concerns non linear continuous systems described by a free motion state space representation such that :

$$(1) \qquad \overset{\circ}{x} = \frac{dx}{dt} = A(x,t)x$$

with
$$\begin{cases} t \in T = [o,+\infty[\\ x \in \mathbb{R}^q,\ x = (x_1,\dots,x_q)^T \text{ state vector} \\ A(x,t) = [\alpha_{ij}(x,t)],\ A(x,t) \in \mathbb{R}^{q\times q} \end{cases}$$

Moreover, the origine x=o is supposed to be the unique equilibrium state. We propose in this paper to study stability of this equilibrium by Liapounov's direct method /2/ based on the particular choice of a "candidate" function.

Let us consider the following special decomposition of evolution matrix coefficient /1/.

$$A(x,t) = \begin{bmatrix} \alpha_{11}(.) & a_{12}(.)f_2^* & \cdots & a_{1i}(.)f_i^* & \cdots & a_{1q}(.)f_q^* \\ a_{21}(.)f_1^* & \ddots & & \vdots & & \vdots \\ \vdots & & \ddots & a_{i-1i}(.)f_i^* & & \vdots \\ a_{i1}(.)f_1^* & \cdots\ a_{ii-1}(.)f_{i-1}^* & & \alpha_{ii}(.) & a_{ii+1}(.)f_{i+1}^* \cdots & a_{iq}(.)f_q^* \\ \vdots & & & a_{i+1i}(.)f_i^* & \ddots & \vdots \\ \vdots & & & \vdots & \ddots & a_{q-1q}(.)f_q^* \\ a_{q1}(.)f_1^* & \cdots & & a_{qi}(.)f_i^* & \cdots\ a_{qq-1}(.)f_{q-1}^* & \alpha_{qq}(.) \end{bmatrix}$$

* $a_{ij}(.) = a_{ij}(x,t)$ are q(q-1) scalar functions from $\mathbb{R}^q \times T$ to $\mathbb{R}$;

* $f_i(x_i)$ $i = 1,\dots,q$, are ϕ scalar functions everywhere defined and integrable on $\mathbb{R}$, continuous and zero at the origine x=o, and whose instantaneous gains are defined by :

$$(3) \qquad \forall i \in \{1,\dots,q\} \quad f_i^*.x_i = f_i(x_i) \qquad \forall x \in \mathbb{R}^q$$

Moreover, all the functions involved in this definition have to be such that matric coefficients $\alpha_{ij}(x,t)$ of A(x,t) are bounded for any bounded values of x.

If we except these bounded-values conditions, matrix A(x,t) (2) constitutesa special decomposition of any general q order matrix : it is obvious, when every instantaneous gain f_i is choosed equal to 1. This remark will allow the generalization of classical stability criteria applications .

II - STABILITY STUDY - GENERAL CONDITIONS.

The non constant proportionality coefficients f_i^* are introduced in order to facilitate the stability analysis by using a specific integral type Liapounov function.

We propose the following function, noted W(x) and defined from the evolution matrix (2) by /2/ :

$$(4) \qquad W(x) = \sum_{i=1}^{q} \int_{o}^{x_i} f_i(u_i)\,du_i$$

Considering the total derivative $\overset{o}{W}(x,t)$, it successively comes out :

$$(5) \qquad \overset{o}{W}(x,t) = \sum_{i=1}^{q} f_i(x_i)\,\overset{o}{x}_i$$

$$(6) \qquad \overset{o}{W}(x,t) = \sum_{i=1}^{q} f_i(x_i)\Big[\sum_{\substack{j=1\\ j\neq i}}^{j=q} a_{ij}(x,t) f_j(x_j) + x_i \alpha_{ii}(x,t)\Big]$$

$$(7) \qquad \overset{o}{W}(x,t) = \sum_{i=1}^{q} x_i f_i(x_i) \alpha_{ii}(x,t) + \sum_{i=1}^{i=q} \sum_{\substack{j=2\\ j>i}}^{j=q} f_i(x_i) f_j(x_j)(a_{ij}(x,t) + a_{ji}(x,t))$$

The sign indetermination of $\overset{o}{W}$, caused by the existence of rectangular terms $f_i(x_i)f_j(x_j)$, is then proposed to be removed by means of the following overvaluation :

$$(8) \qquad \forall (x,y) \in \mathbb{R}^2,\ \forall \omega \in \,]o,+\infty]$$

$$|xy| \le \frac{1}{2}\Big[\frac{x^2}{\omega} + \omega y^2\Big]$$

Therefore the total derivative $\overset{o}{W}(x,t)$ can be overvaluated as :

$$(9) \qquad \forall x \in \mathbb{R}^q,\ \forall t \in T$$

$$\forall \omega_{ij} \in \,]o, +\infty],\ (i,j) \in \{1,\ldots,q\}^2,\ i < j,$$

$$\overset{o}{W}(x,t) \le \sum_{i=1}^{q} x_i f_i(x_i) a_{ii}(x,t) + \sum_{i=1}^{q} \sum_{\substack{j=2\\ j>i}}^{q} \frac{|a_{ij}(x,t)+a_{ji}(x,t)|}{2} \Big(\frac{f_i(x_i)^2}{\omega_{ij}} + \omega_{ij} f_j^2(x_j)\Big)$$

Let us denote :

$$(10) \qquad \forall\ (i,j) \in \{1,\ldots,q\}^2 \quad \begin{cases} \delta_{ij} = \dfrac{1}{\omega_{ij}} & \text{if } j > i \\[2ex] \delta_{ij} = \omega_{ij} & \text{if } j < i \end{cases}$$

It comes out :

$$(11) \qquad \overset{o}{W}(x,t) \le \sum_{i=1}^{q} \Big[x_i f_i(x_i) a_{ii}(x,t) + f_i^2(x_i) \sum_{\substack{j=1\\ j\neq i}}^{q} \frac{|a_{ij}+a_{ji}|}{2}\delta_{ij}\Big]$$

$\overset{o}{W}(x,t)$ is negative or zero if its overvalue is negative or zero. This leads to the following stability conditions :

Theorem 1.

The equilibrium x=o of the system (1) (2) is stable if there exists a domain Ω of $\mathbb{R}^q$ including a neighbourhood of the origine x=o such that :

$$\forall x \in \Omega - \{o\},\ \forall t \in T,\ \forall i \in \{1,\ldots,q\}$$

a) $\displaystyle\int_{o}^{x_i} f_i(u_i)\,du_i > 0 \quad \forall x_i \neq 0$

b) $\displaystyle f_i^* \Big[\alpha_{ii}(x,t) + f_i \sum_{\substack{j=1\\ j\neq i}}^{q} \frac{|a_{ij}(x,t)+a_{ji}(x,t)|}{2}\delta_{ij}\Big] \le o$

If conditions b) are strict, the equilibrium x=o is asymptotically stable.

The introduction of arbitrary parameters δ_{ij} allows the determination of different stability domains. Nevertheless if the use of an overvaluation leads to obtain easy interpretable stability conditions, this is to the prejudice of stability domain precision.

III - GENERALISATION OF SOME CLASSICAL CRITERIA.

In this part we suppose that the diagonal terms of A(x,t) are also decomposed into :

$$(12) \qquad \forall i \in \{1,\ldots,q\}\ \alpha_{ii}(x,t) = a_{ii}(x,t)\, f_i^*$$

Under this hypothesis of decomposition, equation (1) becomes :

$$(13) \qquad \overset{o}{x} = \tilde{A}(x,t)\, F(x)$$

$$\text{where} \quad \begin{cases} \tilde{A}(x,t) = [a_{ij}(x,t)] \in \mathbb{R}^{q\times q} \\[1ex] F(x) = (f_1(x_1),\ldots,f_q(x_q))^T \in \mathbb{R}^q \end{cases}$$

The condition b) in theorem 1 is then :

$$(14) \qquad \forall x \in \Omega - \{o\},\ \forall t \in T,\ \forall i \in \{1,\ldots,q\}$$

$$a_{ii}(x,t) + \sum_{\substack{j=1\\ j\neq i}}^{q} \frac{|a_{ij}(x,t)+a_{ji}(x,t)|}{2}\delta_{ij} \le o$$

Moreover if we choose :

$$(15) \qquad \forall (i,j) \in \{1,\ldots,q\}^2\ \delta_{ij} = 1$$

(14) can be interpreted as the application of Krosembrock criterium /3/ to $\frac{A+A^T}{2}$. When the choice of decomposition is such that :

$$(16) \qquad \forall i \in \{1,\ldots,q\} \quad f_i^* = 1.$$

matrix $\tilde{A}(x,t)$ is equal to A(x,t) and theorem 1 appears as a generalisation of Rosembrock criterium /3/ on $\frac{A+A^T}{2}$.

On another hand, the form (13) allows to leave out the overvaluation on $\overset{o}{W}(x,t)$.

From (7) and (12) it comes out :

$$(17) \qquad \overset{o}{W}(x,t) = F^T(x)\, \frac{\tilde{A}(x,t) + \tilde{A}^T(x,t)}{2}\, F(x)$$

which is a quadratic form of F(x).

So we can state the following stability criterium :

Theorem 2.

The equilibrium x=o of system (13) is stable if there exists a domain Ω of $\mathbb{R}^q$ including a neighbourhood of the origine x=o such that :

$$\forall x \in \Omega - \{o\}, \quad \forall t \in T,$$

α) $\int_o^x F^T(u)du > o$ where $du = (du_1,\dots,du_q)^T \in \mathbb{R}^q$

β) $\tilde{A}(x,t) + \tilde{A}^T(x,t)$ is negative semi-definite /5/

Moreover if $\tilde{A}(x,t) + \tilde{A}^T(x,t)$ is negative definite /5/, then the equilibrium is asymptotically stable.

Ifwe choose here, for every i, $f_i^* = 1$, we find again a generalisation of another classical stability criterium : the Krasovski criterium /4/.

IV - PARTICULAR CASES.

IV.1 - Pseudo-antisymmetrical matrices.

We call "pseudo-antisymmetrical" matrix /6/ a matrix A(x,t) of type (2) such that :

(18) $\quad \forall (i,j) \in \{1,\dots,q\}^2 \qquad a_{ij}(.) = -a_{ji}(.)$

This is to say that, except for the diagonal terms, each column (i) of A(x,t) is proportional to a same non constant term f_i^*, and that the coefficients of proportionality are disposed in an antisymmetric way on both sides of first diagonal line.

For such systems, the choice of the Liapounov candidate function W(x) defined in (4) induces the following expression of $\overset{o}{W}(x,t)$:

(19) $\quad \overset{o}{W}(x,t) = \sum_{i=1}^{q} x_i f_i(x_i) a_{ii}(x,t)$

The overvaluation is not no more necessary and b) and β) conditions of previous theorems are respectively reduced to b') and β') ones :

b') $\quad \forall x \in \Omega - \{o\}, \forall t \in T, \forall i \in \{1,\dots,q\}\ f_i^* \alpha_{ii}(x,t) < o$

β') $\quad \forall x \in \Omega - \{o\}, \forall t \in T, \forall i \in \{1,\dots,q\}\ a_{ii}(x,t) < o$

We remark that the coefficients $a_{ij}(x,t)$ (i,j) are not appearing in the stability conditions a, and (b') or β')), wich are particularly easy to check conditions, with regard to a system that can present a lot of non-linearities.

IV.2 - Pseudo-symmetrical matrices.

In a same way, we call "pseudo-symmetrical matrices" the representations (13) verifying :

(20) $\quad \forall (i,j) \in \{1,\dots,q\}^2 \quad a_{ij}(.) = a_{ji}(.)$

In this case, the reduced matrix A(x,t), obtained from A(x,t) by omitting the terms f_i^*, is symmetrical. Equation (17) becomes :

(21) $\quad \overset{o}{W}(x,t) = F^T(x)\ \tilde{A}(x,t)\ F(x)$

This is particularly easy to interpretate : if the condition (a) of theorem 1 is verified, and if the matrix $\tilde{A}(x,t)$ verifies the linear systems stability conditions for any trajectory, then the stability of A(x,t) is ensured.

Two simplifications are then appearing : on the one hand, the studied non linear model $\tilde{A}(x,t)$ is simplified with regard to the original one A(x,t) ; on the other hand, the stability of this simplified model can be analysed as if it was linear.

Many non linear continuous processes belong to there two particular classes of systems and specially all the second order systems and several cases of Lur'e Postnikov type systems /2/.

V - APPLICATION TO MONOVARIABLE LUR'E POSTNIKOV TYPE SYSTEMS

In this part we will use a specific type of modelisation which can lead to a "pseudo-antisymmetrical" matrix.

Let us consider the free motion of the system described figure 1. It is composed of a linear part governed by a non constant gain $f^*(\varepsilon)$ depending on the input-output difference.

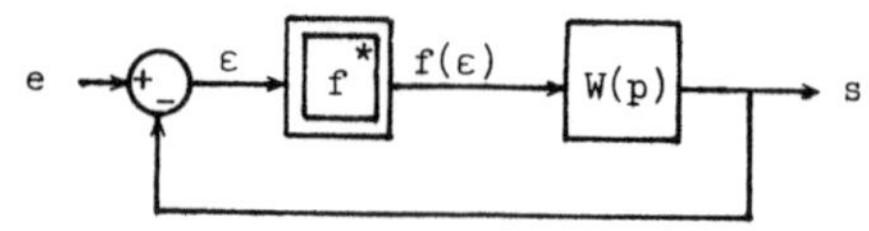

Figure 1.

with $\quad W(p) = \frac{N(p)}{D(p)}$

$$N(p) = \sum_{i=o}^{q-1} \hat{b}_i p^i,\ \hat{b}_{q-1} \geq o$$

$$D(p) = p^q + \sum_{i=o}^{q-1} \hat{a}_i p^i,\ \forall i \in \{o,\dots,q-1\}\ (\hat{a}_i,\hat{b}_i) \in \mathbb{R}^2$$

$$\forall \varepsilon \in \mathbb{R} - \{o\}\ f(\varepsilon) = f^*(\varepsilon).\varepsilon,\ f^*(o) = o$$

Throughout this part, W(p) is supposed to be non degenerated /8/ and $f^*(\varepsilon)$ is denote f^*.

V.1 - Modelisation.

Let us define the polynomial with non constant coefficient $P(\lambda,\varepsilon)$ by (22).

(22) $\quad \forall \varepsilon \in \mathbb{R}\quad P(\lambda,\varepsilon) = D(\lambda) + f^* N(\lambda)$

For the studied systems (figure 1), this polynomial

constitutes, if N(p) and D(p) have no common zero, an invariant of representation /9/. Using this property, previous works /7/ have pointed out that monovariable Lure' Postnikov type systems can be modelised by the state representation (23) :

$$(23)\qquad \overset{\circ}{x} = FE(\varepsilon)x,\quad x^T=[x_{12}x_{11}x_{22}x_{21}\cdots x_{t2}x_{t1}y_{t+1}\cdots y_r\varepsilon]\ \mathbb{R}^q$$

$$FE(\varepsilon) = \begin{bmatrix} \begin{matrix} a_1 & b_1 \\ -b_1 & a_1 \end{matrix} & & & & & 0 & \beta_1^{2*} \\ & & & & & & \beta_1^{1*} \\ & & \ddots & & & & \vdots \\ & & & \begin{matrix} a_t & b_t \\ -b_t & a_t \end{matrix} & & & \beta_t^{2*} \\ & & & & & & \beta_t^{1*} \\ & 0 & & & c_{t+1} & & \alpha_{t+1}^* \\ & & & & & \ddots & \vdots \\ & & & & & c_r & \alpha_r^* \\ 1 & \text{---} & \text{---} & \text{---} & \text{---} & 1 & \gamma^* \end{bmatrix}$$

with $r+t = q$ and $\forall\, \varepsilon \in \mathbb{R}$:

- $\forall i\in\{1,..,t\}$ if $R_i(\lambda) = \prod_{\substack{\ell=1\\ \ell\neq i}}^{t}((\lambda-a_\ell)^2+b_\ell^2)\prod_{\ell=t+1}^{r}(\lambda-c_\ell)$

$$\text{and}\quad A_i = \left[\frac{P(\lambda,\varepsilon)}{R_i(\lambda)}\right]_{\lambda=a_i+ib_i},\quad b_i\neq 0$$

$$\begin{cases} \beta_i^{2*} = \dfrac{1}{2b_i}(\text{Real } A_i - \text{Imag } A_i) \\ \beta_i^{1*} = -\dfrac{1}{2b_i}(\text{Real } A_i + \text{Imag } A_i) \end{cases}$$

$$\forall j\in\{1,..,t\}\ i\neq j \Leftrightarrow a_i\neq a_j \text{ or } |b_i| \neq |b_j|$$

- $\forall i\in\{t+1,..,r\}$ if $R_i(\lambda) = \prod_{\ell=1}^{t}((\lambda-a_\ell)^2+b_\ell^2)\prod_{\substack{\ell=t+1\\ \ell\neq i}}^{r}(\lambda-c_\ell)$

$$\alpha_i^* = -\left[\frac{P(\lambda,\varepsilon)}{R_i(\lambda)}\right]_{\lambda=c_i}$$

$$\forall j\in\{t+1,..,r\}\ i\neq j \Leftrightarrow c_i\neq c_j$$

- $\gamma^* = -\hat{a}_{q-1}-\hat{b}_{q-1}f\ -2\sum_{i=1}^{t} a_i - \sum_{i=t+1}^{r} c_i$

This form is a "thick arrow" matric modelisation/10/,it permits to generalize previous works /11/,/12/,/10/ which use "thin arrow" type representations. The advantage of representation (23) is that it can lead to a pseudo-antisymmetrical one if the nonconstant terms of last column are negatively proportionnal to a same nonconstant term. It is this point of view we are going exploit in the next part.

V.2 - Stability analysis.

Several authors /11/,/12/ have proposed to choose the arbitrary parameters appearing in the main diagonal of the representations like (23) as zeros of a fixed polynomial of degree q or q-1. With this aim in mind, and using the proposed Liapounov candidate function we can state the following theorems.

Theorem 3.

If the system described fig.1 is such that :

$$(24)\ N(p) = K\prod_{i=1}^{t}((p+p_i)^2+r_i^2)\prod_{i=t+1}^{r}(p+p_i)\ ;\ t+r = q\ ;\ K > 0$$

with

$$\forall i\in\{1,..,t\}\begin{cases} (\text{sign } r_i)i\left[\dfrac{D(\lambda)((\lambda+p_i)^2+r_i^2)}{N(\lambda)}\right]_{\lambda=-p_i+ir_i} \in \mathbb{R}^- -\{0\} \\ \text{and } p_i > 0\ ,\ b_i \neq 0 \end{cases}$$

$$\forall j\in\{1,..,t\}\ i\neq j \Leftrightarrow p_i\neq p_j \text{ or } |b_i|\neq|b_j|$$

$$\forall i\in\{t+1,..,r\}\left[\frac{D(\lambda)(\lambda+p_i)}{N(\lambda)}\right]_{\lambda=-p_i} > 0\ ,\ p_i > 0$$

$$\forall j\in\{t+1,..,r\}\ i\neq j \Leftrightarrow p_i\neq p_j$$

Then a sufficient asymptotic stability condition is :

$$\forall \varepsilon\in\mathbb{R}-\{0\}\quad f^* > -\frac{1}{K}\left(\hat{a}_{q-1} - \frac{\hat{b}_{q-1}}{K}\right)$$

Theorem 4.

If the system described fig.1 is such that : $\hat{b}_{q-1} \geq 0$,

$$(25)\ \exists a\in\mathbb{R}, D(p)+aN(p) = \prod_{i=1}^{t}((p+p_i)^2+r_i^2)\prod_{i=t+1}^{r}(p+p_i)(p+p_o)$$

with $t+r = q$ and

$$\forall i\in\{1,..,t\}\begin{cases} (\text{sign } r_i)i\left[\dfrac{N(\lambda)(\lambda+p_o)((\lambda+p_i)^2+r_i^2)}{D(\lambda)+aN(\lambda)}\right]_{\lambda=-p_i+ir_i} \in \mathbb{R}^- -\{0\} \\ \text{and } p_i > 0\ ,\ b_i \neq 0 \end{cases}$$

$$\forall j\in\{1,..,t\}\ i\neq j\quad p_i\neq p_j \text{ or } |b_i|\neq|b_j|$$

$$\forall i\in\{t+1,..,r\}\left[\frac{N(\lambda)(\lambda+p_o)(\lambda+p_i)}{D(\lambda)+aN(\lambda)}\right]_{\lambda=-p_i} > 0\ ,\ p_i > 0$$

$$\forall j\in\{t+1,..,r\}\ i\neq j\quad p_i\neq p_j$$

Then a sufficient asymptotic stability condition is :

$$\forall \varepsilon\in\mathbb{R}-\{0\}\quad f^* > a + \text{Max}\left(0,-\frac{p_o}{\hat{b}_{q-1}}\right)\ \text{ if }\ \hat{b}_{q-1} > 0$$

$$(f^* > a\ ,\ p_o > 0)\quad \text{if}\ \ \hat{b}_{q-1} = 0$$

Proofs of theorems 3 and 4 are given respectively in annexes 1 and 2.

V.3 - Example of application : third order systems.

Interpretation of previous theorems of part V.2 on small order systems leads to easy to check stability conditions. In this section we consider systems described on figure 2 :

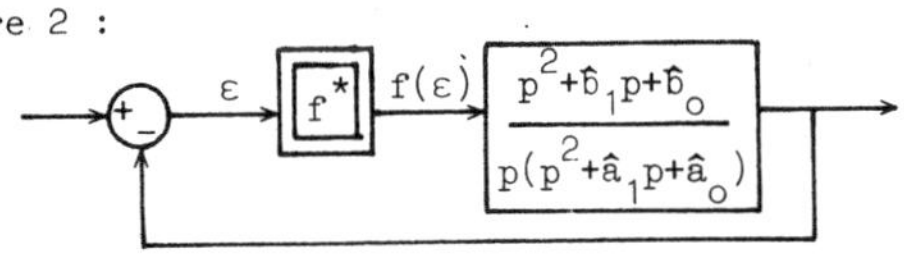

figure 2. with $(\hat{a}_1,\hat{a}_o,\hat{b}_1,\hat{b}_o)\ \mathbb{R}^4, \hat{b}_o\neq 0$

Having a zero pole is not a particular case but can be realized in all configurations by pole assignement technic /8/ if $N(0) \neq 0$. The application of the above parts results allows to state the following properties :

Property 1.

If the system fig.2 is such that : $\hat{a}_1^2 < 4\,\hat{a}_o$ and if

$$\begin{cases} \hat{b}_1 > \hat{a}_1 > 0 \\ \hat{a}\,(\hat{a}_1-\hat{b}_1) = 2(\hat{a}_o-\hat{b}_o) \end{cases}$$

then a sufficient asymptotic stability condition is :

$$\forall \varepsilon \in \mathbb{R}-\{0\}\quad f^* > 0$$

Property 2.

If the system fig.2 is such that : $\hat{b}_1^2 < 4\,\hat{b}_o$ and if

$$\begin{cases} \hat{b}_1^{\,2}-\hat{b}_o^{\,2}-\hat{a}_1\hat{b}_1+\hat{a}_o > 0 \\ \hat{b}_1(3\hat{b}_o-\hat{b}_1^{\,2}+\hat{a}_1-\hat{a}_o) = 2\hat{a}_1\hat{b}_o \end{cases}$$

then a sufficient asymptotic stability condition is :

$$\forall\varepsilon \in \mathbb{R}-\{0\} \quad f^* > 0$$

The proofs of these two properties are immediately gi- given by theorems 3 and 4.

CONCLUSION

Aim of this paper was to associate the stability study of a system and its possible modelisation.

The proposition of a Liapounov candidate function leads to stability conditions which are especially easy to check in the cases of "pseudo-symmetrical" or "pseudo-antisymmetrical" matrices. In the first case, linear systems stability conditions can be applied to a simplified model, obtained from the original one by omitting certain nonlinear multiplicative coefficients. In the second one, the worked out conditions are involving the only diagonal terms.

These results are applied a general class of linear systems with nonlinear feedback.

The main advantage of this work is then to reduce the number of non linear terms involved in the application of stability criteria.

Annexe 1. Proof of theorem 3.

If $N(p)$ has only simple real and complex roots (24), we can choose in (23) ;

$$\forall i\in\{1,..,t\} \qquad a_i = -p_i \ , \ b_i = r_i$$

$$\forall i\in\{t+1,..,r\} \qquad c_i = -p_i \qquad ;$$

consequently the first (q-1) terms of the last column of $FE(\varepsilon)$ are constant and given in (23) by replacing $P(\lambda,\varepsilon)$ by $D(\lambda)$, $R_i(\lambda)$ by $N(\lambda)/((\lambda+p_i)^2+r_i^{\,2})$ if i is in $\{1,..,t\}$, and $R_i(\lambda)$ by $N(\lambda)/(\lambda+p_i)$ if i is in $\{t+1,..,r\}$. Under these conditions : $\gamma^* = (-\hat{a}_{q-1}-Kf^*+\frac{\hat{b}_{q-2}}{K})$.

According to the study we propose the candidate Liapounov function $W(x)$ defined by :

$$W(x) = -\frac{1}{2}\Big(\sum_{i=1}^{t} \Big(\frac{x_{i1}^2}{\beta_i^1}+\frac{x_{i2}^2}{\beta_i^2}\Big) + \sum_{i=t+1}^{r} \frac{y_i^2}{\alpha_i} - \varepsilon^2 \Big)$$

From (23), it comes out, with the condition :

$$\forall i\in\{1,..,t\} \qquad \beta_i^1 = \beta_i^2 = \beta_i \in \mathbb{R}^- -\{0\}$$

the expression of $\overset{\circ}{W}(x)$:

$$\overset{\circ}{W}(x) = \sum_{i=1}^{t} \frac{p_i}{\beta_i}(x_{i1}^2+x_{i2}^2) + \sum_{i=t+1}^{r} \frac{p_i}{\alpha_i}y_i^2 + \gamma\,\varepsilon^2$$

$W(x)$ is a function of Liapounov if are verified conditions established in theorem 3.

Annexe 2. Proof of theorem 4.

If $W(p)$ verifies (25), we can choose in (23) :

$$\forall i\in\{1,..,t\} \qquad a_i = -p_i \ , \ b_i = r_i$$

$$\forall i\in\{t+1,..,r\} \qquad c_i = -p_i \qquad ;$$

consequently, we have in (23) : with $g^* = f^* - a$,

$$\forall i\in\{1,..,r\} \quad \beta_i^2 = \beta_i^2 g^* , \ \beta_i^1 = \beta_i^1 g^* , \ \alpha_i = \alpha_i g^*$$

where $\beta_i^2, \beta_i^1, \alpha_i$ are constant terms given in (23) by replacing $P(\lambda,\varepsilon)$ by $N(\lambda)$, $R_i(\lambda)$ by $(D(\lambda)+aN(\lambda))/((\lambda+p_o)((\lambda+p_i)^2+r_i^{\,2}))$ if i is in $\{1,..,t\}$, and $R_i(\lambda)$ by $(D(\lambda)+aN(\lambda))/((\lambda+p_o)(\lambda+p_i))$ if i is $\{t+1,..,r\}$. Under these conditions γ^* is equal to $-(\hat{b}_{q-1}\, g^* + p_o)$.

Let us propose the candidate Liapounov function of general type proposed in the paper :

$$W(x) = -\frac{1}{2}\Big(\sum_{i=1}^{t} \Big(\frac{x_{i2}^2}{\beta_i^2}+\frac{x_{i1}^2}{\beta_i^1}\Big) + \sum_{i=t+1}^{r} \frac{y_i^2}{\alpha_i} \Big) + \int_0^{\varepsilon} g(\xi)d\xi$$

if we have for every i in $\{1,..,r\}$, the relations :

$$\beta_i^2 = \beta_i^1 = \beta_i < 0 \quad \text{and} \quad \alpha_i < 0 \quad ,$$

it comes out for the eulerian derivative of $W(x)$:

$$\overset{\circ}{W}(x) = \sum_{i=1}^{t} \frac{p_i}{\beta_i}(x_{i2}^2+x_{i1}^2) + \sum_{i=t+1}^{r} \frac{p_i}{\alpha_i}\, y_i^2 - g^*(\hat{b}_{q-1}g^*+p_o)\varepsilon^2$$

So $W(x)$ is a Liapounov function for this system if are verified the conditions of the theorem 4.

REFERENCES.

/1/ZAMBETTAKIS I."Contribution à l'étude de systèmes à non-linéarités multiples - Application aux systèmes électromécaniques"Thèse de Docteur Ingénieur,Lille,n°326, 1983.

/2/LIAPOUNOV A.M."Problème général de la stabilité du mouvement"University Press, Princeton, 1949.

/3/ROSENBROCK H.H."A method of investigating stability" Conf. IFAC, Bâle, 1965.

/4/KRASOVSKI N.N."Certain problem of the theory of stability of motion"Stanford ,Californie,1963.

/5/HAHN W."Theory and application of Liapounov's direct method"Prentice Hall, Englewood Cliffs, 1963.

/6/ZAMBETTAKIS I.,RICHARD J.P.,LAURENT F."Etude des systèmes electromécaniques non-linéaires par la méthode directe de Liapounov"1st IASTED Symp.,Applied Informatics,vol.3,p.169-172,Lille,1983.

/7/ROTELLA F.,RICHARD J.P."Modélisation et synthèse des systèmes monovariables de type Lure'Postnikov"1st IASTED Symp.,Applied Informatics,vol.3,p.173-178, Lille,1983.

/8/POPOV V.M."L'hyperstabilité des systèmes automatiques" Dunod,1973.

/9/RICHARD J.P."Sur la mise en équation d'état des systèmes continus non-linéaires par une méthode de calcul symbolique : définition d'un invariant de représentation"Thèse de Docteur Ingénieur,Lille,n°259,1981.

/10/BENREJEB M."Sur l'analyse et la synthèse de processus hiérarchisés.Application aux systèmes singulièrement perturbés"Thèse de Docteur es Sciences Physiques, Lille,n°479,1980.

/11/BENREJEB M.,BORNE P.,LAURENT F."Sur une application de représentation en flèche à l'analyse des processus" RAIRO Auto.,vol.16,n°2,1982,p.133-146.

/12/ROTELLA F.,ZAMBETTAKIS I.,RICHARD J.P."Application de la méthode du lieu des racines à la synthèse de systèmes non-linéaires"MECO'82,vol.1,p.27-1,Tunis,1982.

EFFICIENT ALGORITHMS FOR FAST FEATURES DETECTION IN BINARY PICTURES

by

N.G.BOURBAKIS, C.A.ALEXOPOULOS and D.SERPANOS
University of Patras
School of Engineering
Dept. of Computer Science
Patras, Greece

Abstract

This paper presents a global consideration of the theme "Methods for fast edge detection of an object", by giving a set of efficient algorithms of picture scanning. The picture scanning algorithms presented are: Raster,Zeta,Raster-zeta, Right brachet, Diagonals,Spirals and Rectangular. The above algorithms provide the advantage of the fast detection of an object, which depends on its position in the examined image. This feature is more significant when the user has information about the position of the object in the picture.

1 - Introduction

On the assumption that a picture is a two-dimensional array nxn, there occurs the problem of the scanning techniques of the picture and their possible applications. A two-dimensional array p(nxn) has a total of n^2 differend points. Then the total of all the possible ways of scanning is $(n^2)!$

The process of detecting an edge of an object which appears in a picture, is important in further processing of the object e.g. extraction of an object, description of an object, encoding of curves, reproduction, classification, pattern recognition etc., which can be implemented in real-time image processing. Scientists involved with this problem so far (1-8) did not cover this subject globally. Instead they gave some partial solutions to it focussing on the further processing.

This paper comes to give a more global view of the theme:"techniques for fast detecting of the edge of an object", giving a set of different techniques of picture scanning like (9):raster scanning, zeta(z) scanning, raster-zeta scanning, right bracket scanning, diagonal scannings, spiral scannings and rectangular scanning.

After the examination of the above scanning techniques there results that depending on the position of the object in the examined picture, there exists a corresponding enhanced technique, which gives the best solution.

An important factor for the detection of an object in the picture is its position on it. This is based on the fact that the user may already have information about the object's position in the picture. Such information results in a transformed picture in which the majority of its information is concentrated in some particular area.

2 - Descriptive Algorithms

In this section the descriptive algorithms of the examined techniques, are briefly presented:

Algorithm: *Raster Scanning* **(RO)**
Input: Set the initial value n;
  Begin:For i:=1, step1 until n do
    Begin For j:=1, step1 until n do
      Scan the P(i,j);
    end;
  end;
Stop

Algorithm: *Zeta Scanning* **(Z)**
Input Set the initial values n,i,j;
  Calculate $L:=\log_2 n$;
  Scan the P(i,j);
  Begin For k:=1, step1 until L do
    Set $i:=i-(2^{k-1}-1)$; $j:=j+1$;
    Realize scanning of (k-1) level;
    Set $i:=i+1$; $j:=j-(2^{k}-1)$;
    Realize scanning of (k-1) level;
    Set $i:=i-(2^{k-1}-1)$ $j:=j+1$;
    Realize scanning of (k-1) level;
  end;
Stop

```
Algorithm: Raster-zeta Scanning  (RZ)
Input:Set the initial value n;
      Begin:For L:=1, step2 until (n-1) do
            Begin:k:=1, step2 until (n-1) do
                  Begin: i:=L, step1, until (L+1) do
                        Begin: j:=k, step1 until (k+1) do
                               Scan the P(i,j);
                        end;
                  end;
      end;
Stop
```

```
Algorithm: Right Bracket Scanning  (RB)
Input:Set the initial values n,i,j;
      Calculate L:=log2n;
      Scan the P(i,j);
      Begin:For k:=1, step1 until L do
            Set i:=i-(2^(k-1)-1); j=j +2^(k-1);
            Realize scanning of (k-1) level;
            Set i:=i+1; j:=j;
            Realize scanning of (k-1) level;
            Set i:=i-(2^(k-1)-1); j:=j-2^(k-1);
            Realize scanning of (k-1) level;
      end;
Stop
```

```
Algorithm:DSVD Scanning
Input:Set the initial value n;
      Begin:For L:=1, step 1 until n do
            if L is odd then
            Begin: For k:=0, step1 until (L-1) do
                   Set i:=L-K; j:=K+1;
                   Scan the P(i,j);
            end;
            else
            Begin:For k:=0, step 1 until (L-1) do
                  Set i:=k+1; j:=L-K;
                  Scan thà P(i,j);
            end;
      end;
      Begin:For L:=1, step 1 until (n-1) do
            if L is odd then
            Begin:For k:=1, step 1 until (n-L) do
                  Set i:=n+1-k; j:=L+k;
                  Scan the P(i,j);
            end;
            else
            Begin:For k:=1, step 1 until (n-L) do
                  Set i:=L+k; j:=n+1-k;
                  Scan the P(i,j):
            end;
      end;
Stop
```

```
Algorithm:DSPR Scanning
Input:Set the initial value n;
      Begin:For k:=1, step 1 until n do
            Scan the P(k,k);
      end;
      Begin: For L:=1, step 1 until (n-1) do
             Begin: For k:=1, step 1 until (n-L) do
                    Set i:=n-L-k+1; j:=n+1-k;
                    Scan the P(i,j);
             end;
             Begin:For k:=1, step 1 until (n-L) do
                   Set i:=L+k;
                   Scan the P(i,k);
             end;
      end;
Stop
```

```
Algorithm:SIS scanning
Input:Set the initial value n;
      Begin:For k:=1, step 1 until n/2 do
            Set i:=k;
            Begin:For j:=k, step 1 until (n-k) do
                  Scan the P(i,j) ;
            end;
            Set j:=n-k+1;
            Begin:For i:=k, step 1 until (n-k) do
                  Scan the P(i,j);
            end;
            Set i:=n-(k+1);
            Begin:For j:=n-k+1, step 1 until (k+1) do
                  Scan the P(i,j);
            end;
            Set j:=k;
            Begin:For i:=n-k+1, step 1 until (k+1) do
                  Scan the P(i,j);
            end;
      end;
Stop
```

```
Algorithm: DQA-scanning
Input:Set the initial value n;
      Scan the p(1,1);
      Begin:For k:=2, step 1 until n do
            Begin:For i:=1, step 1 until n do
                  Scan the p(i,k);
            end;
            Begin:For j:=k-1, step -1 until 1 do
                  Scan the p(k,j);
            end;
      end;
Stop
```

```
Algorithm: SES Scanning
Input: Set the initial value n;
       Begin:For k:=1, step 1 until (n/2) do
             Set j:=(n/2)-k+1;
             Begin:For i:=(n/2)-k+2, step1 utnil (n/2)+k do
                   Scan the P(i,j);
             end;
             Set i:=(n/2)+k;
             Begin:For j:=(n/2)-k+2, step1 until (n/2)+k do
                   Scan the P(i,j);
             end;
             Set:j:=(n/2)+k;
             Begin:For i:=(n/2)+k-1, step1 until (n/2)-k+1 do
                   Scan the P(i,j);
             end;
             Set i:=(n/2)-k+1;
             Begin:For j:=(n/2)+k-1, step1 until (n/2)-k+1 do
                   Scan the P(i,j);

             end;
       end;
Stop
```

3 - Comparison of Scanning Techniques

For the ability of a comparative analysis of the scanning techniques "for the fast detection of an edge", we define 9 possible areas($n^2/4$) in the picture, in which will be the object (fig. 1). The greatest possible amount of positions of an area a^2 in a picture n^2 is: $(n-a+1)^2$

When the examined object lies in one of the 9 positions of a picture (fig.1),then for the above techniques (which are applied for detecting an edge of the object, one fast and other slow, depending on its position), a comparative table 1 is given. This table provides information from which a comparison among the presented techniques is possible. The comparative elements of this table are:

1- The total number of steps of each technique until to cover the whole examined area (TNS).

2- The number of steps that are being realized by the technique until the area of the object is touched for the first time (NSFT).

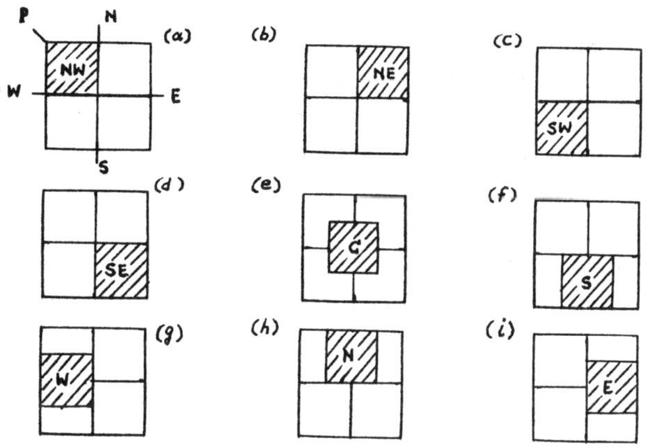

Fig.1: The 9 possible positions of the examined area ($n^2/4$).

TABLE 1

	-NW Area-		-NE Area-		-SW Area-		-SE Area-		-C Area-	
	TNS	NSFT	TNS	NSFT	TNS	NSFT	TNS	NSFT	TNS	NSFT
RO	n(n 1)/2	1	$n^2/2$	n/2+1	$n^2-n/2$	$n^2/2+1$	n^2	$(n^2+n)/2+1$	$(3n^2-n)/4$	$n^2/4+n/4+1$
Z	$n^2/4$	1	$n^2/2$	$n^2/4+1$	$3n^2/4$	$n^2/2+1$	n^2	$3n^2/4+1$	$3n^2/16$	$3n^2/16+1$
RZ*	n(n-2)/2	1	$n^2/2$	n+1	n^2-n	$n^2/2+1$	n^2	$n^2/2+n+1$	$(3n^2-2n)/4$	$n^2/4+n/2+1$
RB	$n^2/4$	1	$n^2/2$	$n^2/4+1$	n^2	$3n^2/4+1$	$3n^2/4$	$n^2/2+1$	$7n^2/8$	$n^2/8+1$
DSVD	$(n^2-2n+2)/2$	1	$(7n^2-6n)/8$	$(n^2+2n)/8+1$	$(7n^2-2n)/8$	$(n^2+6n+8)/8$	n^2	$(n^2+2n)/2$	$(7n^2-4n)/8$	$n^2/8+n/2+1$
DSPR	$(3n^2-4n)/4$	1	n^2-1	3n/2	n^2	(5n-2)/2	$(3n^2-2n)/4$	n/2+1	3n(n-1)/4	n/4+1
SIS	n^2-3	1	n^2-2	n/2+1	n^2	(5n-2)/2	n^2-1	3n/2	n^2	$3n^2/4$
SES	n^2	4	$n^2-n/2$	3	n^2-5n+2	1	$n^2-3n/2$	2	$n^2/4$	1
DQA	$n^2/4$	1	$n^2-3n/2$	$n^2/4+1$	n^2	$(n^2+2n+8)/4$	$n^2-n/2$	$(n^2+2n+4)/4$	$(9n^2-4n)/16$	$(n^2/16+4n+16)/16$

	-N Area-		-W Area-		-E Area-		-S Area-	
	TNS	NSFT	TNS	NSFT	TNS	NSFT	TNS	NSFT
RO	$n^2/2-n/4$	n/4+1	$3n^2/4-n/2$	$n^2/4+1$	$3n^2/4-n/2$	$(n^2+2n+4)/4$	$n^2-n/2$	$(n^2+n+2)/2$
Z	$7n^2/16$	$n^2/16+1$	$5n^2/8$	$n^2/8+1$	$7n^2/8$	$3n^2/8+1$	$15n^2/16$	$9n^2/16+1$
RZ*	$n^2/2-n/2$	n/2+1	$3n^2/4-n$	$n^2/4+1$	$3n^2/4$	$n^2/4+n+1$	$n^2-n/2$	$(n^2+n+2)/2$
RB	$n^2/2$	$n^2/16+1$	$7n^2/8$	$n^2/8+1$	$5n^2/8$	$n^2/2+1$	$15n^2/16$	$n^2/2+1$
DSVD	$23n^2/32-7n/8$	$(n^2+4n+32)/32$	$23n^2/32-5n/8$	$(n^2+12n+32)/32$	$31n^2/32-3n/8$	$(9n^2+20n+32)/32$	$31n^2/32-n/8$	$(9n^2+28n+32)/32$
DSPR	$(15n^2-8n-16)/16$	n/4+1	$15n^2/16-n/2$	n/4+1	$(15n^2-12n-16)/16$	n/2+1	$15n^2/16-n/4$	n/2+1
SIS	n^2-2	n/4+1	n^2	13n/4-2	n^2-1	5n/4	n^2	9n/4-1
SES	$n^2-n/4$	3	$n^2-13n/4+3$	1	$n^2-5n/4+1$	2	$n^2-9n/4+2$	1
DQA	$3n^2/8$	$n^2/16+1$	$9n^2/16$	$n^2/8+1$	$3n^2/4$	$5n^2/16+1$	$15n^2/16$	$3n^2/8+1$

* In the areas N,W,E,S and C, the RZ technique operates for $n>4$.

3.1- Comparative Results

From the elements that gives the table 1 we get the following comparative analysis.

a) NW - area

When the object lies in the NW area of the picture (fig.1a),then the techniques Z,RB,DQA appear to be faster than the other.Fast can also be characterized and the technique DSVD. The most improper technique is SES. This technique scans n^2 picture points to cover the $n^2/4$ points of the examinedarea.

b) NE - area

When the object lies in the NE area (fig.1b), then the techniques RO,RZ are the fastest. The technique RZ covers the same number of points with RO, but it touches the examined area later(n+1 steps). The slowest techniques are SIS,SES,because they cover almost the whole number of picture points to cover the examined area although they touched it very soon.

c) SW - area

Assuming that the object lies in the SW area(fig. 1c), then as the faster techniques appear to be Z, DSVD, as they cover the least points for the covering of the whole area. SES is an efficient technique as it touches the examined area immediately (NSFT=1). The least efficient techniques for this case are SIS,DSPR,RB,DQA ($TNS=n^2$).

d) SE - area

In the SE area the fastest technique for object detection is DSPR. The techniques with the worst conditions are DSVD,RZ,Z.

e) C - area

For the C area (central) of the picture we get that SES is the fastest of all. We presume that from the facts that it touches the area immediately (NSFT=1) and covers the least points($n^2/4$) of the examined area. The worst technique is SIS.

f) N - area

When the object lies in the N area, the most efficient technique appears to be DQA, because it covers whole the examined area in a small number of steps and touches this area very soon. Another fast technique is also characterized the Z. The worst is SIS ($TNS=n^2$).

g) W - area

In the W area the technique DQA is characterized fast and after that the Z. Both techniques cover the same number of points ($NSFT=n^2/8+1$). The worst is SIS.

h) E - area

The most efficient technique appears to be RB and after that the RZ,DQA. The worst one is SIS.

i) S - area

The DSPR technique covers the S area faster and touches it fast too. On the contrary SIS appears to be as the worst technique.

4 - Conclusions

The proposed image scanning techniques could provide with significant advantages(memory,time) the further image processing (features extraction, recognition, reconstruction,...), when the user has some information about the position of the a area. An information of such kind is that "the examined picture was transformed by FFT", which means that almost the total information has been accumulated in the picture middle (9).

References

(1) DAVIS L.S."Understanding Shape:Angles and Sides" IEEE Tran. on Computers,Vol.3-26,No.3,Mar 1977, pp. 236-242.

(2) DAVIS L.S."Shape Matching Using Relaxation Techniques",Proc. of the IEEE Conf. on Pattern Recognition and Image Processing, jun. 1977, pp. 191-197.

(3) TANIMOTO S.L."An Iconic/Symbolic Data Structuring Scheme",Pattern Recognition and Artificial Intelligence, Academic Press, N.Y. 1976, pp. 452-471.

(4) WARNOCK J.E."A Hidden Surface Algorithm for Computer Generated Halftone Pictures",Comp.Sci. Dept.,Univ. of Utah,TR 4-15,Jun. 1969.

(5) FREEMAN H."Techniques for the Digital Computer Analysis of Chain-Encoded Arbitrary Plane Curves",Proc. of the National Electronics Conf., Vol. 17, Oct. 1961, pp. 421-432.

(6) FREEMAN H."Computer Processing of Line-Drawing Images", Computing Surveys, Vol. 6, No. 1, Mar. 1974, pp. 57-97.

(7) FREEMAN H."Analysis of Line-Drawing", Proc. of the NATO Advanced Study Institute on Image Processing, Jun. 1976.

(8) FREEMAN H. and DAVIS L.S."A Corner Finding Algorithm for Chain-Encoded Curves", IEEE Tran. on Computers, Mar. 1977,pp. 297-303.

(9) BOURBAKIS N. " Efficient Image Scanning Algorithms for Picture Processing", submmited for publication, (PhD thesis, Chapter 2, pp. 57-110, Dept. of Computer Science, School of Engineering, Univ. of Patras, Patras, Greece).

A PROCESSOR_PROGRAM FOR A FLOAT_POINT ARITHMETIC PROCEDURES REALIZING IN A MICROPROCESSOR SYSTEM

B. E. Manahilov, M. Avramova, D. Karastojanov

Institute for Industrial Cybernetics and Robotics, BAS, Sofia, Bulgaria

ABSTRACT

The article examins a problem of high-precision calculations realized by a general purpose microprocessor in operating systems for real time process control. The acting algorhythm of a program is described. The program simulates the work of a three address processor with a special instruction set, that is an extension of a standard microprocessor instruction set. The basic arithmetic operations, trigonometric, exponential, logarithmic and special functions, as well as instructions which control calculations logical flow are included. A process-program handles data of any format of binary and decimal number systems and the calculating process itself is performed in a floating point format.

INTRODUCTION

Microprocessor systems for industrial process control today have a comparatively limited computational capabilities. In real time systems, a high speed scanning is required and for this reason the calculational procedures must be written in Assembly language. But robot development puts new problems to the program design. The kinematical control is not sufficient yet, it is necessary to take into consideration the dynamical characteristics and adaptation. The calculations size and complexity are considerably increased. Microprocessors are continuously advancing too. The difficulties in debugging and hardware dependence of the assembler programs make them noneffective. Existing high-level languages cannot be implemented in real-time controlled processes because of their drawbacks. Interpreters have low execution speed, while compilers make debugging difficult and the machine code is noneffective. The advantages of both types were combined into new high-level languages, operating with pseudocodes (1).

The translation of the source program is done by a compiler into pseudo-code rather then directly into machine one. When running on the microprocessor, a special interpreter scans and executes functions defined by the P-code. The software processor described below is an example of a such P-code interpreter.

PROGRAM ARCHITECTURE

The program module simulates a three address processor operation. Information is parted into five blocks, each of them equivalent to different units of a common processor /Fig.1/. Interpreter module itself can be regarded as a microprogram memory. The operating components of the processor are represented by registers, counters and flags /Fig.2/. They can be located in RAM, if hardware resources of the concrete microprocessor are insufficient. Usually a region (30 bytes) of direct addressed memory is used. The processor memory block is defined as an array of words, each containing a variable. Variables are binary 16bit integer or 32bit float_point coded. Data format is as follows: - 8bit exponent in the first byte, represented in 2's complement code (range -127 to +127) and a fraction with nonevident coded sign bit in the most significant 23th bit. This is possible because the fraction is always kept normalized. Machine zero appears when the exponent is equal to -128. The length of the fraction determine calculation accuracy.

The software processor does not operate with the input/output devices directly, but through two buffers with arbitrary structure. These buffers carry out the comunications with the input/output driver programs and other system modules. The data format in the buffers unlike the intermediate results can be various. It is specified in the input/output macroinstructions, which convert data when transferring them

BETWEEN THE BUFFERS AND MEMORY BLOCK. THE MACROPROGRAM, WHICH SUPERVISES THE COMPUTATION EXECUTION IS SEPARATED AS A PROGRAM MEMORY.

ALL BLOCKS DESCRIBED, EXEPT THE OPERATIONAL COMPONENTS ARE ADDRESS INDEPENDENT. ABSOLUTE BUFFER ADDRESSES AND BEGIN ADDRESS OF THE MACROPROGRAM ARE GIVEN BY THE MODULE, THAT INITIATES THE SOFTWARE PROCESSOR TO START THE COMPUTATIONS. THIS PROCEDURE IS ORGANIZED AS A SUBROUTINE CALL IN A STANDARD WAY. THE APPROPRIATE DISTRIBUTION OF THE MEMORY BUFFERS AND BLOCK ORGANISATION OF THE PROCESSOR MAKES POSSIBLE ITS SIMULTANEOUS USE BY THE DIFFERENT SYSTEM MODULES. IN MULTITASKING MODE, THE RECOGNITION OF THE TASKS' REQUESTS ARE DONE BY THEIR NUMBER. A NEW REQUEST FROM THE SAME TASK IGNORES ITS UNFINISHED OLD ONE, IF ANY. EACH TASK REQUEST BREAKS OFF CURRENT EXECUTION. WHEN INITIATING EXECUTION AFTER AN INTERRUPT, THE LAST UNCOMPLETED MACROINSTRUCTION IS REPEATED (3).

PROCESSOR RUNNING

THE FLOW-CHART OF THE MAIN PROGRAM IN THE MODULE IS GIVEN ON FIG.3. AN INITIALIZATION TAKES PLACE WHEN A REQUEST IS RECEIVED AND A NEW STATUS OF THE SOFTWARE PROCESSOR IS ESTABILISHED. IF THE OLD STATUS SHOWS UNCOMPLETED MACROPROGRAM EXECUTION, THE INTERRUPTION FLAG IS SET AND THE CONTROL REGISTERS ARE SAVED IN THE MEMORY BLOCK. THEN THE NEW INFORMATION IS LOADED IN THE REGISTERS AND THE NEW MACROPROGRAM EXECUTION IS INITIATED.

DURING INSTRUCTION FETCH CYCLE, THE MACROPROGRAM COUNTER IS UPDATED ACCORDING TO THE LENGTH OF THE LAST INSTRUCTION EXECUTED.THAT WAY, A CONTROLLED RETRY OF THE MACROINSTRUCTIONS CAN BE DONE. THE ABOVE OPERATIONS ARE CRITICAL FOR THE PROCESSOR'S WORK AND FOR THIS REASON, HARDWARE INTERRUPTS ARE MASKED WHILE THEY RUN.

THE MACROINSTRUCTION'S EXECUTION IS PERFORMED BY MEANS OF A LIBRARY, WHERE THE ROUTINES ARE FUNCTIONALY GROUPED /FIG.4/. THE BASIC CONFIGURATION INCLUDES INPUT/OUTPUT AND ARITHMETIC FUNCTIONS. TO MAKE THE PROGRAM PROCESSOR ADDRESS INDEPENDENT, IT IS NECESSARY TO HAVE THE ROUTINE'S OFFSETS IN THE MACROINSTRUCTION DECODING TABLE. THE FIRST 8 ADDRESSES OF THE TABLE CONTAIN THE OFFSETS OF THE FREQUENTLY USED ROUTINES - ARITHMETIC OPERATIONS WITH FRACTIONS, NORMALISATION AND ERRORS HANDLING.

THE PROCESSOR SUBPROGRAMS RECEIVE DATA IN OPERATIONAL REGISTERS. THEY ARE BASED ON STANDARD WELL KNOWN ALGORHYTHMS, REVISED TO MEET HIGH SPEED AND MINIMUM MEMORY REQUIREMENTS. THE ALGORHITHM OF THE DIVISION PROCEDURE IS GIVEN ON FIG. 5, WHERE DATA AND INDEX REGISTERS ARE USED SIMULTANEOUSLY AS POINTERS AND COUNTERS.

ALL COMPUTATIONAL ERRORS AND DATA STRUCTURE ERRORS, THAT APPEARED DURING EXECUTION ARE REFLECTED IN A SPECIAL ERROR CODE REGISTER. THE ARITHMETIC ERRORS (UNDERFLOW, OVERFLOW, DIVIDE CHECK, ARGUMENT OUT OF RANGE, ECT.) ARE PROCESSED IN A STANDARD FIXUP. IRREGULAR MACROINSTRUCTIONS AND DATA STRUCTURE ERRORS INTERRUPT THE EXECUTION AND THE CONTROL IS RETURNED TO THE CALLING SYSTEM MODULE. THE REQUEST IS MARKED SERVICED.

MACROINSTRUCTION SET

THE MAIN CONSIDERATIONS IN MACROINSTRUCTION CODING ARE EFFICIENCY AND MEMORY EXCESS. EACH MACROINSTRUCTION CONTAINS A ONE BYTE OPERATION CODE, ADDRESSES OF VARIABLES, CONSTANTS, BRANCH OFFSETS OR DATA ADDRESSES. THERE ARE SEVERAL POSSIBLE WAYS TO CODE VARIABLES' ADDRESSES DEPENDING ON THEIR NUMBER: - LESS OR EQUAL TO 64 - ONE BYTE OFFSET FROM THE BEGIN BLOCK ADDRESS, FROM 65 TO 256 - ONE BYTE SUCCESSIVE NUMBER OF THE VARIABLE IN THE BLOCK, AND GREATER THAN 256 - A TWO'S BYTE OFFSET IN THE BLOCK. (4)
THE BRANCH ADDRESSES ARE IN RELATIVE ADDRESSING FOR ADDRESS INDEPENDENCE.THE INPUT/ OUPUT INSTRUCTIONS USE BUFFER OFFSETS AS WELL AS ABSOLUTE ADDRESSES TO ACCESSDATA AND CONSTANTS.

OPER V1 V2 V3	(V1) OPER (V2) - V3
FUNC V1 V2	FUNC (V1) - V2
TEST V LABEL	IF TRUE, BRANCH TO LABEL
SUBR NAME	BRANCH TO SUBROUTINE NAME
RTRN	RETURN FROM SUBROUTINE
INP/OUT V F N P	OFFSET ADDRESS BEGINNING FROM THE VARIABLE INPUT/OUTPUT A NUMBER OF CONSEQUTIVE DATA WITH FORMAT AND POSITION OF THE POINT
LOAD CONSTANT V	LOAD THE CONSTANT INTO VARIABLE SPECIFIED
ASMB	THIS MACROCOMMAND PERMITS ASSEMBLY INSTRUCTION TO BE INCLUDED IN THE MACROPROGRAM. RETURN OF THE CONTROL IS REALIZED BY THE ASSEMBLY INSTRUCTION "BRANCH TO SUROUTINE" IN

	The special entry point of the processor module
END	last macroprogram instruction. The request is served and control is returned to the caller.

CHARACTERISTICS

Input/output macroinstructions convert data in binary and decimal number systems with fixed or floating point. Binary data length must not exceed 7 bytes, with sign byte, bit or in two's complement code. The point is located to a bit. Decimal data can be in packed (BCD) or unpacked (ASCII) code with maximum length of 10 bytes. The point is located to a byte or exists in the number. Internal variable range is from 2^{-127} to 2^{+127} Accuracy, speed and range of the P-interpreter memory requirements depend on the fraction length and microprocessor used. The alteration of these characteristics has an exponential form. Concrete

- accuracy - 24 bits fraction or 7 decimal digts
- speed - average 1000 multip/sec MC 6800
  1400 multip/sec Int 8086
- memory - 3K for MC 6800, 2.6K for Int 8086 including SQR,SIN,COS,TAN,ATAN,LN, LOG, EXP,X!Y functions (5,6)

HIGH-LEVEL LANGUAGE

Calculation procedures are programmed in a high-level language similar and compatible with BASIC. The source program is debugged by a sequence of tests and corrections. During the test, the program is line by line scanned and macroinstructions, generated by the operators, are executed by the software processor module. Data is input in decimal number system from the keyboard. The creation of and the corrections in the source program are held by an editor with its own control comands. After a complete tracing, the program is translated by a compiler module. Real structure of the input/output buffers and data formats can be given in two ways - in interactive mode or by language declarations. As a result, an object code of the macroprogram is created, which can be stored in PROM or periferal device(7).

The complex of the language modules, usually works with the operation system of the user design devices, but it can work autonomously as well. The time for interpretation of the source program operators is about 40% of the total execution time.

CONCLUSIONS

The main advantage of such products is the considerable time reduction in software design. It is not necessary to rewrite old programs when changing the accuracy and speed of computations or hardware equipment, because of the P-code independence. Only the interpreter module has to be translated by nonexpert programmers. The routines library of the software processor can be filled up with a number of special blocks, tailored to a concrete realisation. Including macroinstructions for editting and symbol processing allow high-level language program modules to be written in P-code, which makes them hardware free too. At the end, a comparison with BASIC shows a gain of five per one increased productivity of the presented language in off-line run and two per one - with on-line FORTRAN object code (8).

REFFERENCES

1. Teshler L., High-level languages for machine control, Machine design, vol 52, Nr 20, September 1980, p. 113 - 125
2. Knuth D., The Art of Computer Programming, Vol 2, Seminumerical algorithms, Addison-Wesley, Reading Mass.,1969
3. Skates J., Distributed Control Systems Applied to Batch Processing, Control & Instrumentation, August 1981, p. 22
4. Manahilov B., Computational Programs for Linear and Circular Interpolation, Internal rep., ISOMATIC labs, March 1981
5. HP issue, Programming Guide for Personal Computers, January 1979, p. 145 - 232
6. Perile M., Cordic Technique Reduces Trigonometric Functions Look up, Computer Design, June 1971, p. 72
7. Manahilov B., K. Popov, A Language to Describe Computations, Int. Conf., Industrial Robot Control Problems, Varna, October 1981
8. Manahilov B., M. Avramova, High-level Language for Microprocessor Calculations Automation, Anual Report, IICR-BAS, vol 19, 1982

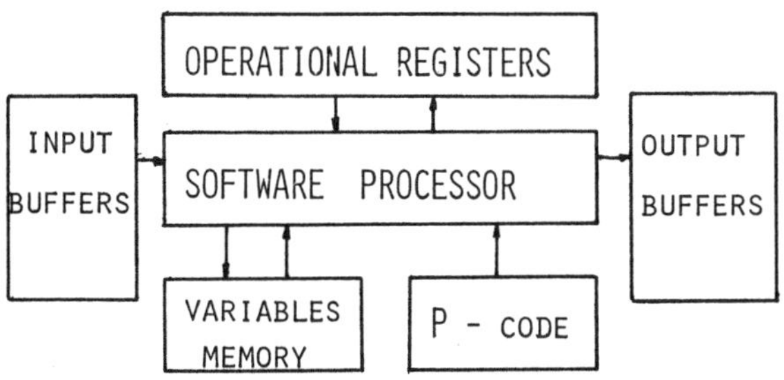

Fig. 1

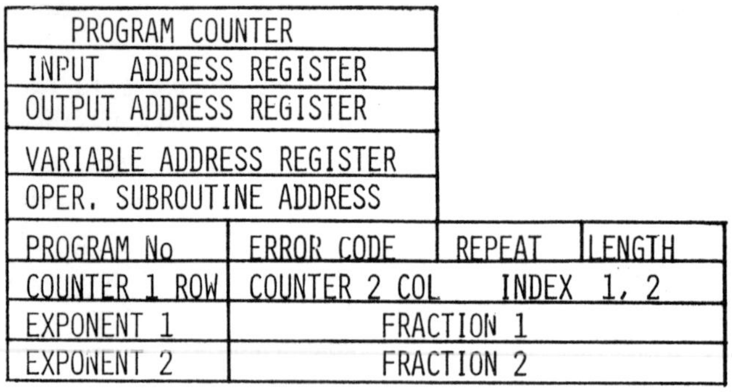

FIG. 2

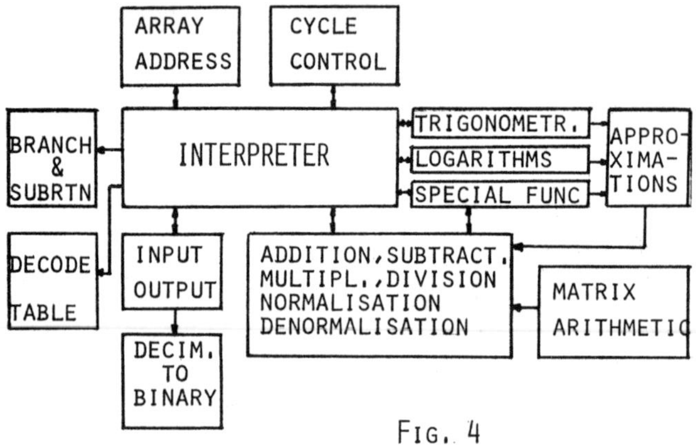

FIG. 4

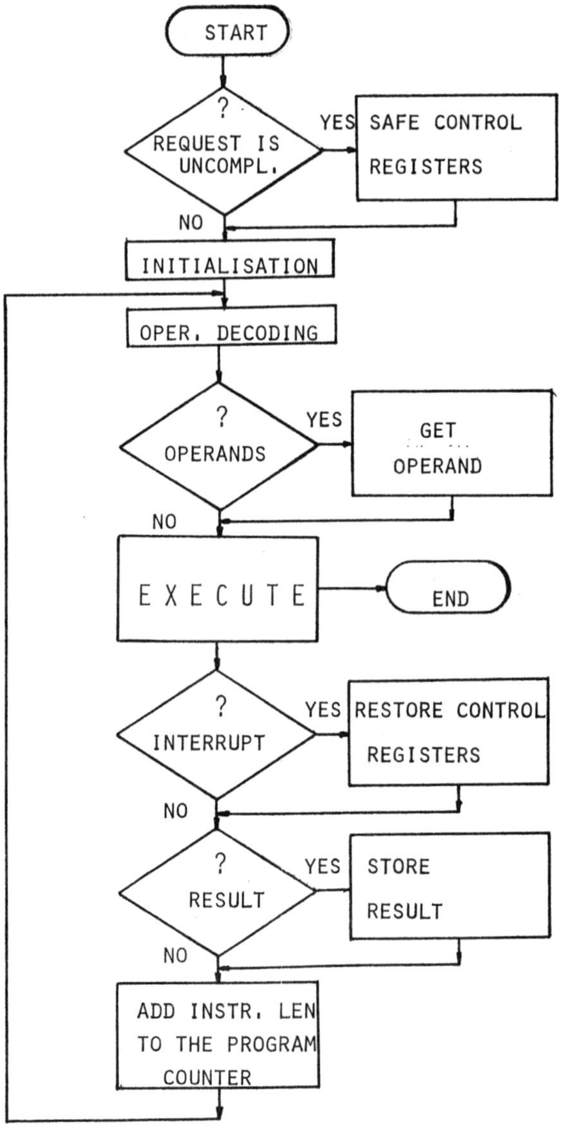

FIG. 3

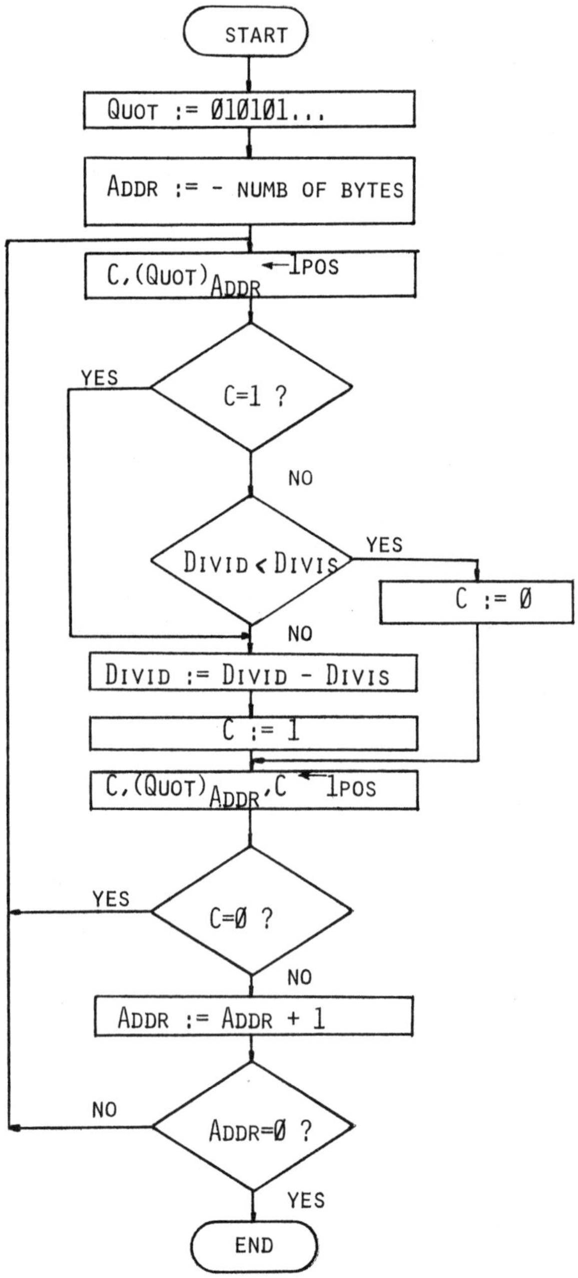

FIG. 5

APPLICATION DE LA REGRESSION ORTHOGONALE A L'IDENTIFICATION D'UNE CHAINE DE MESURES COMPLEXE DE TYPE CASCADE

M. ROBERT, J.L. NOIZETTE, J. RAGOT, T. DISANO

LARA - CRAN - ERA 905
Rue du Doyen Marcel Roubault
54501 VANDOEUVRE CEDEX

Tél. (8) 351.43.71

RESUME

Les auteurs appliquent une méthode de moindres carrés étendue (régression orthogonale multiple) à l'identification de processus linéaires et non linéaires interconnectés de façon cascade ; simultanément, l'algorithme proposé donne les estimations vraisemblables des variables d'état du processus à partir de leur mesure. Une chaîne complexe de mesure de densité par rayonnement gamma a été identifiée selon ce principe.

INTRODUCTION

La plupart des systèmes physiques sont caractérisés par un certain nombre de variables mesurables ; chercher un modèle qui représente ce processus consiste à déterminer des relations entre ces variables.

Cette détermination suppose au préalable de répondre à un certain nombre de questions :

- Quelle est la structure du modèle ?
- Quelles sont les "propriétés" des erreurs qui affectent les mesures ?
- Quelle méthode d'estimation choisir ?

A ces questions on ne peut apporter que des éléments de réponse. Le choix de la structure d'un modèle relève, soit du domaine de l'analyse et de la mise en équation des systèmes, soit du domaine purement heuristique où, bien souvent, la détermination de la structure du modèle se fait de façon itérative, compte tenu des résultats successifs obtenus. La première approche utilise ce que l'on nomme un modèle de connaissance et se fonde en général sur les lois physiques qui décrivent un processus. La seconde approche utilise ce que l'on nomme un modèle de type boîte noire dont la structure n'a aucune justification physique. Dans la plupart des cas l'utilisateur essaie de combiner les avantages et les inconvénients de ces deux approches.

Les caractéristiques et les propriétés des erreurs de mesure conditionnent le traitement à faire subir aux mesures ; on conçoit aisément l'importance de savoir si, parmi les erreurs de mesure, certaines sont accidentelles, d'autres systématiques (biais) ou d'autres encore à caractère aléatoire. Dans ce dernier cas on est souvent conduit à poser l'hypothèse que ces erreurs sont des réalisations de variables aléatoires dont la fonction densité de probabilité est connue ou partiellement connue.

Le choix d'une méthode d'estimation est lié aux hypothèses sur la nature des erreurs de mesure ; les plus courantes sont la méthode des moindres carrés, celle du maximum de vraisemblance et celle du maximum a priori.

Le but de cette communication est de montrer l'application d'une méthode de moindres carrés étendue (régression orthogonale) à l'identification de processus linéaires et non linéaires interconnectés de façon cascade. Une première partie rappelle le principe de la régression orthogonale ; une deuxième partie développe ce type de régression aux systèmes cascade et en justifie l'importance à propos d'une chaîne de mesure complexe.

1. LA REGESSION ORTHOGONALE

Le système physique sur lequel on effectue N observations est caractérisé par n variables. On note W la matrice des mesures, W^* celle des grandeurs vraies. Les grandeurs vraies sont soumises à des contraintes de modèle notées :

$$G(W^*, \theta) = 0$$

où θ^* sont les paramètres de forme du modèle.

Dans le cas général on adopte comme structure des différentes grandeurs :

$$W = \begin{bmatrix} W_1 \\ \vdots \\ W_i \\ \vdots \\ W_N \end{bmatrix} \quad \text{où } W_i = \begin{bmatrix} W_{i1} \\ \vdots \\ W_{ij} \\ \vdots \\ W_{in} \end{bmatrix}$$

$$G = \begin{bmatrix} G_1 \\ \vdots \\ G_i \\ \vdots \\ G_N \end{bmatrix} \quad \text{où } G_i = \begin{bmatrix} G_{i1} \\ \vdots \\ G_{ij} \\ \vdots \\ G_{in} \end{bmatrix} \tag{1}$$

Ainsi W_{ij} représente l'état de la $j^{ème}$ variable ($j = 1 \ldots n$) pour l'observation i ($i = 1, \ldots, N$) et G_{ij} représente la $j^{ème}$ contrainte modèle ($j = 1 \ldots n$) toujours pour l'observation i ; W et G sont respectivement les vecteurs d'observation (dimension $N \cdot n$) et de contrainte (dimension $N \cdot m$) ; enfin, le vecteur des paramètres θ est de dimension l.

Le problème que nous nous proposons de résoudre est :

$$(\hat{W}, \hat{\theta}) = \underset{W^*,\, \theta^*}{\text{Arg min}}\ \Phi(W^* - W) \quad \text{sous } G(\hat{W}, \hat{\theta}) = 0 \tag{2}$$

où Φ est une fonction quadratique telle que :

$$\Phi(W^* - W) = \| W^* - W \|^2_{h^{-1}}$$

Par opposition à la régression ordinaire où l'estimation privilégie la distribution des erreurs de mesure à certaines variables, la régression orthogonale utilise une hypothèse moins restrictive où toutes les variables mesurées sont entachées d'erreur. Le problème défini en (1) est équivalent à rendre extrémal le Lagrangien :

$$\mathcal{L} = \left\| W^* - W \right\|^2_{H^{-1}} + \lambda^T G\,(W^*, \theta^*) \qquad (3)$$

par rapport à W^*, θ^* et .

Les conditions de stationnarité au premier ordre s'expriment :

$$\mathcal{L}_{\hat W} = H^{-1}(\hat W - W) + G^T_{\hat W}(\hat W, \hat\theta)\,\lambda = 0$$

$$\mathcal{L}_{\hat\theta} = G^T_{\hat\theta}(\hat W, \hat\theta)\,\lambda = 0 \qquad (4)$$

$$\mathcal{L}_{\lambda} = G\,(\hat W, \hat\theta) = 0$$

Dans le cas général et en raison du caractère non linéaire du modèle G la résolution des équations (4) ne peut se faire que de façon itérative ; à cet effet on peut linéariser G autour d'une estimation $\hat W$, $\hat\theta$ de W^*, θ^*.

Soit $\tilde G$ la contrainte approchée définie par :

$$\tilde G\,(\hat W + \delta\hat W, \hat\theta + \delta\hat\theta) = G\,(\hat W, \hat\theta) + G_{\hat W} \cdot \delta\hat W + G_{\hat\theta} \cdot \delta\hat\theta \qquad (5)$$

avec

$$G_W = \frac{\partial G}{\partial W^T} \qquad (nm \,.\, rr)$$

$$G_\theta = \frac{\partial G}{\partial \theta^T} \qquad (nm \,.\, 1)$$

Le Lagrangien (3) est remplacé par :

$$\mathcal{L}(\delta\hat W, \delta\hat\theta, \lambda) = \left\| \hat W + \delta\hat W - W \right\|^2_{H^{-1}} + \lambda^T \tilde G\,(\hat W + \delta\hat W, \hat\theta + \delta\hat\theta) \qquad (6)$$

et les conditions de stationnarité au premier ordre s'écrivent :

$$\mathcal{L}_{\hat W} = H^{-1}(\hat W + \delta\hat W - W) + G^T_{\hat W}\cdot\lambda = 0 \qquad (7)$$

$$\mathcal{L}_{\hat\theta} = G^T_{\hat\theta}\,\lambda = 0 \qquad (8)$$

$$\mathcal{L}_{\lambda} = G + G_{\hat W}\cdot\delta\hat W + G_{\hat\theta}\cdot\delta\hat\theta = 0 \qquad (9)$$

Comme H est régulière on déduit de (7) :

$$\delta\hat W = W - \hat W - H\,G^T\lambda \qquad (10)$$

Puis λ s'obtient à partir de (9) et (10) :

$$\lambda = A^{-1}\left[G_{\hat\theta}\,\delta\hat\theta + G + G_{\hat W}(\hat W - W) \right] \qquad (11)$$

avec

$$A = G_{\hat W}\, H\, G^T_{\hat W}$$

Enfin, le report de (11) dans (8) exprime $\delta\hat\theta$:

$$\delta\hat\theta = -(G^T_{\hat\theta}\, a^{-1}\, G_{\hat\theta})^{-1}\, g^T_{\hat\theta}\, A^{-1}\left[G + G_{\hat W}(\hat W - W) \right] \qquad (12)$$

En résumé, les équations (10), (11) et (12) définissent les estimations $\delta\hat\theta$ et $\delta\hat W$ qui génèrent les nouvelles solutions :

$$\hat\theta + \delta\hat\theta \quad \text{et} \quad \hat W + \delta\hat W$$

La reprise du cycle de calcul est décidée selon le taux de décroissance du critère .

Remarque : Dans le cas d'un critère additif séparable par rapport aux observations le problème (2) se formule souvent :

$$(\hat W_I, \hat\theta) = \text{Arg min} \left\| W_i^* - W_i \right\|^2_{H^{-1}}$$

$$\textbf{sous} \quad G_i\,(W_i^*, \hat\theta^*) = 0 \qquad (13)$$

et sa solution s'en trouve allégée.

La résolution du problème général (2) et même du problème "additif séparable" (13) reste très complexe en raison de la dimension et de la nature non linéaire des équations mises en jeu. Dans la pratique des simplifications notables interviennent dans les cas suivants :

- modèles uniquement couplés par une partie des variables W,
- modèles uniquement couplés par une partie des paramètres $\hat\theta$,
- modèles bilinéaires par rapport aux variables et aux paramètres,

pour lesquels des structures de résolution hiérarchisée s'avèrent performantes (2), (3).

Dans le cas des systèmes cascade de tels algorithmes peuvent être mis en oeuvre dont certains, quand ils s'appliquent à des modèles bilinéaires, se caractérisent par des solutions quasi analytique.

2. IDENTIFICATION D'UNE CHAINE DE MESURE CONSTITUE DE SOUS-SYSTEMES EN CASCADE

2.1. Description de la chaîne de mesure (figure 1)

Cette chaîne est destinée à la mesure de distribution granulométrique ; elle comprend entre autres éléments un densimètre à rayons gamma constitué de deux éléments :

- l'ensemble émetteur, ou source radioactive, permettant la production de rayonnement gamma par désintégration d'un élément de césium 137. Cette source radioactive enfermée dans un container en plomb à une intensité de 500 millicuries.
- l'ensemble récepteur, ou chambre d'ionisation qui permet de détecter les rayons émis par la source et atténuer par les particules dont la densité est à étudier ; le rayonnement reçu par la chambre d'ionisation est donc fonction de la densité du produit. Cette chambre d'ionisation est maintenue à une température de 45°C. Le rayonnement pénétrant le gaz produit des paires d'électrons (-) ions (+), les électrons sont attirés par l'électrode centrale (Fil conducteur) et, de ce fait, vont créer un courant i.

Le courant recueilli étant faible, de l'ordre de 10^{-8} à 10^{-13} A, un préamplificateur transforme cette information délivrée sous haute impédance en une tension V_{s1} délivrée sous basse impédance.

Le but de l'étage suivant est de délivrer une tension V_{s2} directement proportionnelle à la densité en utilisant la propriété que possèdent certains semi conducteurs, à savoir, pour un transistor de type NPN, le courant de collecteur est lié à la tension base - émetteur par une loi logarithmique. Cette tension V_{s2} est amplifiée en une tension V_{s3} à l'aide d'un amplificateur de gain K_4, puis filtrée. Un additionneur - soustracteur permet d'effectuer le réglage de la valeur de la densité minimale qu'il est possible de mesurer. Dans notre application, cette densité minimale correspond bien entendu à celle de l'eau pure.

L'étage suivant, constitué d'un amplificateur de tension à gain variable, permet de fixer la densité maximale mesurable (réglage de la plage de mesure). Cette densité maximale équivalente à 90 % de la pleine échelle correspond au signal délivré par le densimètre lorsqu'un échantillon de 5 kg de sable de Fontainebleau est maintenu en suspension par brassage à l'aide de la pompe. L'amplification finale est réalisée à l'aide de deux étages : le premier délivre une tension V_c applicable à un galvanomètre situé sur la face avant de l'appareil, l'étage suivant assure une conversion tension - courant afin d'obtenir un signal i_N normalisé 4 - 20 mA, directement proportionnel à la densité mesurée.

2.2. Identification de l'élément non linéaire

On note X et Y les vecteurs de dimension n des variables d'entrée et de sortie de l'élément non linéaire (préamplificateur logarithmique). Le Lagrangien (3) s'explicite ici :

$$\mathcal{L}^1 = \left\| \hat{X} - X \right\|^2_{h_X^{-1}} + \left\| \hat{Y} - Y \right\|^2_{h_Y^{-1}} + \lambda^T . F(\hat{X}, \hat{Y}, a, b) \quad (14)$$

avec $$F(\hat{X}, \hat{Y}, a, b) = \hat{Y} - a\, L\hat{X} - b \quad (15)$$

où $L\hat{X}$ est le vecteur dont les composantes sont les logarithmes des composantes du vecteur $\hat{X}$, où est le vecteur de dimension n dont toutes les composantes sont égales à 1.

Le Lagrangien (14) est stationnaire si :

$$\mathcal{L}^1_{\hat{X}} = h_X^{-1} (\hat{X} - X) + F^T_{\hat{X}} (\hat{X}, \hat{Y}, a, b)\lambda = 0 \quad (16)$$

$$\mathcal{L}^1_{\hat{Y}} = h_Y^{-1} (\hat{Y} - Y) + \lambda = 0 \quad (17)$$

$$\mathcal{L}^1_{\lambda} = \hat{Y} - a\, L\hat{X} - b = 0 \quad (18)$$

$$\mathcal{L}^1_{a} = - \lambda^T L\hat{X} = 0 \quad (19)$$

$$\mathcal{L}^1_{b} = - \lambda^T = 0 \quad (20)$$

Ce système non linéaire peut être résolu par une méthode hiérarchisée utilisant la relaxation suivante :

Un premier niveau de calcul estime, à l'étape k+1, la solution $\hat{X}$ en utilisant l'équation (16) selon un algorithme à itération directe :

$$\hat{X}_{k+1} = X - \left[F^T_{\hat{X}} (\hat{X}, \hat{Y}, a, b) . \lambda \right]_k \quad (21)$$

Un deuxième niveau de calcul estime la solution a, b, $\hat{Y}$, par simple résolution du sytème linéaire (17) à (20) dans lequel $\hat{X}$ prend la valeur élaborée au premier niveau. Ces deux niveaux de calcul sont repris jusqu'à convergence des suites $(\hat{X}, \hat{Y}, \;, a, b)_k$ vers leurs limites respectives.

2.3. Identification des éléments linéaires

En notant X_l $(l = 1, \ldots, m+1)$ les vecteurs de dimension n des variables de la chaîne de mesure (figure 1), le Lagrangien (3) s'explicite :

$$\mathcal{L}^2 = \sum_{l=1}^{m+1} \left\| \hat{X}_l - X_l \right\|^2_{h_l^{-1}} + \sum_{j=1}^{m} \lambda_j^T F_j (\hat{X}, a, b) \quad (22)$$

avec $$F_j (\hat{X}, a, b) = \hat{X}_{j+1} - a_j \hat{X}_j - b_j \quad (23)$$

Il est stationnaire si :

$$\mathcal{L}^2_{\hat{X}_1} = h_1^{-1} (\hat{X}_1 - X_1) - \lambda_1 a_1 = 0 \quad (24)$$

$$\mathcal{L}^2_{\hat{X}_j} = h_j^{-1} (\hat{X}_j - X_j) - \lambda_j a_j + \lambda_{j-1} = 0 \quad (25)$$

$$\mathcal{L}^2_{\hat{X}_{m+1}} = h_{m+1}^{-1} (\hat{X}_{m+1} - X_{m+1}) + \lambda_m = 0 \quad (26)$$

$$\mathcal{L}^2_{\hat{a}_j} = \lambda_j^T \hat{X}_j = 0 \quad (27)$$

$$\mathcal{L}^2_{b_j} = \lambda_j^T = 0 \quad (28)$$

A partir des relations (24), (25), (28) on montre facilement que les valeurs moyennes des mesures et des estimations sont identiques. Si Y et $\hat{Y}$ représentent les valeurs centrées des mesures X et des estimations $\hat{X}$, le modèle se simplifie sous la forme :

$$\hat{Y}_{j+1} - C_{j+1} \hat{Y}_1 = 0 \qquad j = 1, \ldots, m \quad (29)$$

avec $$C_{j+1} = a_j - a_{j-1} \ldots a_1$$

Au Lagrangien (3) est substitué le critère :

$$\Phi = \sum_{l=1}^{m+1} \left\| C_l \hat{Y}_1 - Y_l \right\|^2_{h_l^{-1}} \quad (30)$$

où par convention $C_1 = 1$.

La minimisation de Φ par rapport à C_l et $\hat{Y}_1$ est un problème de régression orthogonale développée en (6).

En posant :

$$\bar{C}_l = C_l\, h_l^{-1/2}$$

$$\bar{Y}_l = Y_l\, h_l^{-1/2}$$

$$C = (\bar{C}_1\, \bar{C}_2 \ldots \bar{C}_{m+1})^T \quad (31)$$

$$W = (\bar{Y}_1\, \bar{Y}_2 \ldots \bar{Y}_{m+1})$$

$$\hat{W} = (\hat{Y}_1\, \hat{Y}_2 \ldots \hat{Y}_{m+1})$$

on montre que le vecteur C des paramètres vérifié :

$$\left(W^T W - \frac{C^T W^T W C}{C^T C} I\right) C = 0 \quad (32)$$

où I est la matrice unité de rang m+1.

Dans ces conditions le critère résiduel vaut :

$$\Phi_R = \text{Trace } W^T W - \frac{C^T W^T W C}{C^T C} \quad (33)$$

L'examen de (32) et (33) montre que C est le vecteur propre de $W^T W$ correspondant à sa plus grande valeur propre.

2.4. Identification de la chaîne complète

La chaîne complète résulte de la concentration de la chaîne non linéaire et de la chaîne linéaire (figure 1).

Les traitements algébriques concernant l'estimation des paramètres et des grandeurs précèdemment effectués restent valables à condition de satisfaire aux conditions de "continuité" :

$$\hat{Y} - \hat{X}_1 = 0 \qquad (34)$$

$$\mathcal{L} = \mathcal{L}^1 (\hat{X}, \hat{Y}, a, b, \lambda) + \mathcal{L}^2 (\hat{X}_1, a_1, b_1, \lambda_1) + \rho^T (\hat{Y} - \hat{X}_1) \qquad (35)$$

où est le vecteur des paramètres de Lagrange associé à la contrainte (34).

Il se structure suivant une forme additive séparable :

$$\mathcal{L} = \mathcal{L}^1 (\hat{X}, \hat{Y}, a, b, \ ) + \rho^T \hat{Y} + \mathcal{L}^2 (\hat{X}_1, a_1, b_1, \ _1) - \rho^T \hat{X}_1 \qquad (36)$$

Si les paramètres de Lagrange ρ étaient connus $\mathcal{L}^1$ et $\mathcal{L}^2$ seraient indépendants. Cette remarque conduit à une optimisation de selon un algorithme hiérarchisé à deux niveaux :

- le niveau supérieur estime de façon itérative les paramètres selon, par exemple, un algorithme de type gradient :

$$\rho^{k+1} = \rho^k + \Delta . (\hat{Y} - \hat{X}_1)^k$$

où k est l'indice de l'itération courante et Δ est le vecteur des pas d'itération.

- le niveau inférieur estime, d'une part, $\hat{X}$, $\hat{Y}$, λ, a, b selon l'algorithme défini en 22 et, d'autre part, X_1, a_1, b_1, λ_1 selon l'algorithme défini en (23).

3. RESULTATS

3.1. Campagnes de mesure

Elles ont été réalisées en injectant à l'entrée de la chaîne de mesure (à la place de la chambre d'ionisation) des signaux d'amplitudes variables issus d'un générateur de courant. Le relevé des tensions aux différents points de la chaîne d'amplification est réalisé automatiquement par un minicalculateur muni d'une chaîne d'acquisition analogique. Les mesures obtenues sont converties, compte tenu de leurs échelles respectives, puis transférées dans un fichier qui est traité statistiquement selon la procédure précèdemment exposée.

3.2. Résultats et interprétation

Le tableau 1 est relatif à 15 mesures effectuées en régime stable sur les 8 voies de la chaîne de mesure ; on le comparera au tableau 2 des estimées qui vérifient les équations des modèles dont les paramètres sont donnés dans le tableau 3.

Pour des campagnes de mesure plus importantes on peut exprimer de façon significative la sensibilité et la variance des estimations obtenues ; des tests statistiques ont été mis en oeuvre pour donner les intervalles de confiance des paramètres des modèles. D'autre part la répétition de campagnes de mesures autours de différents points de fonctionnement de la chaîne de mesure a permis de valider les structures de modèles choisis.

3.3. Conclusion

Soulignons une nouvelle fois la nécessité d'utiliser une méthode globale d'estimation des paramètres des modèles de la chaîne de mesure : le couplage entre les différents modèles de cette chaîne d'une part, la répartition des erreurs de mesure qui affectent l'ensemble des variables mesurées d'autre part, rendent peu adéquates l'utilisation des régressions multiples qui privilégient à certaines variables la répartition des erreurs.

Comme premier résultats de l'identification nous disposont d'un modèle représentatif de l'ensemble de la chaîne de mesure ; cette chaîne est donc étalonnée et elle peut être utilisée comme capteur dans un processus plus complexe de conduite automatique.

Le modèle obtenu peut être utilisé dans le but d'établir un diagnostic sur le fonctionnement de la chaîne de mesure. Le minicalculateur connecté à cette chaîne reçoit en temps réel les mesures des tensions aux différents points de la chaîne ; ces mesures, sous réserve de test de validité, sont injectées dans les modèles ; des tests statistiques permettent de comparer le fonctionnement de la chaîne réelle avec son modèle, de détecter des pannes éventuelles et de les localiser.

BIBLIOGRAPHIE

1. J. RAGOT - Contribution à l'extension de la méthode des moindres carrés. Application à l'équilibrage de bilans industriels. Thèse de Docteur ès-Sciences, Nancy, mai 1980.

2. G. KRZAKALA - Contribution à l'extension du calcul hiérarchisé. Thèse de Docteur ès-Sciences, Nancy, octobre 1981.

3. M. ROBERT - Conception et réalisation d'un granulomètre à sédimentation. Thèse de 3ème Cycle en Génie Electrique, option Automatique. Nancy, novembre 1982.

4. J.L. NOIZETTE - Conduite d'un granulomètre à sédimentation par microprocesseur. D.E.A. de Génie Electrique, Nancy, octobre 1981.

5. A. RAULT, D. GAUME, M. VERGE - Identification en temps réel et détection de défauts sur processus industriels. 1rst IASTED Symposium on Applied Informatics, Lille, France, 15-17 March 1983.

6. J. RAGOT, M. AUBRUN - Application de la régression orthogonale sous contrainte linéaire à un problème d'équilibrage de bilan-matière. Revue de Statistique Appliquée, 1982, vol. 90, n° 2, p. 45-56.

X	$Y = X_1$	X_2	X_3	X_4	X_5	X_6	X_7
1503	659	7255	7241	112	2320	2310	3740
1528	649	7182	7172	43	878	878	2622
1440	673	7446	7436	302	6147	6118	6738
1401	683	7558	7553	419	8530	8491	8598
1464	668	7353	7343	214	4272	4248	5268
1484	664	7304	7294	161	3271	3251	4487
1538	649	7158	7148	19	375	375	2231
1479	664	7314	7299	175	3500	3486	4672
1513	654	7226	7216	92	1816	1806	3354
1474	664	7333	7329	200	4013	3999	5073
1455	668	7397	7387	253	5161	5141	5966
1420	678	7490	7480	351	7119	7084	7500
1440	673	7431	7421	292	5869	5844	6528
1508	659	7236	7226	102	2026	2011	3515
1406	678	7529	7514	385	7856	7817	8076

Tableau 1 - Mesures

$a = -0.8015$		$b = 799.5$	
$a_1 = 12.2149$		$b_1 = -781.6$	
$a_2 = 1.0032$		$b_2 = -34.3$	
$a_3 = 0.9867$		$b_3 = -7032.1$	
$a_4 = 20.3887$		$b_4 = -30.6$	
$a_5 = 0.9952$		$b_5 = 0.4$	
$a_6 = 0.7854$		$b_6 = 1933.3$	

Tableau 3 - Paramètres des modèles

$\hat{X}$	$\hat{Y}$	$\hat{X}_1$	$\hat{X}_2$	$\hat{X}_3$	$\hat{X}_4$	$\hat{X}_5$	$\hat{X}_6$	$\hat{X}_7$
1502.5	657.8	657.8	7253.8	7242.9	114.9	2312.6	2302.0	4741.3
1528.1	651.9	652.0	7182.7	7171.6	44.5	877.2	873.4	2619.2
1437.3	673.3	673.4	7443.7	7433.4	302.9	6145.2	6116.4	6737.0
1397.9	682.9	683.0	7561.7	7551.9	419.8	8528.0	8487.9	8599.6
1467.7	666.0	665.8	7351.0	7340.5	211.2	4274.5	4254.6	5274.8
1485.4	661.8	661.7	7301.1	7290.4	161.7	3266.9	3251.8	4487.2
1537.1	649.9	650.0	7157.8	7146.6	19.9	375.1	373.7	2226.8
1481.2	662.8	662.7	7312.8	7302.2	173.4	3504.6	3488.4	4673.0
1511.6	655.7	655.8	7229.3	7218.4	90.7	1818.0	1809.8	3354.6
1473.1	664.7	664.8	7338.3	7327.7	198.6	4018.3	3999.6	5074.5
1453.8	669.3	669.4	7394.8	7384.4	254.5	5149.2	5135.1	5966.3
1420.5	677.4	677.3	7491.9	7481.9	350.7	7119.4	7086.0	7498.6
1441.1	672.3	672.3	7430.3	7420.1	289.7	5876.3	5848.8	6526.9
1507.4	656.7	656.7	7239.5	7228.7	100.8	2025.2	2016.0	3516.6
1408.4	680.3	680.3	7528.3	7518.3	386.6	7852.5	7815.6	8071.6

Tableau 2 - Estimations

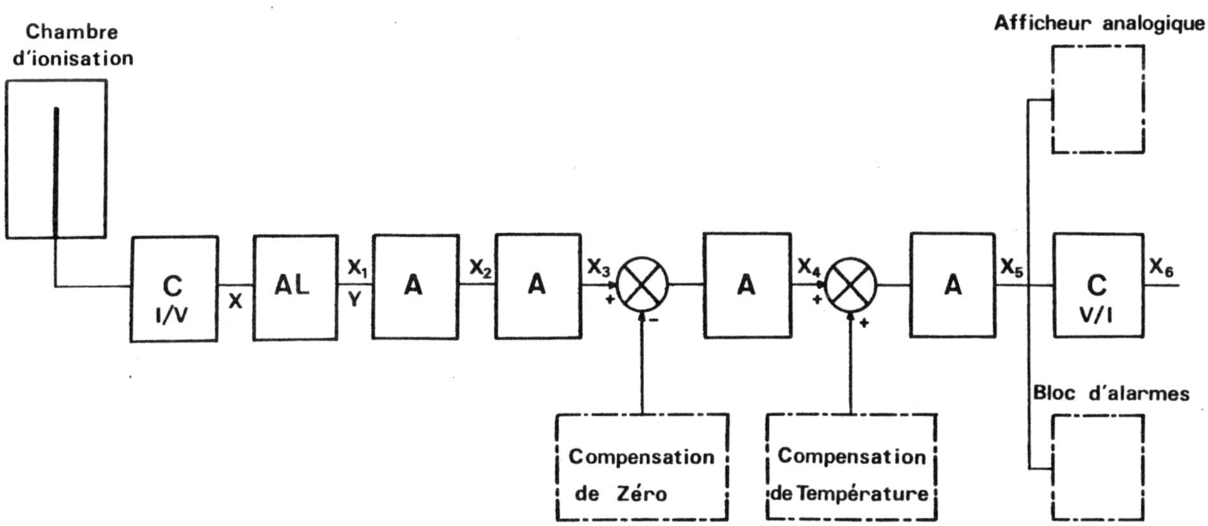

Figure 1 : Chaine de mesure densimétrique

AL – Amplificateur logarithmique

A – Amplificateur

C – Convertisseur

ABSOLUTE LOCATION OF A MOBILE ROBOT WITH A SCANNING LASER RANGE FINDER

L. MARCE, M. JULLIERE, H. PERRICHOT, G. UTJES, C.J. ZHAO

I.N.S.A. (Institut National des Sciences Appliquées)
L.A.T.E.A. (Laboratoire d'Application des Techniques Electroniques Avancées)

35043 RENNES CEDEX FRANCE - Tel. : (99)36.48.30

ABSTRACT

The absolute location of a mobile robot can be computed from the recognition of the obstacles. We present a method using the range data provided by a scanning laser range finder and eventually from a color sensor. We divide the free space for the robot into visibility areas characterized mainly by the number of vertices of obstacles sensed by the range finder in a cylindrical polyhedral world.

INTRODUCTION

We are interested in the autonomous control of mobile robots such as CART (1). The experimentation with the electrical vehicle VESA (Vehicule Electrique Semi-Autonome) (2) results in the idea that for the guidance of a robot and the planning of its path, the absolute location of the robot is a problem of great importance and still opened for further research. The odometry (3) gives only a relative position and the use of beacons (4) is reduced when they are occluded by obstacles.

That's why are developing for VESA 3 , a VESA's daughter, a scanning laser range finder (5) able to detect the vertices of cylindrical polyhedral obstacles and to provide their robocentric coodinates. The robot is also equipped by an optical sensor (6) made of photosensitive cells and filters, allowing the determination of the color of flat surfaces.

The question is to find the absolute location of the robot in a cylindrical polyhedral colored world, the features of which are known (figure 1) from a robocentric colored view such as on the figure 2.

PRINCIPLE OF THE LOCATION

In image processing for missile guidance (7), the correlation between the sensed view and restored reference map data, is used to locate the missile. The correlation is maximum when matching occurs. This principle can be used for our application, but with several important differences : a camera gives an image of pixels with several grey levels, when we have only in our case a boundary ; furthermore, there is none occlusion of the landscape by obstacles, because the view is taken vertically by the missile.

Besides the correlation algorithms require much computing time and a very large memory. So we choose another method which can be divided in three parts :

a) First the coordinates of the vertices of the obstacles are estimated from the range data.

b) The number of vertices sensed by the range finder and eventually the colors of the obstacle sides identified by the optical color sensor allow a rough estimate of the location of the vehicle in a map divided in visibility areas.

c) After a labelling of the vertices, the location becomes more accurate when we try to match the ranges of vertices to the robot between the scanning view and the prestored map.

a) Determination of the coordinates of the vertices

From the range laser data, we construct a model of the environment, estimating the coordinates of the obstacle vertices (figure 3).

It is not necessary to know the color of the sides, but the feature extraction is easier with it.For this extraction the changes (in sign and amplitude) of the polar robocentric ranges between five successive scanning laser data are computed .There are only few possible combinations.

b) Areas of visibility

The use of the scanning telemeter allows the division of the free space for the movement of the robot in visibility areas , characterized by the number of vertices seen from the robot.

So on figure 4, where the hatched part represents a triangular obstacle, the free space can be divided in 6 areas belonging to 2 classes : first class A for which the number of vertices is 2 (areas 2,4,6), and class B for which the number of vertices is 3 (areas 1,3,5).

Thus we can locate approximatively the robot, if we know the number of vertices sensed by it; but the relation between the areas and the number of vertices is not bijective in the above example. The classification is much less ambiguous when the number of obstacles is growing. So if the previous triangle is in a room, the corners of which the robot can sense, the method gives the map of the figure 5 with 23 areas. The number of areas grows rapidly with the number of obstacles.

If the robot can detect the color of each wall and each side of obstacles supposed uniformly painted, the number of identical areas characterized now by the number of vertices and colors of surfaces, diminishes very quickly. In this example all the areas are distinct. The relation between the areas and their characteristics is bijective ; hence we have a means to locate the robot in an unique visibility area. This method does note use the values of the ranges and the angles provided by the scanning laser.

The visibility areas are determined by segments belonging to lines meeting at least two vertices of obstacles, or a vertex and a corner of the room. These lines can involve several segments said valid if they delimit a visibility area, invalid in the other hand. We found several rules to determine the validity of segments.

On the figure 6 are represented the visibility areas for two obstacles. If we paint the obstacles and the walls, all the areas are distinct.

c) Accurate location of the robot

Once we determined the area of visibility where the

robot is fixed, we can match in the aggregate the subset of vertices of the reference map to the subset of sensed vertices, scanned by the laser range finder ; the robocentric coordinates of these vertices are known.

We must now determine the individual matching vertex by vertex ; it is necessary to compute accurately the location of the robot ; this matching is obtained by minimization of the following criterion(figure 7) :

$$\sum_{I,J}\left[\sqrt{(X_{I,i}-X_{I+1,i})^2+(Y_{I,i}-Y_{I+1,i})^2}-\sqrt{\rho^2_{J,j}+\rho^2_{J+1,j}-2\rho_{J,j}\rho_{J+1,j}\cos(\alpha_{J,j}-\alpha_{J+1,j})}\right]^2$$

for every possible matching of vertices.

The knowledge of the color of surfaces makes easier this labelling. This minimization is also used to choose the good area between several areas with same characteristics.

AN EXAMPLE OF LOCATION

For the environment of the figure 1, we obtain the visibility map of figure 8. In its actual position the robot senses 13 vertices. It computes two possible areas without using color (area 37 and area 66). After the labelling it retains the area 66 and gives the computed coordinates of the robot :

$X_c = 78 \qquad Y_c = 67,$

when the real coordinates are :

$X_r = 79 \qquad Y_r = 64.$

DISCUSSION

This method of location is interesting, because the visibility map is determined only once for a given environment.The search of the areas of location is very fast. It does not run if some vertices are not sensed or if there are noisy vertices. From the actual experimentation of the scanning laser range finder, we conclude that only the first type of defect is possible. In this case, we can improve the labelling of the vertices by means of the robocentric range data for the sensed vertices. From two successive vertices of a same obstacle, the length of the edge of the obstacle can be computed and matched with all the prestored edge lengths of every obstacle. If the lengths are accurate enough, the location area of the robot can be found. After labelling the location is improved by matching the ranges only for the sensed obstacles. If there are several possible edges, we retain only the best one.

CONCLUSION

We developed a method of location for which the major quality of the scanning laser range finder is the ability to sense the vertices of polyhedral cylindrical obstacles. In case of bad sensing, it is the precision of measurements that allows the computing of the location.

The knowledge of the color of the edges of obstacles speeds up the algorithms, but is not necessary.

REFERENCES

(1) H.P. MORAVEC. Obstacle avoidance and navigation in the real world by a seeing robot rover. Ph. D. Thesis, Standford University, sept. 2, 1980.

(2) L. MARCE, M. JULLIERE, H. PLACE, H. PERRICHOT. A semi-autonomous remote controlled mobile robot. The Industrial Robot, 1980, 7, 4, pp. 232-235.

(3) M. JULLIERE, L. MARCE, H. PLACE. Capteurs de navigation pour un robot mobile. Congrès AFCET/Automatique. Nantes, 27-29 oct. 1981, pp. 367-374.

(4) M.H.E. LARCOMBE. Mobile robots for industrial use. The Industrial Robot, 1979, 6,2, pp. 70-76.

(5) S. MONCHAUD, R. PRAT. Couplage d'un laser et d'une caméra solide pour la reconnaissance d'objets polygonaux en robotique.Opto 82, Nov., Paris, pp. 179-182.

(6) S. MONCHAUD, R. PRAT. Télémétrie acousto-optique pour robot mobile. Etat de la Robotique en France, 1982, Tome 1 , Hermes Publishing, pp. 201-207.

(7) Thomas F. WIENER. Image processing for missile guidance. Proceedings of the S.P.I.E., Vol. 2338, july 29 - august 1, 1980, San Diego, California.

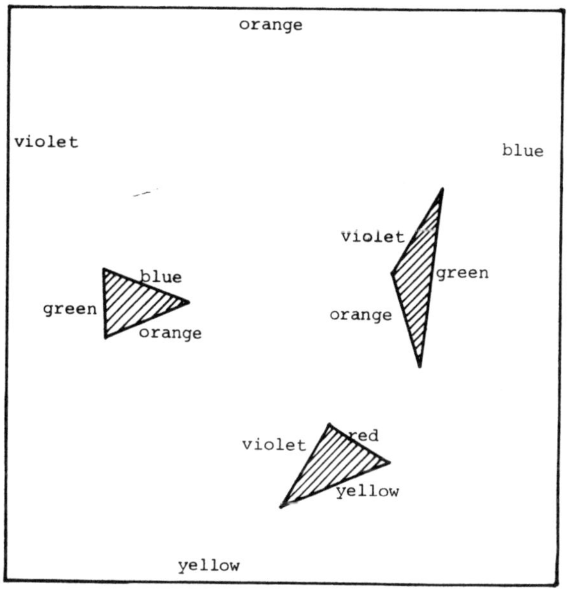

FIG. 1. AN EXAMPLE OF POLYHEDRAL WORLD

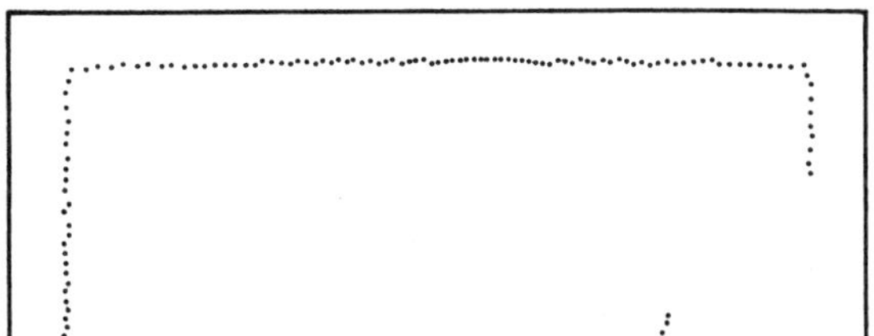

FIG.2. ROBOCENTRIC VIEW OF THE PREVIOUS ENVIRONMENT

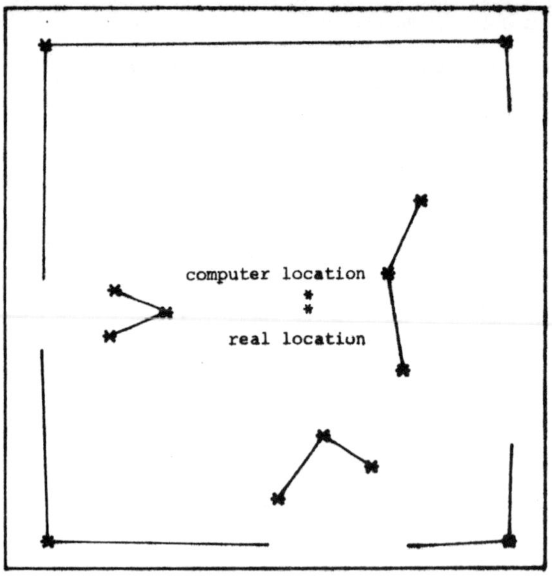

FIG. 3. ESTIMATED VERTICES

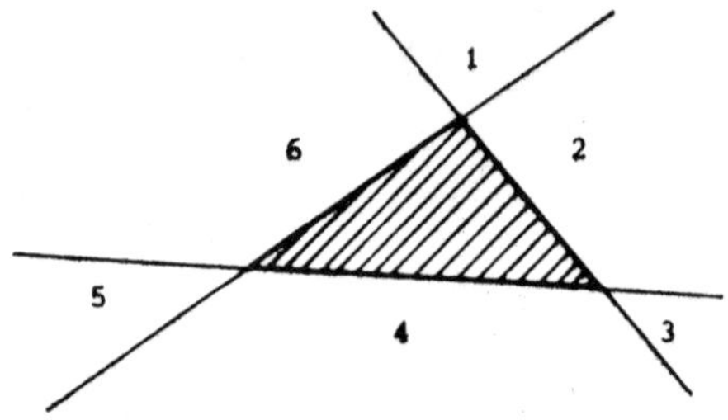

FIG. 4. EXAMPLE OF VISIBILITY AREAS

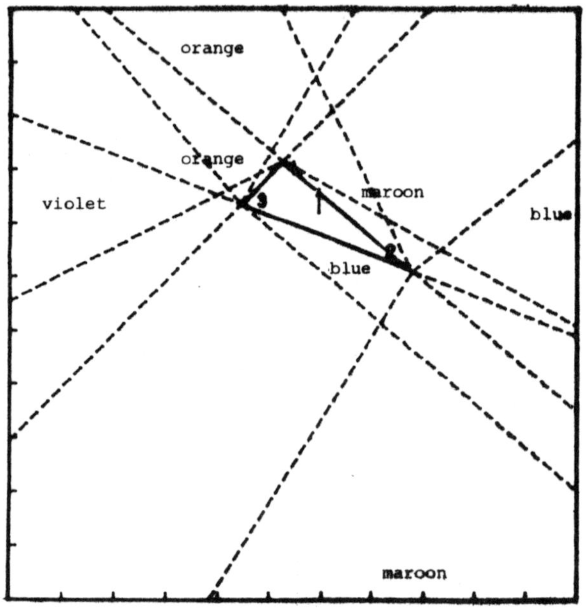

FIG.5. ONE TRIANGLE IN A ROOM

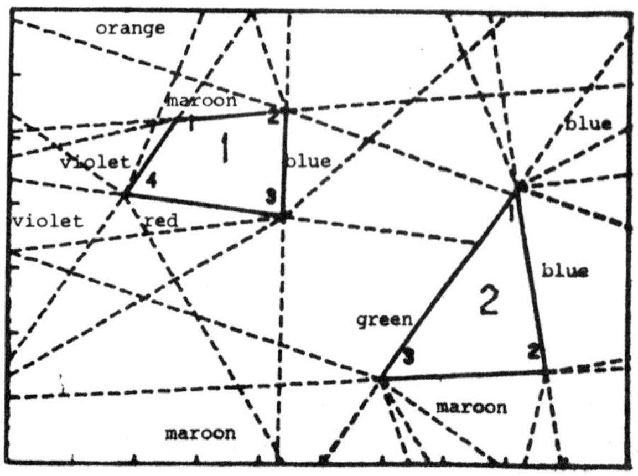

FIG.6. TWO OBJECTS IN A ROOM (59 areas)

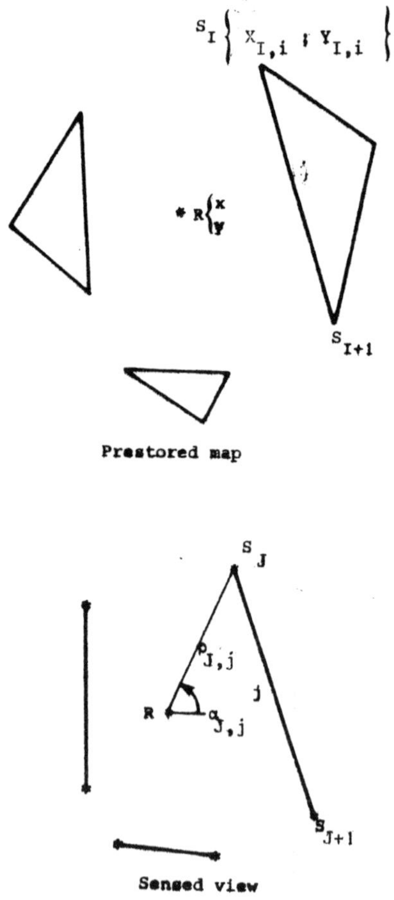

FIG.7. MATCHING FOR ACCURATE LOCATION

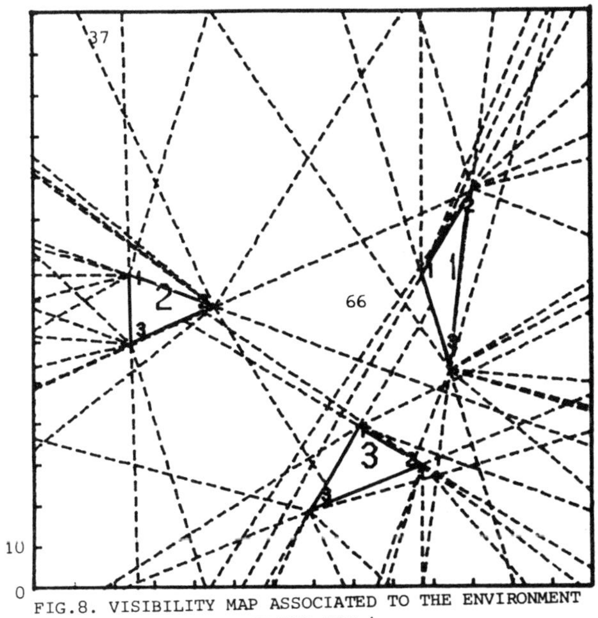

FIG.8. VISIBILITY MAP ASSOCIATED TO THE ENVIRONMENT OF THE FIG.1.

ASSERVISSEMENT DE VITESSE D'UN MOTEUR SYNCHRONE DIPHASE ACTIONNANT UN AXE DE ROBOT

-=-=-=-=-=-=-=-=-=-=-=-=-=-=-=-

C. BERGMANN ; T. MARILIER ; C. SOL

Laboratoire d'Electrotechnique et Robotique

ECOLE NORMALE SUPERIEURE DE L'ENSEIGNEMENT TECHNIQUE

61 avenue du Président Wilson - 94230 CACHAN

-=-=-=-=-=-=-=-=-=-=-=-=-=-=-=-

RESUME

Pour actionner un axe de robot nécessitant des couples dynamiques de plusieurs centaines de mN et des vitesses de rotation de quelques rad/s, nous proposons un système moteur synchrone-réducteur alimenté par un convertisseur à thyristors.

Nous utilisons un moteur diphasé à aimant permanent (3kW nominal, 8kW crête). Ce moteur est commandé en tension à travers un convertisseur à thyristors à modulation de largeur d'impulsion. De ce fait nous obtenons une commande en couple adaptée à la robotique et l'asservissement de vitesse entièrement numérique que nous avons réalisé ne dépend que d'un seul capteur. Ce capteur, incrémental, associé à un réducteur "harmonic drive" sert également à l'autopilotage du moteur.

La commande est élaborée à partir d'un microprocesseur de 16 bits et d'une logique câblée. Le programme effectue le calcul en temps réel de deux fonctions sinusoïdales en quadrature et assure un rapport tension/fréquence constant.

Nous présentons des résultats obtenus en asservissement de vitesse ; le moteur est chargé par un axe de robot incluant une inertie variable. Ces résultats sont exploités en faisant intervenir un modèle de l'ensemble moteur-réducteur+axe en régime variable. L'objectif est de permettre la prédétermination du fonctionnement de l'actionneur en asservissement de position.

INTRODUCTION

L'utilisation de microprocesseurs de plus en plus puissants associés à l'électronique de puissance permet d'envisager dans le domaine de la robotique, la commande de moteur à commutation électronique performant.

Ces moteurs apportent d'énormes améliorations sur les actionneurs traditionnels : le couple massique (0,3MN/Kg) est doublé par rapport aux moteurs à collecteur ; l'entretien est minimum (absence de balais) ; l'inertie thermique est beaucoup plus grande (les enroulements à refroidir se trouvent au stator).

Dans cet article, nous présentons la conception et la réalisation par microprocesseur d'un système de commande d'onduleur à thyristors en P.W.M. alimentant un moteur synchrone à aimant permanent (samarium cobalt) autopiloté, destiné à la motorisation d'un axe en robotique.

DESCRIPTION DE L'ETUDE

Les moteurs à aimants permanents présentent des réactances de commutation relativement faibles ; on peut admettre que si nous alimentons la machine en tension, avec une loi sinusoïdale telle que $V\sin(\omega t) = -N d\Phi/dt$, nous imposons le flux.

Le samarium cobalt ayant un champ coercitif très important, le moment magnétique de l'aimant $|\vec{M}|$ est constant. Dans l'hypothèse d'un champ à répartition spatiale sinusoïdale, en imposant l'angle de déphasage entre M et le champ tournant statorique (autopilotage), nous imposons le couple moteur $\vec{C} = \vec{M} \wedge \vec{B}$. La machine est ainsi commandée en couple.

PARTIE PUISSANCE

L'alimentation de puissance est constituée d'un pont redresseur PD3 commandé ; il assure la protection du montage contre toutes pertes de contrôle de l'onduleur. Il maintient, à l'aide d'un asservissement, la tension de sortie constante quelque soit le courant débité.

Caractéristiques : 220V/30A.

L'onduleur est un onduleur de tension à thyristors, fonctionnant en commutation forcée (P.W.M.). L'extinction est assurée par un thyristor auxiliaire et un circuit LC. Une branche de l'onduleur est donc constituée de 2 thyristors principaux, 2 thyristors auxiliaires et un circuit LC. Le moteur étant diphasé, nous avons 4 branches ; les caractéristiques par branche sont les suivantes : 200V/20A commutation 1kHz. (Voir figure 1).

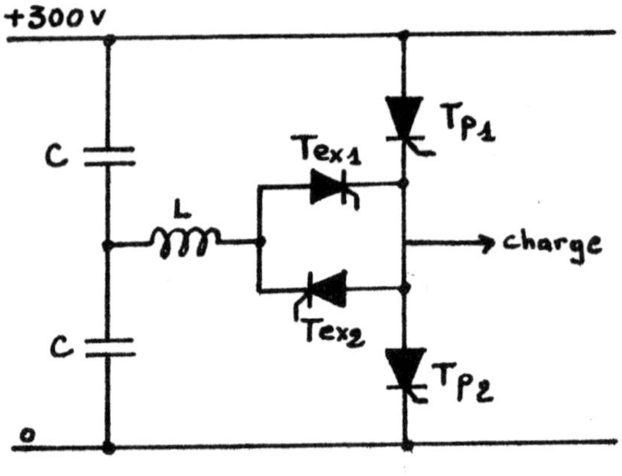

FIGURE 1.

La durée de conduction d'un thyristor principal est imposée par un nombre numérique de 8 bits $0 < N < 256$. Pour garder le pouvoir de coupure de la branche, il faut créer un temps mort de 100µs entre la conduction de 2 thyristors principaux (temps de blocage de Tp1 et inversion de la charge du condensateur).

Cette commande est réalisée en logique câblée T.T.L. pour décharger le microprocesseur de cette tâche.(Voir f.2)

ELABORATION DU NOMBRE N

La machine est commandée en tension, nous devons reconstituer à partir de la tension continue, une tension sinusoïdale échantillonnée à Fe = 1kHz (la limite supérieure est imposée par l'onduleur).

Le calcul de l'échantillon N peut se faire de 3 manières :

- le développement limité d'une fonction sinus par approximation polynomiale.

- la tabulation.

- une série récurrente à deux termes.

1) - La méthode par approximation polynomiale conduit à résoudre deux polynomes du sixième degré. Cette méthode est très bien adaptée pour une calculatrice relativement lente (basic) mais inadaptée pour l'utilisation en robotique.

2) - Tabulation des valeurs : soit N phases considérant une fréquence de découpage f variable de manière que f/F = cte, f fréquence d'échantillonnage.

$$Vi(KT) = a\ E \sin 2\pi \left(fKT + \frac{i-1}{N}\right)$$

Soit $a = \lambda f$ $\quad \lambda$ = cte d'où un rapport cyclique

$$\frac{T_i}{T}(KT) = \frac{1}{2}\left[1 + \lambda f \sin 2\pi\left(fkT + \frac{i-1}{N}\right)\right]$$

donc le terme $\frac{1}{2}\lambda f T \sin 2\pi fkT$ est à tabuler.

Si $fT = cte = \frac{1}{\alpha}$, si l'on choisit en plus α multiple entier de N le terme devient :

$$\frac{1}{2}\ \frac{\lambda}{\alpha}\ \sin 2\pi\left(\frac{k'}{\alpha}\right)$$

L'inconvénient pour les fréquences basses (faible fréquence d'échantillonnage F = λf) peut être limité en modifiant le coefficient λ, il suffit de modifier dans le même rapport le coefficient multiplicateur. Il est préférable que ces modifications soient quantifiées par un multiple de 2 pour des facilités de calcul binaire.

3) - Le terme général d'une suite récurrente est de la forme : $U_n = A \sin(\omega t + b)$, si l'on utilise la relation de récurrence à deux termes de la forme :
$U_n = 2\cos(a).U(n-1) - U(n-2)$, (voir démonstration en annexe).

Il suffit de définir : $U_o = A\sin(b)$ et $U(1) = A\sin(a+b)$, b étant le déphasage et $a = 2\pi.F/Fe$, Fe fréquence d'échantillonnage et f fréquence désirée donc vitesse du moteur.

En fait $a \ll 1$, on prendra b = 0 d'où $U_o = 0$ et $U(1) = A.a^2$. Le rapport V/f doit être constant, il faut donc poser $U(1) = A.a^2$. Pour obtenir une tension sinusoïdale d'amplitude A.a de fréquence F échantillonnée à 1000Hz il suffit d'effectuer une multiplication et une soustraction par échantillon (1ms).

$U(n) = K.U(n-1) - U(n-2)$; K n'est calculé qu'à l'initialisation $K = 2 - a^2$.

Pour le calcul de la fonction déphasée de 90° (moteur diphasé) il suffit de poser à l'initialisation :

$U(o) = A.a^2$ et $U(1) = Aa^2.(1 - a^2/2)$

et l'on prendra le même algorithme de calcul que précédemment.

Le calcul de la fonction sinus et cosinus prendra environ un temps de calcul de 330µs par échantillon donc un temps d'utilisation du microprocesseur d'environ 33 %. (Voir ci-joint la conversion analogique figures 2 et 3).

INTERFACE DE COMMANDE

La commande de chaque branche est réalisée par une carte dont le schéma synoptique est représenté ci-dessous:

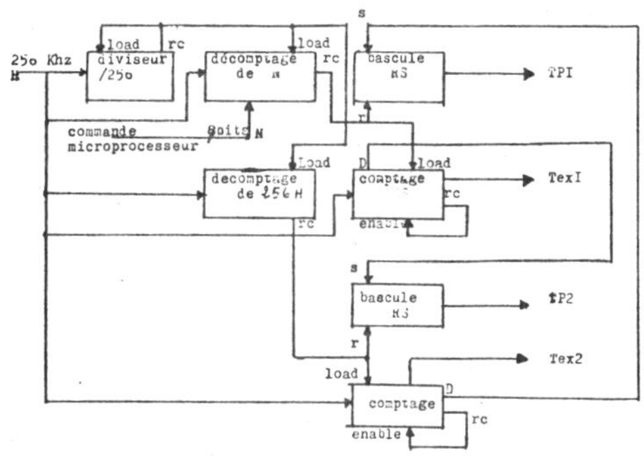

DIAGRAMME TEMPOREL

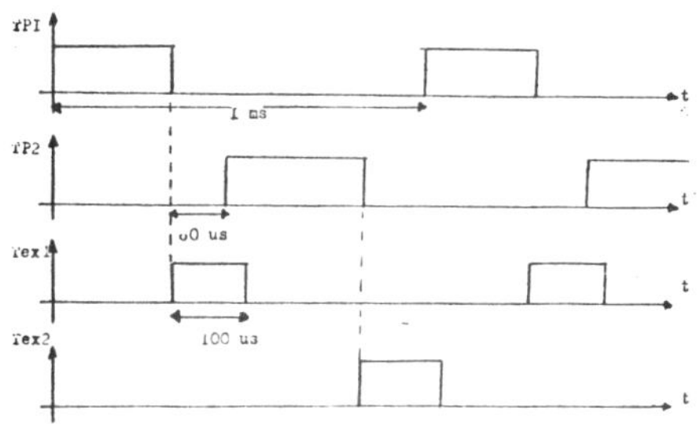

FIGURE 2.

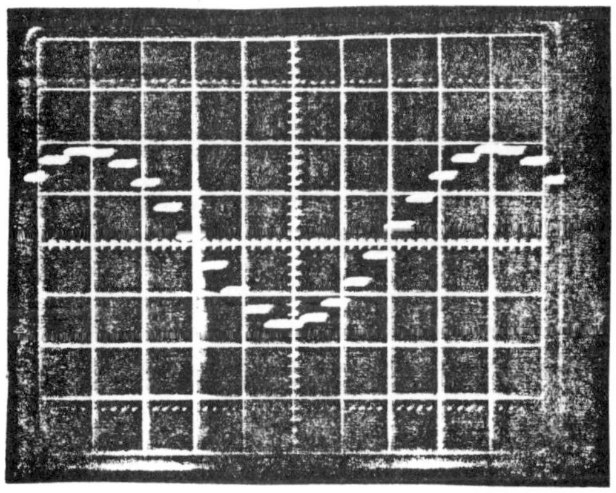

FIGURE 3 - f = 25Hz

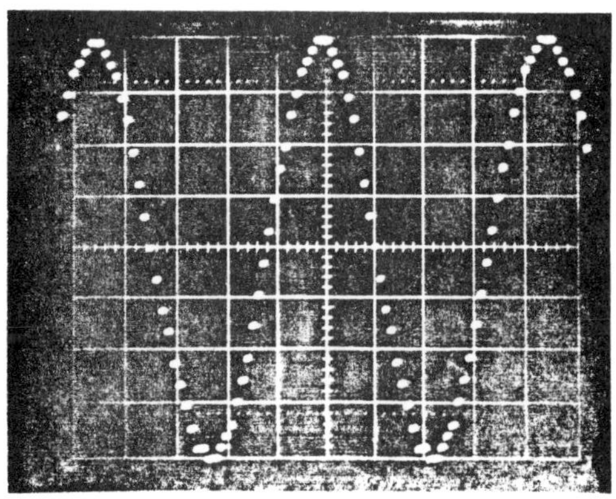

FIGURE 4 - f = 50Hz

ASSERVISSEMENT DE VITESSE

La vitesse est donc échantillonnée quatre fois par période de signal d'où un retard important à la commande. La vitesse affichée n'évolue pas entre deux échantillons nous avons ainsi un bloqueur d'ordre "0".

Transformée de Laplace :

$$B = \frac{1}{p} - \frac{1}{p} e^{-\tau p}$$

Régime harmonique :

$$B = 2e^{-\frac{j\omega\tau}{2}} \frac{1}{\omega} \left[\frac{e^{\frac{j\omega\tau}{2}} - e^{-\frac{j\omega\tau}{2}}}{2j} \right]$$

$$\phi = - \frac{2\pi f}{2} \frac{1}{4f} = - \frac{\pi}{4}$$

Nous devons utiliser un réseau correcteur dérivé pour compenser ce déphasage. (Voir figure 5).

$$V_{S(t)} = \frac{1}{B} \left[V_{e(t)} + \tau_e \frac{dV_e}{dt} - \tau_S \frac{dV_S}{dt} \right]$$

$$\frac{dV_e}{dt} = \frac{V_{e(n)} - V_{e(n-1)}}{T_e}$$

$$V_{S(n)} = \frac{1}{B} \left[V_{e(n)} + \frac{\tau e}{T_e} (V_{e(n)} - V_{e(n-1)}) - \frac{\tau S}{Te} (V_{S_n} - V_{S(n-1)}) \right]$$

$$H_{(p)} = \frac{1 + {}^{\tau e}/Te - r\,{}^{\tau e}/Te}{B + {}^{\tau S}/Te - r\,{}^{\tau S}/Te} = \frac{(1 + {}^{\tau e}/Te) \left[1 - r \frac{{}^{\tau e}/Te}{1 + {}^{\tau e}/Te} \right]}{(B + {}^{\tau S}/Te) \left[1 - r \frac{{}^{\tau S}/Te}{B + {}^{\tau S}/Te} \right]}$$

Soit la fonction :

$$H_r = 4 \frac{1 - 0{,}9\, r}{1 - 0{,}6\, r}$$

Soit :

$$V_{S(n)} = 4\, V_{e(n)} - 3{,}6\, V_{e(n-1)} + 0{,}6\, V_{S(n-1)}$$

Après essai, nous pouvons faire une variation de vitesse 1/30 (de 3Hz à 100Hz) la limite supérieure est imposée par la fréquence d'échantillonnage de l'onduleur et la limite inférieure est imposée par la précision du calcul (12 bits) dans l'état actuel du microprocesseur. Nous avons ainsi constitué l'asservissement de vitesse d'un moteur synchrone autopiloté avec un seul capteur incrémental assurant ainsi un taux de fiabilité maximum.

CONCLUSION

Nous avons mis au point ce type de commande en couple adéquat pour l'utilisation d'un asservissement de vitesse en robotique.

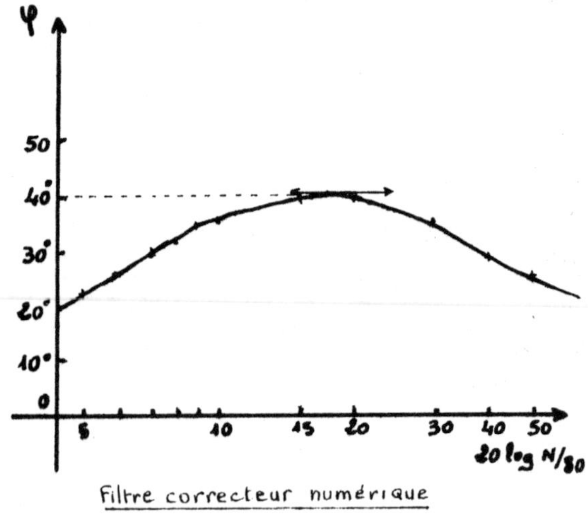

FIGURE 5

ANNEXE

Soit une relation de récurrence linéaire à 2 termes:

$u_{(n+1)} = P\, u_n + Q\, u_{n-1}$ avec P et Q réels non nuls

on cherche des solutions de $u_{(n+1)}$ de la forme :

$u_n = k_1 R^n + k_2 S^n$

Soit $k_1 R^{n+1} + k_2 S^{n+1} = P(k_1 R^n + k_2 S^n) + Q(k_1 R^{n-1} + k_2 S^{n-1})$

$R^2 - PR - Q = 0$ considérons $P^2 - 4Q < 0$

$S^2 - PS - Q = 0$ $R = ae^{j\alpha}$ $S = ae^{-j\alpha}$

avec $a^2 = -1/Q$

$2a \cos\alpha = P$

d'où $u_n = k_1 a^n e^{jn\alpha} + k_2 a^n e^{-jn\alpha}$

$u_o = k_1 + k_2$ $u_o = A \sin \beta$

$u_1 = k_1 a e^{j\alpha} + k_2 a e^{-j\alpha}$ $u_1 = aA \sin (\alpha + \beta)$

Soit $k_1 = + Ae^{j\beta}/2j$

$k_2 = - Ae^{-j\beta}/2j$

d'où $u_n = Aa^n \sin (n\alpha + \beta)$

Posons $a = 1 \longrightarrow u_o = A \sin \beta \longrightarrow$

$u_1 = A \sin (\alpha + \beta)$

$$u_n = (2 \cos \alpha)\, u_{n-1} - u_{n-2}$$

BIBLIOGRAPHIE

- Hacheurs et onduleurs autonomes de l'Institut National Polytechnique de TOULOUSE.

- C. CONRATH
  Etude et réalisation d'un onduleur à thyristors. Thèse Docteur Ingénieur 3ème cycle.

DISCRIMINATION DES ORIENTATIONS DE PIECES
POUR L'ALIMENTATION AUTOMATISEE FLEXIBLE

S. JACQMART
Centre de Recherche - Microsystèmes et Robotique
E N S M M - 25030 BESANCON CEDEX - FRANCE

RESUME :

Cet article propose un algorithme de détermination d'orientation de pièces industrielles, rapide bien que programmé sur un microprocesseur 8 bits.

L'approche peut être résumée par :

- une réduction de l'ordre d'indétermination de la position de la pièce par un dispositif mécanique. De ce fait, la pièce prend un nombre fini N d'états lorsqu'elle est immobilisée dans un trièdre,
- une caméra matricielle fixe observant la pièce dans le trièdre,
- un auto-apprentissage lorsque les images des N états d'équilibre ont été mémorisées. Il fournit :
  . les adresses des points image caractéristiques de tous les états,
  . les valeurs des seuils à appliquer aux intensités numériques obtenues pour ces points de façon à binariser leurs réponses,
  . les vecteurs de référence ayant pour composantes les valeurs binaires obtenues en seuillant les intensités numériques observées aux points caractéristiques,

une phase d'identification de l'état pris par la pièce consistant à rechercher un état parmi N. L'algorithme de reconnaissance effectue une comparaison composante par composante du vecteur de l'état à identifier avec les vecteurs de référence obtenus précédemment.

La durée de traitement lors de l'identification dépend du nombre d'états de la pièce dans le trièdre ; et pour une pièce présentant huit états d'équilibre, la reconnaissance est effectuée en moins de 200 millisecondes.

INTRODUCTION :

L'utilisation de robots en remplacement de l'opérateur humain pour assembler des pièces en petites et moyennes séries exige que, lors de leur alimentation, les composants d'un assemblage soient positionnés et orientés précisément à des cadences élevées (de l'ordre d'une pièce par seconde). Notre but consiste à développer un système d'alimentation automatique programmable qui, à partir de pièces à l'état de vrac volumique, fournit au robot d'assemblage les pièces, une à une, dans l'orientation désirée et à une place définie. De plus, ce dispositif doit pouvoir s'adapter très rapidement à l'alimentation de nouvelles pièces industrielles aux formes variées.

Comme le montre la figure 1, nous avons décomposé le problème d'alimentation à partir du vrac en trois étapes : la phase d'extraction du vrac et de préorientation, la phase de détermination de l'orientation de la pièce et la phase de réorientation et de positionnement final de la pièce. Des prototypes ont déjà été développés dans cette voie de recherche, comme par exemple ceux qui consistent à placer une caméra TV [1] , [2] ou un appareil opto-électronique [3] au-dessus de la piste d'un bol vibrant qui distribue les pièces, ou à utiliser un pousseur qui sépare les pièces en sortie de bol vibrant et les amène à vitesse constante devant un module de vision constitué par deux rubans de fibres optiques se rejoignant en un seul sous une caméra ligne [4] , ou encore à isoler et à bloquer la pièce après sa sortie de bol vibrant pour analyser son orientation [5]. Cependant, les temps de cycle de ces dispositifs sont trop longs pour la robotique légère, une part importante de ce temps de cycle étant consacré au traitement de l'image dans les prototypes [1], [2] , [3] .

Dans cet article, nous nous limiterons à la description d'un dispositif rapide et peu coûteux d'identification de l'orientation de la pièce. Dans un premier temps, nous exposerons les hypothèses et les contraintes de ce système. Puis nous décrirons l'architecture du dispositif de vision et nous formulerons les algorithmes d'apprentissage et d'identification des pièces. Enfin, nous concluerons sur les performances du système.

HYPOTHESES :

Nous observons des pièces tridimensionnelles isolées.

Nous supposons que les pièces à alimenter prennent un nombre fini d'états stables lorsqu'elles sont placées dans un trièdre.

De plus nous supposons que lorsque la pièce est dans son k ième état d'équilibre stable, l'orientation et la position de la pièce (représentée par exemple par les coordonnées du centre de gravité données dans un repère lié au trièdre) sont toujours les mêmes. Ceci traduit la répétabilité du processus mécanique de préorientation.

D. Andrée et A. Wernersson [6] ont eu une approche similaire. En effet, la méthode proposée n'effectue pas l'extraction des contours et des descripteurs (périmètre, surface...) associés à la pièce ; elle repose sur l'obtention de points caractérisant chaque état d'équilibre de la pièce. Mais notre méthode se différencie :

- par la facilité de réalisation de l'apprentissage. En effet, l'opérateur doit seulement positionner la pièce dans le trièdre, dans tous les états stables qu'elle est susceptible de prendre. Les capacités de l'opérateur ne sont pas utilisées pour guider la détermination des points caractéristiques. Notre méthode effectue donc un auto-apprentissage.
- par la prise en compte des défauts du capteur (points manquants), des poussières sur l'objectif lors de l'apprentissage, car ils apparaissent sur les images de chaque état d'équilibre et ne permettent donc pas de caractériser ces états.

- par la rapidité du traitement lors de l'identification de l'état, inférieur à 200 ms pour une pièce présentant au plus 8 états d'équilibre.

DESCRIPTION DU DISPOSITIF

Les principaux composants de ce dispositif sont une caméra CCD TN 2200 GE et un système microinformatique articulé autour d'un microprocesseur Z 80. Les traitements de l'apprentissage se font sur des images 64 x 64 pixels à 16 niveaux de gris stockées dans une mémoire de 32 K octets, par accès direct.

PRINCIPE DE L'APPRENTISSAGE

Après avoir inventorié tous les états d'équilibre que la pièce pouvait prendre dans un trièdre, l'opérateur positionne la pièce successivement dans chacun de ses états stables dans le trièdre. Le système acquière alors une image de chaque état échantillonnée spatialement par la caméra, code l'intensité lumineuse reçue par chaque pixel sur 4 bits (16 niveaux de gris) et mémorise les matrices images correspondant à chaque état. Puis travaillant sur les matrices images non binarisées, l'algorithme d'apprentissage détermine les q points P_i, $(i \in \{1,\dots,q\})$ qui caractérisent tous les états et les q seuils locaux S_i, $(i \in \{1,\dots, q\})$ qui leur sont associés. Ceci permet d'obtenir le vecteur V_k de référence du kième état défini par

$$V_k = \begin{pmatrix} v_1 \\ \cdot \\ \cdot \\ v_i \\ \cdot \\ \cdot \\ v_q \end{pmatrix} \text{ où } \forall\, i \in \{1,..,q\} \left[\begin{array}{l} v_i = 0 \text{ si } I_k(P_i) < S_i \\ \\ v_i = 1 \text{ si } I_k(P_i) \geqslant S_i \end{array} \right.$$

avec $I_k(P)$, l'intensité lumineuse quantifiée obtenue au point P de l'image discrétisée pour le kième état d'équilibre stable.

CRITERE DE SELECTION DES POINTS CARACTERISTIQUES

Nous définissons comme point caractéristique un point permettant de distinguer un ou plusieurs états de la pièce des autres états possibles tout en s'affranchissant au maximum des différences entre pièces (tolérances dimensionnelles, états de surface), des défauts de positionnement et des parasites (lumineux, électriques...). Il s'agit donc de mettre en place un critère de sélection des points caractéristiques qui ait les propriétés suivantes :

- il doit être très peu sensible aux variations des images de chacun des états, c'est-à-dire que les points caractéristiques et les seuils associés déterminés lors d'un premier apprentissage et appliqués à des apprentissages successifs, doivent fournir des vecteurs de référence identiques à ceux obtenus lors du premier apprentissage.
- il doit minimiser le nombre de points caractéristiques pour que la phase de reconnaissance soit la plus rapide possible.
- il doit être rapide à exécuter, afin que la phase d'apprentissage ne dure que quelques minutes.

A chaque point image A est associé un nuage des états d'équilibre constitué par l'ensemble

$$E_A = \{ I_k(A) \;/\; k \in \{1, \dots, N\}\} .$$

En ce point, les états vont être classifiés selon une hiérarchie ascendante indicée.

La proximité entre états est mesurée par la métrique définie par :

$$\begin{array}{l} I_k(A) \in E_A \\ I_j(A) \in E_A \end{array} \quad d(I_k(A),\, I_j(A)) = |I_k(A) - I_j(A)|$$

Le critère d'agrégation choisi est celui du minimum, car il permet de mesurer les plages de niveaux de gris non utilisés par les différents états au point A. De telles plages sont indispensables pour s'affranchir des variations des images. En outre, elles permettent de placer un seuil S_A dans une zone de niveaux de gris non occupés de façon à séparer les états en deux groupes.

Aussi, étant donné deux classes C_1 et C_2, la distance interclasse est :

$$\delta(C_1,C_2) = \min \{d(I_j(A),\, I_k(A)),\; I_j(A) \in C_1,\; I_k(A) \in C_2\}$$

L'algorithme général s'écrit alors :

étape 1 : on considère des classes $C_k = \{I_k(A)\}$

étape 2 :
- on agrège les classes C_k et C_j qui minimisent le critère $\delta(C_k,C_j)$
- on supprime les classes C_k et C_j et on les remplace par une seule $C_k \oplus C_j$.

étape 3 :
- on calcule les nouvelles distances entre classes

$$\delta(C_i,C_k \oplus C_j) = \frac{1}{2}\, \delta(C_i,C_k) + \frac{1}{2}\, \delta(C_i,C_j) - \frac{1}{2}\, |\delta(C_i,C_k) - \delta(C_i,C_j)|$$

pour $i \neq k$ et $i \neq j$

- puis on retourne à l'étape 2, (NP-2) fois.

L'arbre de classification ainsi obtenu permet de caractériser le point A par :

- le nombre de plages de niveaux de gris non utilisés,
- la taille de ces plages, indiquée par l'indice de la classification $\delta(C_i,C_j)$.

Nous appèlerons point caractéristique d'ordre n, un point P présentant n plages de niveaux de gris non utilisés dont la taille est supérieure à 3 niveaux de gris (limite déterminée expérimentalement pour s'affranchir des variations des images).

Un exemple de point caractéristique d'ordre 2 est donnée figure 2.

ALGORITHME D'APPRENTISSAGE

Dans une première étape, nous associons à chaque point de l'image son ordre n.

Dans une seconde étape, nous considérons l'ensemble des points caractéristiques d'ordre 1, comme ensemble de base. Ces points ne présentant qu'une seule plage de taille suffisante, le seuil local qui peut leur être associé est unique et est situé au milieu de la zone inoccupée séparant les états en deux groupes, c'est-à-dire à la distance maximale de chaque groupe. Nous déterminons alors les vecteurs V_k de référence et recherchons les nombres de différences entre ces vecteurs pris 2 à 2 en les comparant composante par composante et enfin, nous déterminons le nombre minimal de différences obtenues.

Tant que ce nombre minimal sera inférieur à un seuil SM fixé empiriquement, nous effectuerons la troisième étape qui consiste à déterminer les états k_1 et k_2 les moins différenciés, à rechercher pour ces états les points caractéristiques d'ordre supérieur à 1 où il existe une zone de niveaux de gris inoccupés entre I_{k_1} et I_{k_2} et à ajouter à l'ensemble de base les points caractéristiques où, de plus, la taille de la zone est maximale. Puis nous redéterminons les vecteurs de référence et les nombres de différences pour comparer le nombre minimal au seuil SM et éventuellement itérer cette étape.

Pour le moment, nous n'avons pas éliminé les points caractéristiques redondants d'ordre 1, car expérimentalement les résultats obtenus quant au nombre de points caractéristiques ne justifiaient pas un alourdissement de l'algorithme d'apprentissage.

PHASE DE DETECTION D'ETAT

Lorsque la pièce est dans un état d'équilibre dans le trièdre, une image de la pièce est prise, digitalisée sur 256 niveaux de gris et mémorisée.

La première étape de l'identification consiste à extraire de cette image les intensités numériques observées aux points caractéristiques, à binariser leurs valeurs à l'aide des seuils locaux et à stocker les résultats qui constituent le vecteur V_X de l'état à identifier.

La seconde étape consiste à appliquer le critère de décision. L'état inconnu sera identifié au kième état si le nombre de réponses différentes entre son vecteur V_X et le vecteur V_k est tel que :

(1) $||V_X - V_k||^2 < SD$

où : $||.||$ représente la norme euclidienne classique.

SD représente le nombre maximal de réponses différentes admissibles, au-delà duquel la pièce est rejetée comme étant dans un état inconnu.

(2) $\forall \underset{i \neq k}{i} \in \{1,\ldots, N\} \quad ||V_X - V_i||^2 > ||V_X - V_k||^2 + M$

où : M représente une marge de sécurité.

CONCLUSIONS - PERFORMANCES

Expérimentalement, nous avons mesuré les propriétés du critère de sélection des points caractéristiques. Ainsi, nous avons observé que pour deux apprentissages effectués sur la même pièce, 90% des points caractéristiques déterminés lors du premier apprentissage donnaient les mêmes vecteurs de référence pour les deux apprentissages. Le critère de sélection est donc peu sensible aux variations des images. Par contre, les seuils locaux varient avec l'éclairement de la scène et nous devrons réaliser une auto-adaptation des seuils locaux à l'éclairement ambiant.

Par ailleurs, la durée nécessaire pour mémoriser les images des divers états de la pièce et pour exécuter l'algorithme d'apprentissage n'excède pas quinze minutes, ce qui est compatible avec les temps nécessaires pour reprogrammer les autres éléments du système d'assemblage flexible.

Quant à la phase d'identification, sa durée T est donnée en millisecondes par la loi :

$$T \simeq \frac{q}{8} + \frac{Nq}{50}$$

où N représente le nombre d'états d'équilibre stables de la pièce,

q représente le nombre de points caractéristiques.

Il faut y ajouter la durée de prise de l'image c'est-à-dire au maximum 100 ms (en tenant compte de la durée de synchronisation et d'intégration de la lumière).

Donc, nous avons développé un système de vision programmable et rapide qui répond aux exigences d'un dispositif d'alimentation flexible pour machine d'assemblage. Cependant, pour des raisons de taille mémoire, nécessaire lors de la phase d'apprentissage, notre système se limite à des pièces présentant au maximum huit états d'équilibre stables.

REFERENCES

[1] J.W. HILL, Programmable bowl feeder design based on computer vision, Assembly Automation, Vol.1, n° 1, Nov. 1980, pp. 21-25

[2] J. HARTLEY, Picking parts from a bowl feeder with image sensing, Sensor Review, Vol.1, n° 1, January 1981, pp. 30-31

[3] U. STRATEMEIER, Optoelektronische Lageerkennung von Werstücken mit geringen Merk malsunterschieden, Proceedings of the 8 th International Symposium on Industrial Robots, Stuttgart, May-June 1978, pp.734-743

[4] W.B. HEGINBOTHAM, D.F. BARNES, D.R. PURDUE, D.J. LAW, Flexible assembly module with vision controlled placement device, Proceedings of the 11th International Symposium on Industrial Robots, Tokyo, October 1981, pp.479-488

[5] T. SUZUKI, M. KOHNO, The flexible parts feeder which helps a robot assemble automatically, Assembly Automation, Vol.1, n° 2, February 1981, pp.86-92

[6] D. ANDREE, A. WERNERSSON, Linear vision for finding the orientation of parts : Learning procedures, Proceedings of the 2 nd International Conference on Robot Vision and Sensory Controls, Stuttgart, Novemver 1982, pp.147-158

[7] S. JACQMART , Contribution à l'étude de systèmes flexibles d'alimentation en composants pour robots d'assemblage légers, Thèse Docteur-Ingénieur (à paraître).

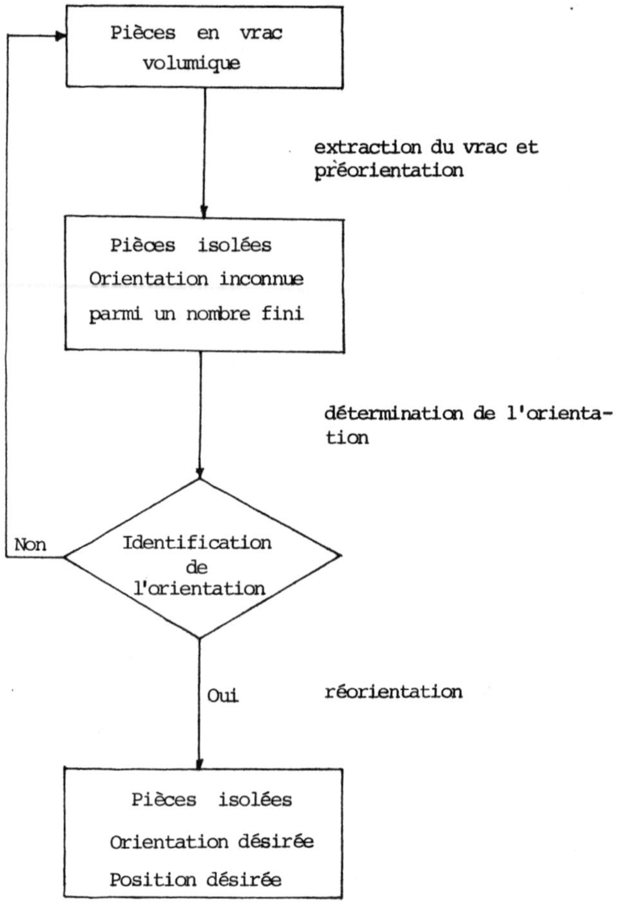

Figure 1 : Synoptique du dispositif d'alimentation

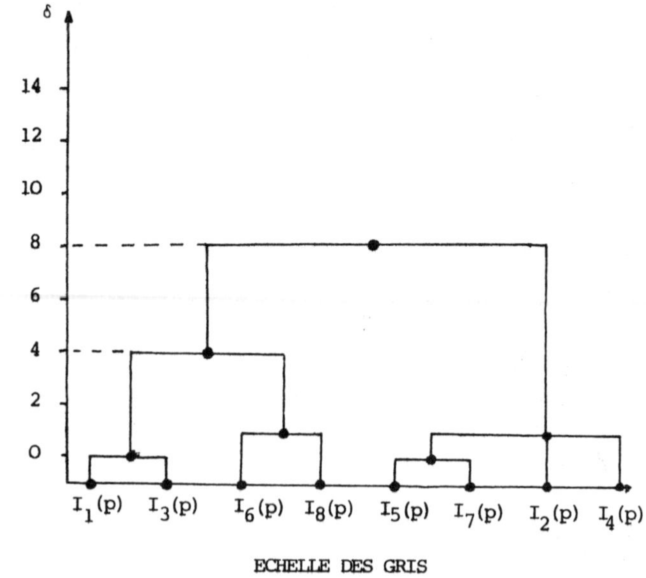

Figure 2 : Exemple d'arbre de classification d'un point caractéristique d'ordre 2

Cas d'une pièce présentant 8 états d'équilibre dans le trièdre.

Soit le point image p décrit par le tableau de donnée:

$$\mathcal{I}(p) = \left[0 \quad 14 \quad 0 \quad 15 \quad 13 \quad 4 \quad 13 \quad 5 \right]$$

où l'élément de la colonne k est $I_k(p)$.

Nous pouvons remarquer que l'agrégation des états débute avec une distance inter classe δ nulle, ce qui met en évidence les états qui ne pourront pas être différenciés à ces points.

Le point p est caractérisé par :

- deux plages de niveaux de gris non utilisés
- de taille 4 et 8

Il est caractéristique d'ordre 2 et peut séparer :

- les états {1,3} des états {2,4,5,6,7,8} avec un seuil de valeur 2 ;

- les états {1,3,6,8} des états {2,4,5,7} avec un seuil de valeur 9 .

Computer Aided Design of Robust Control Algorithms for Robot Manipulators

R. DUMAS, M. LEBORGNE, J.J. BORELLY, B. ESPIAU
I.R.I.S.A.

Av. du Général Leclerc
35042 RENNES CEDEX

ABSTRACT

This paper is devoted to modelling and simulation of an industrial 6 degrees of freedom manipulator: the SCEMI Universal Robot.

Our aim is to simulate and study as deeply as possible the control of this manipulator in order to prepare the implementation of a new control algorithm.

The model of the robot is of the dynamic type with rigid links. The inertia matrices of the links have been computed using a simple shape parts decomposition taken directly from the blueprints.

The simulation includes a model of the payload. Modelled actuators are DC motors controlled by current drives with torque limitations. Different kind of frictions are taken into account and real velocity measurement noises are simulated.

The control algorithm uses a reference model and nonlinear adjustable gains, the variation of which is based upon the error between the reference model and the robot. This kind of control does not need a full knowledge of the manipulator, and allows to use a simplified model. This control scheme has been theorically proved stable and robust. Compare to the classical dynamic control algorithms, the computationnal burden is lower and the implementation is possible using a multimicrocomputers controller.

INTRODUCTION

La nécessité d'améliorer les performances des robots manipulateurs, tant en vitesse qu'en précision et répétabilité, a motivé de nombreuses études qui ont abouti sur deux types de commande: les commandes dites dynamiques et les commandes adaptatives. Toutefois lors des simulations de ces algorithmes de commande, ne sont pas toujours pris en compte un certain nombre de facteurs perturbateurs, tels que les frottements, les bruits de mesure et les limitations des actionneurs. Notre propos est de tester l'efficacité de certains types de commandes en présence de ces perturbations. Pour ce, nous avons choisi d'étudier un robot industriel, le "Robot Universel SCEMI", afin d'en obtenir un modèle aussi complet que possible à partir des plans et des spécifications techniques.

I - MODELISATION

1 - Structure du manipulateur

Le "Robot Universel SCEMI"(Fig. 1) est un robot manipulateur industriel à six degrés de liberté en rotation.

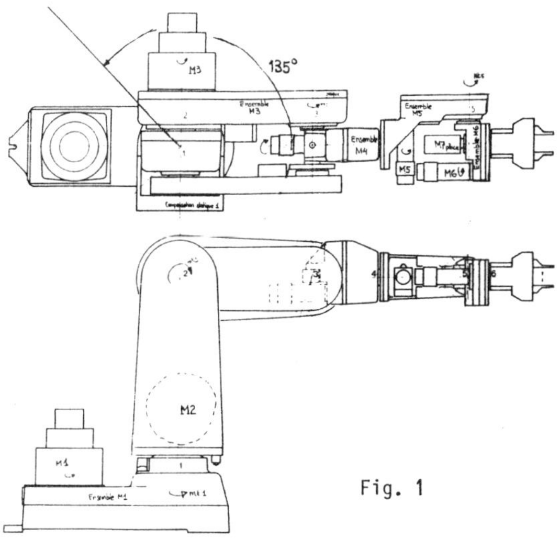

Fig. 1

Ses actionneurs sont des moteurs électriques à courant continu et la transmission des couples moteurs aux articulations se fait par l'intermédiaire d'un train d'engrenages. (rapport de réduction ≈80).
Le modèle mécanique retenu est classique dans la littérature et a été obtenu à partir des équations de Lagrange.
Un programme de calcul symbolique a été réalisé pour obtenir les équations explicites.

L'identification des caractéristiques géométriques et inertielles du manipulateur est basée sur l'exploitation des plans du robot. Un logiciel a été développé permettant, à partir d'une décomposition en éléments simples (cylindres, parallélépipèdes... etc), d'obtenir la position du centre d'inertie et le tenseur d'inertie de tout corps complexe. Les inerties propres des chaînes de réducteurs ont été prises en compte et il est intéressant de noter que leurs valeurs se sont avérées être de l'ordre de grandeur de celles des bras du robot par rapport à leurs axes de rotation.

Un modèle de la charge transportée a été ajouté permettant de prendre en compte des variations d'inertie réalistes.

2 - Les actionneurs

Les algorithmes de commande étudiés impliquent une commande en couple du manipulateur et par conséquent une commande en courant des moteurs. Cette dernière est réalisée par des alimentations à découpage, adaptées au moteur de chaque axe. Une étude en simulation du fonctionnement des variateurs a montré que l'on pouvait négliger l'influence du découpage et de la constante de temps électrique de l'ensemble moteur/self induction. Seules sont retenues les contraintes liées aux limitations des variateurs.
(± 63 Volts et ± 6 Ampères).

Le modèle du moteur prend en compte la limitation de couple liée à la vitesse.

3 - Les capteurs

a - mesure de position

La mesure de position est réalisée par codeur optique incrémental fournissant une mesure dont le bruit de quantification, de l'ordre de 10^{-4} radian, peut être considéré comme négligeable.

b - mesure de vitesse

Le manipulateur est équipé d'origine de génératrices tachymétriques montées sur les axes des moteurs. Ce type de mesure de la vitesse est fortement bruité. Le signal délivré présente un taux d'ondulation de l'ordre de 5% à une fréquence fonction de la vitesse de rotation du moteur. Le modèle est le suivant:

$$\dot{\theta}_2 = \dot{\theta}_1(1 + 0,05 \sin (31\theta)) + W(o, \sigma)$$

$\dot{\theta}_1$: vitesse réelle du moteur

$\dot{\theta}_2$: vitesse mesurée par la génératrice tachymétrique

W : bruit blanc centré d'écart-type σ

Les simulations ont montré que de telles perturbations induisaient pour l'algorithme choisi des oscillations indésirables de la commande (Fig. 6).

Nous avons donc envisagé une mesure de vitesse par comptage de la durée des créneaux du codeur optique de position, ce qui donne une image de l'inverse de la vitesse. Le bruit de mesure dépend de la précision angulaire du codeur; un étalonnage peut être nécessaire.

Dans ces conditions la précision minimum espérée est de l'ordre de 10^{-3} à vitesse maximale.

Notons toutefois qu'un tel dispositif ne permet pas la mesure correcte de vitesses inférieures à une valeur limite dépendant de la fréquence d'échantillonnage.

4 - Les frottements

Le modèle simulé inclus des frottements visqueux et secs. En l'absence d'informations plus précises nous avons retenu pour les frottements secs:

- à vitesse nulle: 25% du couple nominal
- à vitesse non nulle:15% " " "

- Les valeurs numériques retenues pour les modèles de moteurs, capteurs de mesure et frottements ne doivent être considérées que comme des ordres de grandeur réalistes. En effet, les valeurs réelles sont difficilement accessibles à l'expérimentation et sujettes à variations.

II - ALGORITHMES DE COMMANDE

Les algorithmes de commande le plus couramment étudiés se divisent en deux grandes catégories: ceux de types autoadaptatifs et ceux de types dynamiques. Les méthodes de commande autoadaptative ignorent le modèle physique du manipulateur et sont censées identifier en temps réel des paramètres supposés lentement variables. Pratiquement, la dynamique de variation des paramètres est sensiblement égale à celle du manipulateur. A l'opposé, les commandes dites dynamiques supposent une modélisation parfaite du manipulateur. En dehors de la lourdeur de mise en oeuvre, cette commande s'est révélée peu robuste aux erreurs entre modèle et système réel (par exemple: changement de charge, retard d'application de la commande "optimale", butées).

La démarche adoptée ici, permet de prendre en compte les erreurs de modélisation et a fait l'objet d'une étude théorique approfondie en (7) . La commande proposée est du type à modèle de référence.

Si le modèle du manipulateur est

$$(A(q)\ddot{q} + B(q,\dot{q}) + G(q) + H(q,\dot{q}) = \Gamma \qquad (2.1)$$

où $\ddot{q}, \dot{q}, q$: vecteurs accélération, vitesse et positions généralisés.

$A(q)$: Matrice d'inertie.

$B(q,\dot{q})$: Vecteur des forces centrifuges et de Coriolis.

$G(q)$: Vecteur des forces de gravité

$H(q,\dot{q})$: Vecteur des frottements

Γ : Couples ou forces appliqués aux articulations.

la commande appliquée devra permettre de suivre les trajectoires d'un modèle de référence de la forme:

$$U_r = +Kpr\ (q_c-q_r) - Kvr\ \dot{q}r \qquad (2.2)$$

$$U_r = \ddot{q}r \qquad (2.3)$$

où $\ddot{q}r, \dot{q}r, qr$: vecteurs accélération, vitesse, position du modèle de référence

Kpr, Kvr : gains en position et en vitesse

U_r : commande de référence.

La commande s'exprime comme suit:

$$\Gamma = \hat{A}(q)\ (U_r - Kp(q,\dot{q})\varepsilon - Kv(q,\dot{q})\ \dot{\varepsilon}) + \hat{B}(q,\dot{q}) + \hat{G}(q) + \hat{H}(q,\dot{q}) \qquad (2.4)$$

où $\varepsilon = q - qr$: écart en position (2.5)

$\dot{\varepsilon} = \dot{q} - \dot{q}r$: écart en vitesse (2.6)

$Kp(q,\dot{q})$: gain en position Scalaire variable

$Kv(q,\dot{q})$: " " vitesse " "

$\hat{A}(q)$, $\hat{B}(q,\dot{q})$, $\hat{G}(q)$ et $\hat{H}(q,\dot{q})$ sont des matrices choisies par l'utilisateur, plus ou moins distantes de, respectivement $A(q)$, $B(q,\dot{q})$, $G(q)$ et $H(q,\dot{q})$ suivant la finesse de la modélisation retenue.

Il est alors montré en (7) l'existence d'une fonction $k(q,\dot{q})$ positive telle que les conditions (2.7) et (2.8) garantissent des écarts en position et vitesse uniformément bornés.

$$kv(q,\dot{q}) > k(q,\dot{q}) \qquad (2.7)$$

$$kp(q,\dot{q}) = \mu\ kv(q,\dot{q}) \qquad (2.8)$$

La fonction $k(q,\dot{q})$ dépend essentiellement des caractéristiques du manipulateur et des erreurs de modélisation.

Il faut noter que l'étude théorique citée ne prend pas compte les limitations physiques des actionneurs. Il en résulte des difficultés pratiques dans le réglage des gains et l'utilisation d'un modèle de référence. En effet, si l'on désire se placer dans le cadre de l'étude théorique, nous sommes amenés à choisir des commandes Ur du modèle de référence qui évitent de se heurter aux limitations sur la commande U du manipulateur. Ce faisant, les performances en vitesse obtenues se trouvent être inférieures au potentiel du manipulateur.

III - SIMULATIONS

a - Caractéristiques des résultats présentés

Chaque cas de figure comporte l'indication:
- de la charge transportée
- des gains utilisés

Les fig.2-3-4-5-7 concernent un mouvement aller et retour des 3 degrés de liberté.
La fig.6 concerne un mouvement en rotation du 1er degré.

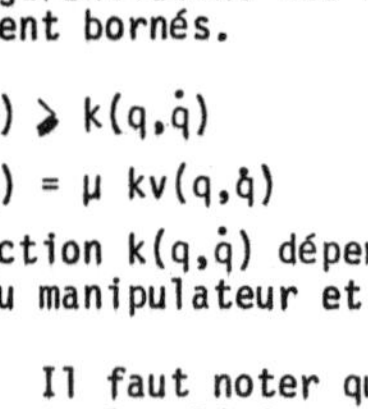

mvt fig. 2-3-4-5-7

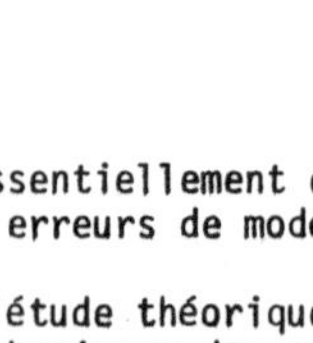

mvt fig. 6

Les courbes correspondent, de haut en bas, à :
- la commande(m.N.) calculée en trait plein, réellement appliquée en trait pointillé.
- la position(degré) du robot en trait plein, du modèle de référence en trait pointillé.
- la vitesse(degré/s) du robot en trait plein, du modèle de référence en trait pointillé.

L'échelle en Y est l'amplitude avec cadrage automatique, les paramètres affichés étant: ymin: valeur minimale et pay: le pas.

b - Résultats

- Fig. 2-3 : Commande dynamique. On utilise le modèle exact du manipulateur et les gains sont réglés de façon à obtenir un temps de réponse de 0,9 secondes. Les limitations des actionneurs sont prises en compte. On peut constater que les performances sont compatibles avec la puissance des moteurs.

- Fig. 4 : Commande avec modèle de référence, utilisant un modèle simplifié du robot manipulateur. Le modèle de référence est réglé de façon à retrouver les trajectoires de la fig. 2 . Les limitations des actionneurs sont prises en compte.

On peut noter un assez bon suivi dans la première partie du mouvement. Des phénomènes de rebroussement apparaissent dans la deuxième partie, correspondant à une anticipation trop importante des couplages compte tenu des limitations des commandes réelles.

- Fig. 5 : Cas identique à fig. 4 , mais les limitations des actionneurs sont supprimées. Il apparaît que les phénomènes indésirables ont disparu.

- Fig. 6 : Commande avec modèle de référence utilisant le modèle exact du manipulateur. Les bruits de mesure de la vitesse d'une génératrice tachymétrique sont simulés . On voit que ce genre de bruit induit sur la commande des perturbations assez importantes.

- Fig. 7 : Commande avec modèle de référence utilisant un modèle simplifié du manipulateur. Les frottements sont pris en compte. Ceux-ci induisent un affaiblissement de la dynamique du manipulateur compensable par une augmentation des gains.

IV - CONCLUSIONS

Nous avons illustré, à travers ces quelques exemples les problèmes pouvant être rencontrés lors de la mise en oeuvre d'algorithmes de commande de manipulateurs proposés récemment.

L'utilisation d'un modèle de référence s'est avérée bénéfique sur le plan de la robustesse de la commande vis à vis des erreurs de modélisation. Toutefois elle suppose un niveau de bruit de mesure suffisamment faible.

En ce qui concerne les limitations des actionneurs l'utilisation de ce type de commandes ne semble pas permettre l'exploitation maximale des possibilités du manipulateur. Elles n'en demeurent pas moins intéressantes dans les applications où le manipulateur est piloté en temps réel à partir d'informations extéroceptives. Le domaine de l'utilisation des gains variables reste à explorer.

BIBLIOGRAPHIE

(1) ALDON M.J. "Elaboration automatique de modèles dynamiques de robots en vue de leur conception et de leur commande", Thèse 3e cycle. USTL Montpellier 1982

(2) COIFFET P." Les Robots - Modélisation et commande" Hermès 1981.

(3) IBARRA-ZANNATHA J.M. "Sur la commande auto-adaptative des robots manipulateurs" Thèse de Docteur-Ingénieur, IRISA , 1982.

(4) LIEGEOIS A., ALDON M.J. "Génération et programmation automatiques des équations de Lagrange des robots manipulateurs", Rapport de Recherche INRIA N°32,1980

(5) PAUL R. " Advanced Industrial robot control systems" 1st Report, Purdue University, West Lafayette, 1978

(6) RENAUD M."Contribution à la modélisation et à la commande dynamique des Robots-Manipulateurs" Thèse d'Etat, U.P.S. Toulouse 1980.

(7) SAMSON C. "Commande non linéaire robuste des Robots Manipulateurs", IRISA, Rapport Interne N°184, 1983

(8) SAMSON C., IBARRA-ZANNATHA J.M. "Sur la commande des manipulateurs", Colloque CNRS, Belle-Isle, 1982

Fig. 2 Mt = 0 Kg kp=18.8 kv=8.7

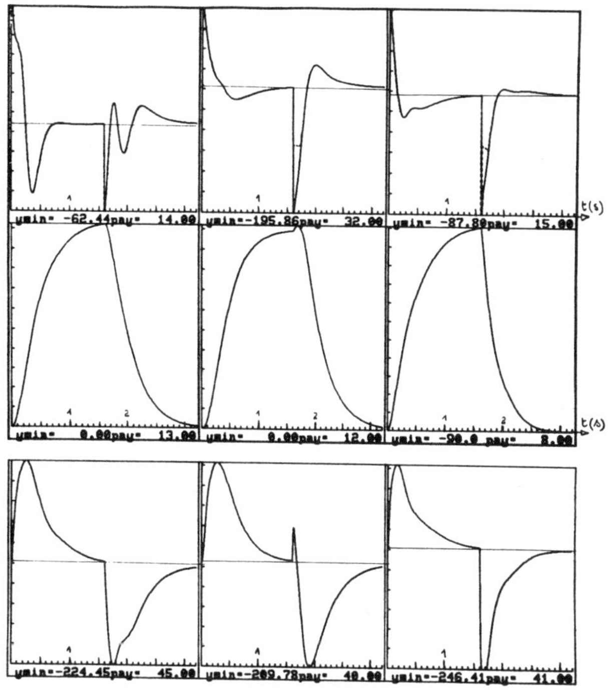

Fig. 3 Mt = 2 Kg kp=18.8 kv=8.7

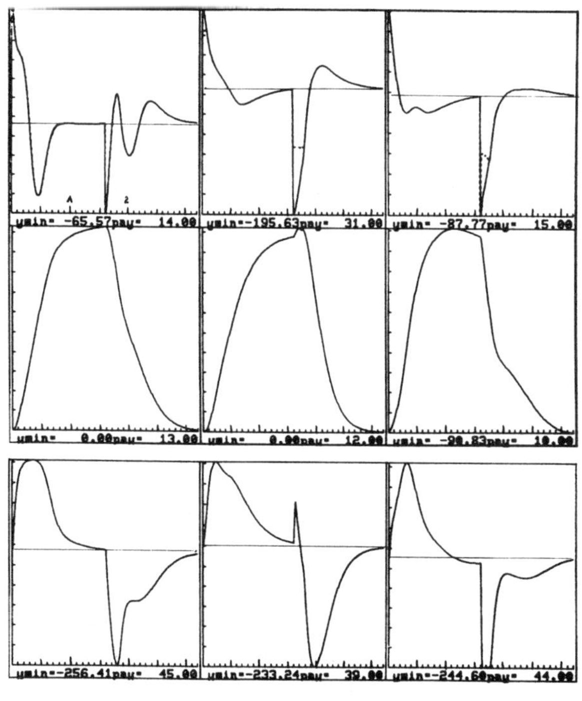

Fig.4 Mt= 0 Kg kp 400 200 180
kv 50 25 25

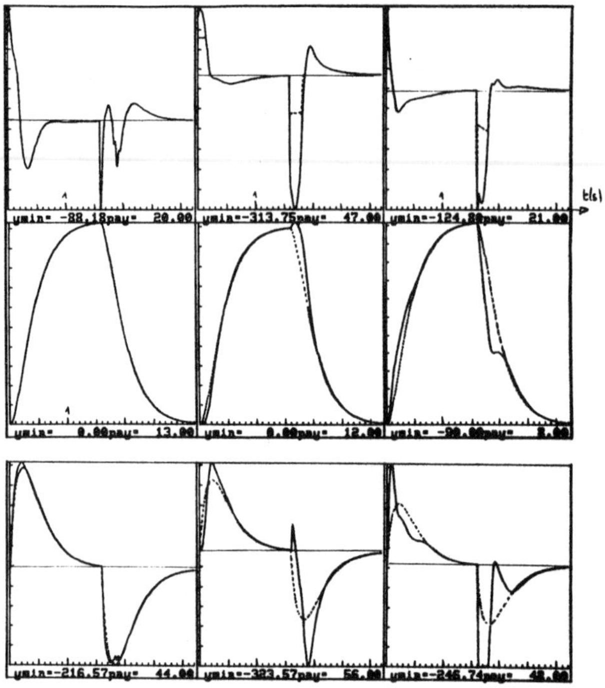

Fig. 6 Mt= 0 Kg Kp 600
Kv 60

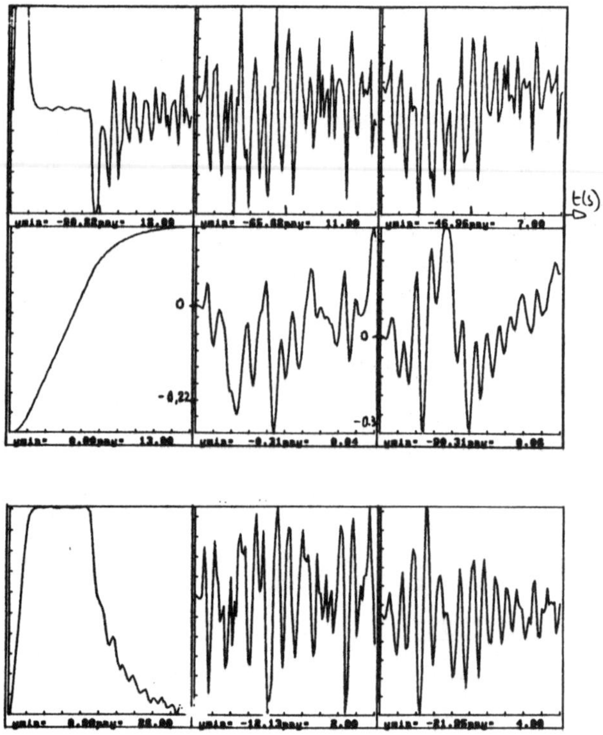

Fig. 5 Mt= 0 Kg kp 400 200 180
kv 50 25 25

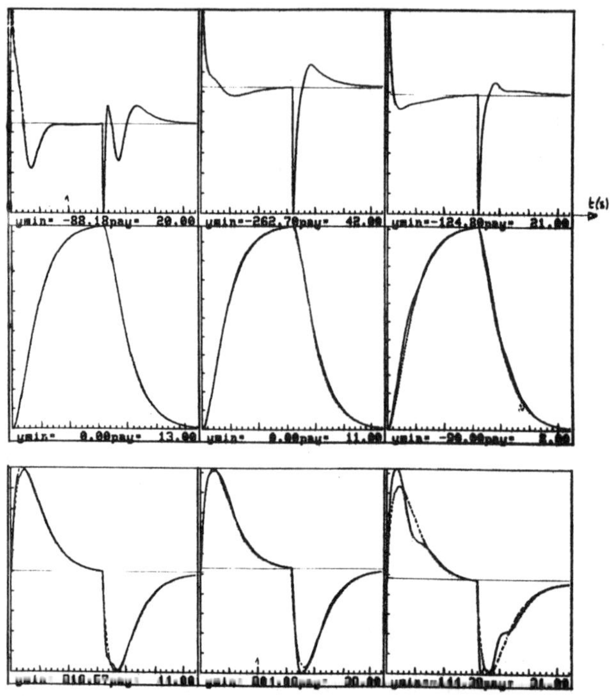

Fig. 7 Mt= 0 Kg kp 600
kv 60

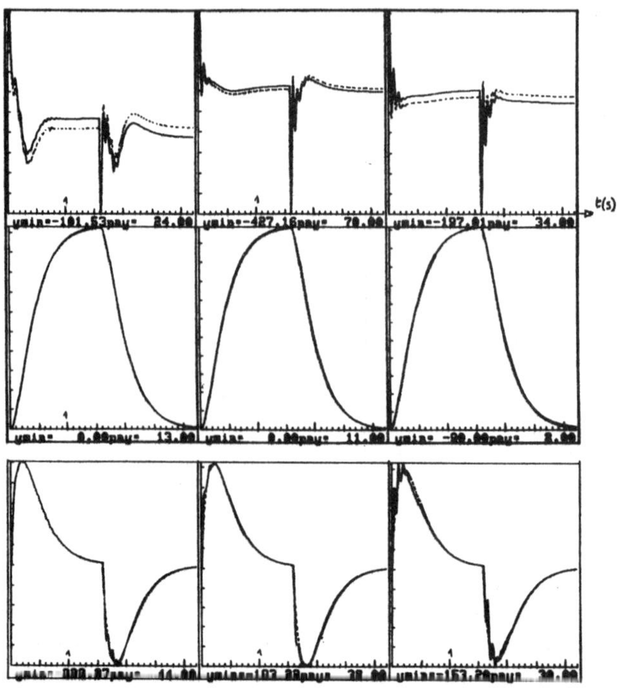

ANALYSE D'IMAGES 2 D. EN VUE DE LA DESCRIPTION D'OBJETS 3 D.

par MM. KERKENI N.*, LEROI M.**, BOURTON M.*** -

Laboratoire d'Informatique, Robotique et Reconnaissance des Formes

UNIVERSITE DE VALENCIENNES ET DU HAINAUT-CAMBRESIS
Le Mont-Houy - 59300 VALENCIENNES - FRANCE

RESUME

La procédure de description d'objets, proposée ici, se développe en 2 phases :

1. Traitement de la vue perspective 2 D. fournissant les points importants permettant d'aboutir à une forme ponctuelle descriptive de l'objet.

2. Reconstitution tridimensionnelle de l'objet à partir de cette description par inversion de la perspective.

Par manque de place, le second point ne sera qu'esquissé alors que, pour le point 1, les deux phases imbriquées seront développées, à savoir :

- Détection de la zone de transition par repérage directionnel multiple et fonction d'appartenance.

- Réduction ponctuelle de la forme après "amincissement".

Le résultat fournit une description directionnelle du contour de l'objet 3 D.

MOTS-CLES : Robotique, extraction de contour, description d'objet 3 D. ...

INTRODUCTION

L'implantation d'un organe de vision sur un robot nécessite la mise au point de logiciels très performants (en temps notamment) en matière de traitement d'image. Pour ce faire, il faut absolument établir des procédures rapides d'extraction de contour et de détermination de l'orientation (et du positionnement) d'objets en vue de leur préhension par un manipulateur par exemple.

L'algorithme présenté dans cette communication permet d'effectuer les deux procédures en simultanéité en établissant un contour directionnel.

I - PRINCIPES DU FILTRAGE DIRECTIONNEL

La limite entre deux régions est caractérisée par une zone de transition d'épaisseur variant entre 3 et 5 pixels. L'orientation de cette zone peut être définie par une direction de filtrage (cf. fig.1) en quadrature avec la direction de l'orientation.

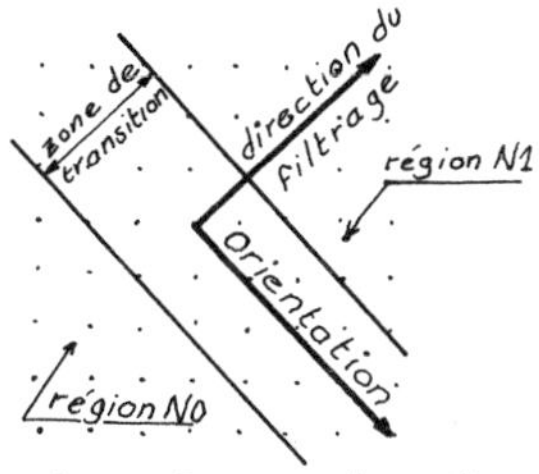

Fig-1- Zone de transition

I.1 - Les Directions Filtrées

Afin d'améliorer la précision de la direction détectée en un point, sur la base des quatre directions premières de Freeman, 12 orientations sont retenues dont les codes sont :

a^2, ab, ba, b^2, bc, cb, c^2, cd, dc, d^2, d(-a), (-a)d.

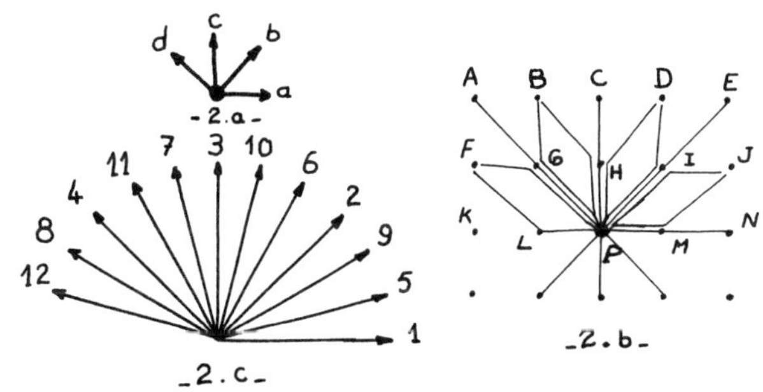

Fig-2- Diverses limites de régions

Toutefois, afin d'assurer une meilleure garantie à la direction, ces orientations sont étendues à une troisième primitive en doublant le premier élément du code soit finalement :

a^3, a^2b, b^2a, b^3, b^2c, c^2b, c^3, c^2d, d^2c, d^3, $d^2(-a)$ et $(-a)^2d$.

I.2 - Estimation Directionnelle d'une Transition

I.2.1- Soit D, l'ensemble de directions d_k :

$$D = \{d_k\}_{k\in[1,12]} = \{a^3, b^3, c^3, d^3, a^2b, b^2c, c^2d, d^2(-a), b^2a, c^2b, d^2c, (-a)^2d\}.$$

ordonné de telle sorte que les éléments d_k et d_{k+2} soient orthogonaux.

Par ailleurs, il faut remarquer, que, pour un pixel P(i,j), la direction en ce point doit être étudiée dans D, sur une fenêtre 5x4, chaque direction d_k, étant estimée dans une fenêtre 4x4 (cf. fig.3) en considérant par exemple les points P(i,j), P(i+1,j+1), P(i-1,j-1) et P(i-2,j-2) pour la direction d_4.

Observant qu'une transition de direction d_k implique de fortes variations, selon d_{k+2}, l'estimation E_k de la direction d_k est obtenue par :

$$E_k = \sum_{\lambda=1}^{4} C_{k,\lambda} \quad \text{où :}$$

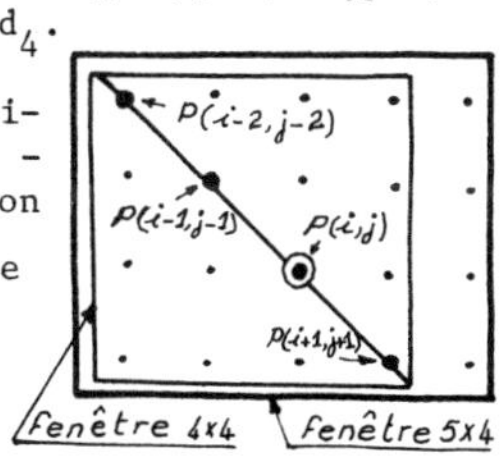

Fig-3- Direction 4

* : Chercheur , ** : Assistant , *** : Professeur.

$$\cdot\ C_{k,1} = \begin{cases} 1 & \text{si } \left| V(i,j) - V(i-i_k, j-j_k) \right| > \theta \\ 0 & \text{sinon} \end{cases}$$

$$\cdot\ C_{k,2} = \begin{cases} 1 & \text{si } \left| V(i,j) - V(i-2i_k+\alpha_k, j-2j_k+\beta_k) \right| > \theta \\ 0 & \text{sinon} \end{cases}$$

$$\cdot\ C_{k,3} = \begin{cases} 1 & \text{si } \left| V(i+i_k, j+j_k) - V(i-i_k, j-j_k) \right| > \theta \\ 0 & \text{sinon} \end{cases}$$

$$\cdot\ C_{k,4} = \begin{cases} 1 & \text{si } \left| V(i+i_k, j+j_k) - V(i-2i_k+\alpha_k, j-2j_k+\beta_k) \right| > \theta \\ 0 & \text{sinon} \end{cases}$$

La pente est alors définie par $d_k = \left| 3j_k - \beta_k \right| / (3i_k - \alpha_k)$ en précisant les valeurs de i_k, j_k, α_k et β_k par le tableau de la figure 4.

k	1	2	3	4	5	6	7	8	9	10	11	12
$I=\{i_k\}$	-1	-1	0	1	-1	-1	0	1	-1	0	1	1
$J=\{j_k\}$	0	1	1	1	0	1	1	1	1	1	1	0
$\alpha\ \{\alpha_k\}$	0	0	0	0	0	-1	-1	0	0	1	1	0
$\beta=\{\beta_k\}$	0	0	0	0	-1	0	0	1	1	0	0	1
pente	0	-1	∞	1	-1/3	-3/2	3	2/3	-2/3	-3	3/2	1/3

Fig.4 - Tableau des paramètres de direction

I.2.2- Exemple d'estimation pour une transition parfaite

La figure ci-dessous illustre la procédure d'estimation directionnelle

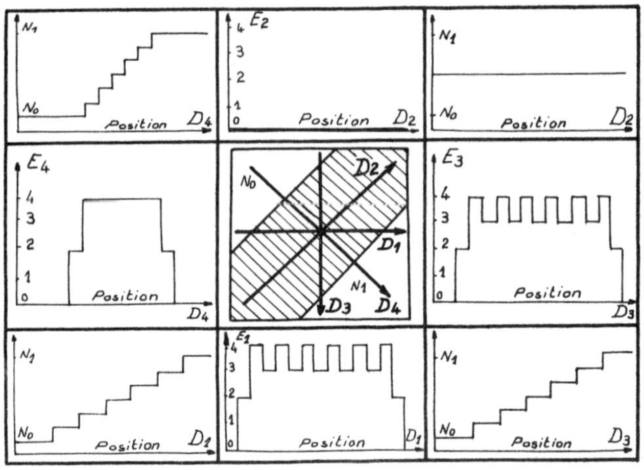

Fig.5 - Estimation (E_k) et variation de niveau de gris selon les 4 directions (D_k) représentées.

Il apparaît donc que, sur $\{E_k\}_{k\in[1,4]}$, pour une transition parfaite, l'une des quatre valeurs d_k est maximale, l'une des directions orthogonales d_{k+2} ou d_{k-2} est nulle, les deux autres d_{k+1} et d_{k-1} sont égales.

II - DESCRIPTION DIRECTIONNELLE

II.1 Détermination du Contour Directionnel

Si, pour une transition parfaite, la détermination de la direction d_k est simplifiée comme le montre la figure 5, par contre, pour une transition moins nette, c'est-à-dire bruitée, il convient d'apprécier, ce qui est rendu possible par la différence existant entre les estimations des directions secondaires d_{k+1} et d_{k-1} soit ΔE_{k2} ($\Delta E_{k2} = |E_{k+1} - E_{k-1}|$) qui doit être minimale mais également par ΔE_{k1} ($\Delta E_{k1} = |E_k - E_{k+2}|$) qui doit être maximale.

Ainsi en généralisant, un point est réputé appartenir au contour de direction d_k si la fonction d'appartenance est maximale, fonction d'appartenance définie par :

$$A(p) = E_p^2 - \mathrm{Min}(\Delta E_{k1}, \Delta_{k2})$$

$$\text{où } \begin{cases} p = k \text{ si } \Delta E_{k1} \gg \Delta E_{k2} \text{ et } E_p = \mathrm{Max}(E_k, E_{k+2}) \\ p = k+1 \text{ si } \Delta E_{k2} \gg \Delta E_{k1} \text{ et } E_p = \mathrm{Max}(E_{k+1}, E_{k-1}) \end{cases}$$

On notera que la fonction a été rendue quadratique par rapport à E_k pour accentuer le caractère décisionnel du critère, le bruit apprécié par ΔE_{ki} permettant de réduire selon son importance la qualité de l'appartenance du point au contour.

II.2 - Amincissement du Contour Directionnel

Malgré la nature quadratique de la décision d'appartenance au contour, celui-ci reste, bien souvent, "épais" et nécessite une procédure d'érosion. Celle utilisée dans l'algorithme recherche, dans l'image directionnelle du contour, des maxima locaux par le biais d'une exploration par fenêtre 3x3.

Ainsi le contour, étant défini par un seul point, se présente sous la forme d'un ensemble de points caractérisés par une direction.

6.a.

6.b.

Fig. 6. Image initiale

Fig. 7. Estimation d'Appartenance

Fig. 8. Contours directionnels "Amincis"

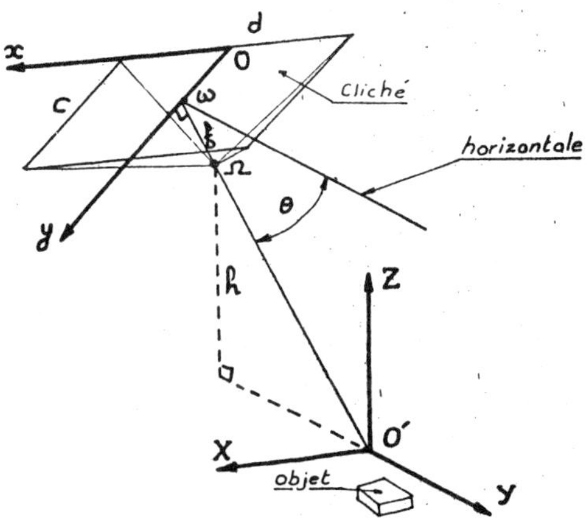

Fig. 9. Referentiels de prise de vue

III - RECONSTITUTION TRIDIMENSIONNELLE DE L'OBJET

Etant dans l'impossibilité de développer dans le détail les procédures mises en oeuvre, ne seront présentées, dans ce paragraphe, que les principes généraux de leur développement en précisant de plus que la méthode n'est actuellement opérationnelle que pour des objets polyédriques.

III.1 - Recherche des Points Multidirectionnels

En théorie, pour ce type d'objets, les limites de régions sont des segments de droites, ce qui, au niveau du contour directionnel, se traduit par une suite alignée de directions toutes identiques, pour peu que cette direction figure dans la base de définition (cf. fig.8.) sinon deux directions définissent le contour.

En pratique, la largeur du contour signifie plusieurs directions, mais avec une certaine dominante.

Les limites d'une portion équidirectionnelle fournissent les extrémités de l'arête et une direction affectée à ces sommets.

Par exploration de l'ensemble des limites de régions sont progressivement déterminées des points affectés des directions des arêtes afférentes.

Ces points multidirectionnels suffisent pour reconstituer l'objet en ses trois dimensions par inversion de perspectives.

III.2 - Procédures d'Inversion

Connaissant les paramètres de la prise de vue, à savoir :

- La focale f de l'objectif,
- Les dimensions (c x d) du cliché,
- La hauteur h du centre optique de l'appareil,
- L'orientation θ par rapport à l'horizontale de l'axe optique,

il est possible de déterminer dans le référentiel O'X Y Z de l'univers de l'objet, la position d'un point P connu par sa localisation dans le référentiel O x y du cliché au moyen de formules assez complexes :

$$X = K\,x\,(y_p - y_b)\,y_v \quad [(y_p - y)\,(y_v - y_b)]$$

$$Y = 2\,K\,f\,y \,/\, [\cos\theta\,(c + 2\,f\,tg\,\theta - 2\,y)] - Z\,[(2\,f - c\,tg\,\theta)\,/\,(2\,f\,tg\,\theta + c)]$$

$$Z = h - h\cos^2\theta\,(y_p - y_c)\,(2\,f - c\,tg\,\theta)\,/\,(2\,f\,tg\,\theta + c)\,[f - \sin\theta\cos\theta\,(y_p - y_c)]$$

avec $y_p = \frac{c}{2} - f\,tg\,\theta$

$y_v = c/2 - f\,tg\,\theta$

$$K = 2\,h\,/\,(2\,f\sin\theta - \cos\theta)$$

$$y_b = y_p - f\,(h-Z)\,(2\,f\,tg\,\theta + c)\,/\,[h\cos^2\theta\,(2\,f - c\,tg\,\theta) + (h-Z)\sin\theta\cos\theta\,(2\,f\,tg\,\theta + c)].$$

III.3 - APPLICATION

Afin d'illustrer l'ensemble des méthodes présentées, les divers traitements sont appliqués à une même image.

Les paramètres de prise de vue sont les suivants :

- Dimension du cliché : c = 24 mm
  d = 36 mm
- Focale de l'objectif : f = 50 mm
- Hauteur de prise de vue : h = 60 mm
- Orientation de l'axe optique : θ = 20°

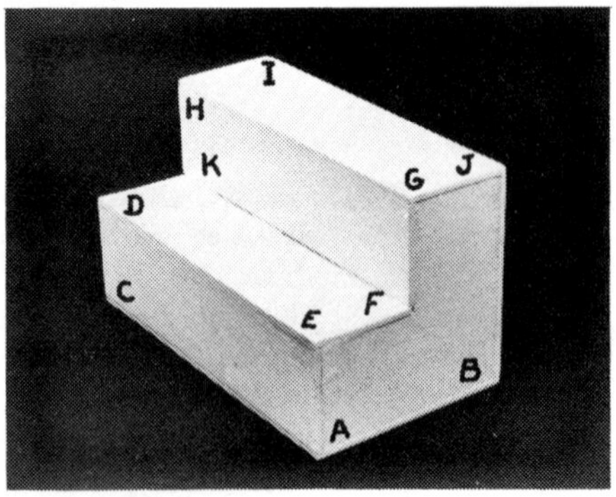

Fig. 10 - Photographie de l'objet

Le tableau ci-dessous présente les dimensions réelles et calculées des différentes arêtes :

Arêtes	AB	AC	CD	AE	DK	EF	FG	KH	BJ	HI	IJ
Dimensions calculées (cm)	16,6	31	9	8,8	8	8,2	9	8,1	17,1	10	32
Dimensions réelles (cm)	17	30	9	9	8	8	9	8	17	9	30

Fig. 11 - TABLEAU RECAPITULATIF

La Fig. 12 présente l'orientation de l'objet à travers la vue de dessus dans le référentiel (O'X , O'Y).

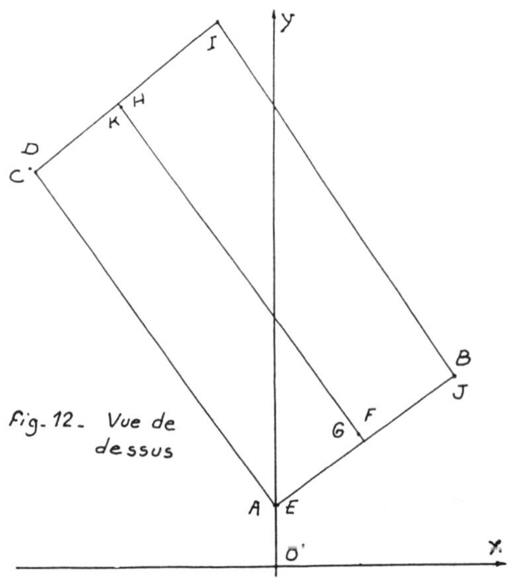

Fig. 12 - Vue de dessus

CONCLUSION

En matière de télémanipulation, la "lecture" tridimensionnelle de l'objet d'application de la tache s'impose avant toute automatisation de l'opération.

L'ensemble des traitements présentés permet de résoudre ce problème sans mise au point de matériels spécifiques 3 D.

La précision atteinte rend opérationnelle cette vision 3 D. pour les robots mobiles en matière d'observation de leur environnement.

En téléopération, la précision étant décroissante au fur et à mesure que le point "lu" s'éloigne de l'organe de visée, il conviendrait d'opérer par approches successives.

BIBLIOGRAPHIE

/1/ M.C KEE J.W. and AGGRAWAL J.K.

"Finding the edges of the surfaces of three-dimensional curved objects by computer" - Pattern Recognition - pp. 25 - 52 (1975).

/2/ ANDREWS H.C.

Tutorial in "Digital Image Processing" I E E 1978.

/3/ ASFAR L.

"Recherche des contours dans une image LANDSAT" 3ème Congrès A F C E T - NANCY, Sept. 81.

/4/ HSUN K. LIU

"Two-and Three Dimensional Boudary Détection" Computer Graphics and Image processing 6, 123-124 (1977).

/5/ JAYAM K. UDUPA, SARGUR N.SRIHARI.

"Boundary Détection in Multidimensions" I E E P A M I - 4 - 1 (1982)

...oooOooo...

SIMULTANEOUS PATH-FORCE-TORQUE-TEACH-IN FOR ROBOTS - THE SENSOR PROGRAMMING CONCEPT

Heindl, Johann and Dr. Hirzinger, Gerhard

Deutsche Forschungs- und Versuchsanstalt fuer Luft- und Raumfahrt e.V.
(DFVLR) Institut fuer Dynamik der Flugsysteme, D-8031 Wessling/Obb.

ABSTRACT

A method is presented that allows a very immediate programming of robots either for paths only or at the same time for forces-torques to be exerted onto the robot's environment. It avoids all the difficulties inherent with off-line or CAD programming caused by uncertainties in the geometry of robot and material to be processed. The method is based on either a space-fixed or robot-mounted force-torque sensor ball via which the human hand commands robot motions in six degrees of freedom. The robot follows the human hand commands as long as it has no contact with its environment. If however it touches a work-piece or the environment, a second force-torque-sensor in the robot's wrist takes care that the robot continues moving only so far that it just exerts the forces or torques given by the human hand. Path and forces/torques are stored and used for automatic execution later. The method is very useful for any kind of teaching robots machining processes as deburring or assembly.

I. INRTRODUCTION

The best way to teach a robot's motion is still a point of controversary discussions. Former keyboards where each key represented a joint have been replaced now by keyboards where a key represents translation in or rotation around a cartesian axis with fixed speed. But there remains awkwardness if the robot is to be taught some oblique or curved motion. Joy-sticks as available with the new ASEA-robots seem to be a step in the proper direction, though they do not allow, for example, combined translational and rotational teach-in. On the other side there are programming languages and CAD-off-line techniques that until now have not yet found very wide applications in industry. One of the reasons is that a robot as well as its environment are subject to variations in mechanical conditions e.g. variable loads or tolerances that destroy precomputed geometries.

Our approach is based on showing the robot in a very direct way what to do. The first steps of our development using a 6-dimensional force-torque sensor-knob by which the human hand tells the robot where to go were outlined in [1]. In this paper the main aim is to describe how by using an additional force-torque sensor in the robot's wrist the robot moves as before, however if it touches the environment or some work-piece, it tries to exert just the forces-torques as given by the human hand. Let us briefly review first the basic path-only teach-technique, the theoretical investigation of which was given in [1].

II. THE BASIC SENSOR-BALL-TEACH-TECHNIQUE

The principles can be taken from fig. 1, 2 and 3. A teach-person grips the six-dimensional force-torque sensor integrated into a plastic hollow ball that fits into a human hand (we call it a sensor ball, see fig. 1). The measuring center of the sensor approximately coincides with the hand's center of attack. The sensed force vector $\underline{f}^i$ in inertially fixed cartesian space is interpreted by some static or dynamic relation as translational velocity command vector $\dot{\underline{x}}$ and the torque vector $\underline{t}^i$ as rotational velocity command $\dot{\underline{\phi}}$. In all the figures the lower indices "hum", "rob" indicate that the forces /torques are exerted by the human hand or sensed by the robot respectively, while the upper indices "i", "eff" indicate the inertially fixed or the end-effector-(tool)-based coordinate system.

Two alternatives are obvious

a) A sensor-ball is fixed on ground or on a teachboard (fig. 2). The forces/torques $\underline{f}_{hum}$, $\underline{t}_{hum}$ are either interpreted in the inertially fixed cartesian system or in a tool-based system. Only this latter case requires multiplication by the orientation matrix θ to arrive at $\underline{f}^i_{hum}$, $\underline{t}^i_{hum}$, but in fig. 2 we assume that the interpretation is inertial.

b) A sensor-ball is mounted at the robot's hand and allows to pull and turn the robot as wanted (fig. 3). The forces/torques $\underline{f}^{eff}_{hum}$, $\underline{t}^{eff}_{hum}$ in the sensor-based coordinate system (which coincides with the end-effector system when using an appropriate calibration matrix) have first to be multiplied by the orientation matrix $\underline{\theta}$ to arrive at $\underline{f}^i_{hum}$, $\underline{t}^i_{hum}$.

In both cases one human hand controls all six degrees of freedom of a robot, the joint motors being servoed via the cartesian transformation and appropriate control laws.

Concerning the closed-loop dynamic relations involved pure proportional control may yield a stable system made up by human and robot dynamics [1]. However essentially better teach-in results are attained if one uses at least a simple PD-controller or even better a state space-controller. We found that it is sufficient to use the magnitudes of force and torque vector only as control input and split up the control output into the directions given by the force and torque components. So we have to deal with only two single-input-single-output control loops. Control law details however are omitted here.

III. THE SIMULTANEOUS PATH-FORCE-TORQUE-TEACH-STRUCTURE

In addition to the arrangement outlined in section II we use now an additional 6-dimensional force-torque-sensor as wrist sensor between the robot's last joint and the end-effector. For simplicity we assume 6 axis-robots here but the technique is applicable with modifications to other configurations too. In our experiments wrist sensor and sensor ball are of exactly the same principle (the DFVLR-sensor concept was outlined e.g. in [2]), though the mechanical size of the sensor ball in general is smaller than that of the wrist sensor.

We now use the force-torque addition $\Sigma \underline{f}^i$, $\Sigma \underline{t}^i$ between the sensor ball and the wrist sensor to derive the robot commands; again static or dynamic control laws are involved, and again the two alternatives, i.e. sensor-ball inertially fixed (fig. 4) or robot mounted (fig. 5) are possible. It immediately comes clear that as long as the robot can move freely, we have the situation as outlined in section II. If however the robot has contact with its environment the forces-torques in the robot's hand increase with negative sign compared to those of the human hand so that by $\Sigma \underline{f}^i$, $\Sigma \underline{t}^i \rightarrow \underline{0}$ the robot finally comes to rest exerting now the same forces/torques on its environment as given by the human hand. But note that the summation of forces/torques holds for each component. Thus depending on the teach-information the robot may move in certain directions, while it has already come to rest in perpendicular directions where it tries to exert the teacher's forces or torques (see Mason's concept of "artificial" and "natural" constraints [4]). Contour following is a good example herefore. In case the sensor-ball is inertially fixed (fig. 4), only the wrist-sensor-data have to be transformed into the inertial frame, while in case of the robot-mounted sensor ball, the two force-torque sensor-data which by appropriate calibration are both given in the end-effector frame have to be transformed into the inertial frame (in fig. 5 this is done before summation for clearness, but by proper calibration the sensor data may be added directly before transforming the sum into the inertial frame).

Of course if the forces and torques the robot has to exert exceed that of a human hand, then in case of contact with the environment the human hand values may be scaled by some factor > 1.

IV. INHERENT PROBLEMS

IV.1. THE FORCE-TORQUE-BIAS

For the robot-wrist-sensor it has to be taken in account that in case the end-effector's or tool's weight is not negligible, the sensor "nulled" in some reference orientation would show up forces/torques in different orientations due to the changing gravitational load of the end-effector. Thus we compensate the gravitational forces/torques the robot has to expect in different orientations by mapping the present orientation into the expected forces/torques sensed in a short learning or training phase and subtracting these expected bias forces/torques from those actually sensed to arrive at the real machining forces/torques. Details of this associative mapping scheme are not discussed in this paper.

IV.2. THE INTERPRETATION OF TORQUES

Remember that the main idea of the structures in fig. 4 and 5 is to assure that the robot exerts only those forces-torques onto its environment that are given by the human teacher.

However with the simple addition technique outlined in section III it is obvious that a problem indicated in fig. 6 might occur. As the robot's tool center does not coincide with the sensor's measuring center, reaction torques may occur due to reaction forces (let us call them bias torques) and would tend to rotate the robot's hand if these torques are not counterbalanced by the human hand. Two solutions come into mind immediately:

a) if one is not interested in teaching torques (e.g. in grinding problems) then one may just disregard (i.e. set to zero) the wrist sensor's torques.

b) if one is interested however in teaching torques too, then the bias torques $\underline{t}_b$ may be either computed by $\underline{t}_b = \underline{f} \times \underline{r}$ if the vector $\underline{r}$ from tool tip to the sensor's center in tool-based coordinates is known or again by an associative mapping generated in a training phase. These torques associated now with the forces sensed may be subtracted from the really sensed torques in order to arrive at the torques the robot exerts onto its environment.

IV.3. EVALUATION OF STORED INFORMATION

There are a lot of ways to make use of the informations gained during the teach-in process. The most obvious one is to store all the cartesian commands that are issued by the controller in the teach-phase and to recall them in the operational phase, the starting point being arbitrarily shiftable. In our lab we are using a sampling period of 30 msec with an ASEA IRB6 robot, an Intel 8086/87 combination performing on-line coordinate transformation and control law realization.

Increase in motion speed may be attained in different ways e.g. simply by adding n successive commands.

If machining forces/torques are of interest the technique outlined in section III allows to store the forces/torques as sensed by the robot's wrist (and commanded by the human hand) together with the motion commands. It depends on the special task how this information is evaluated later on in the operational phase. Simple examples are:

- the robot tries to sense the same forces/torques as sensed in the teach phase by using these forces-/torques as nominal values, deviations from which are used to correct (by addition) the stored motion commands via the same control law structure as used in the teach phase (note that instead of the human hand's forces/torques the negative values of the stored forces/torques are fed into the loop now)
- the robot tries to keep prescribed magnitudes of the force/torque vectors (e.g. constant), using the forces/torques of the teach phase only as directional information.
- the robot tries to null the forces/torques even though in the teach-phase there may have been reaction forces/torques. Again this may be accomplished by taking the stored motion commands and correcting them by the output of the same control law as used in the teach-phase, the sensor ball's values being replaced now by zero values.

An example for the last case might be the assembly of parts, where by iterated execution and storing the latest motion commands the robot learns to do the job better and better. Indeed self-learning robots are one main background topic of the technique proposed here, where the human operator shows the robot the task to be performed in a very direct way and then the robot tries to repeat and at the same time improve its performance.

V. CONCLUSION

Techniques are proposed that allow a very direct teach-in of robot motions combined with sensory patterns. The method using a force-torque-sensor-ball is particularly obvious in the case of teaching reaction forces/torques but may be extended to other sensory information.

The techniques are applicable to many teach problems as occur in grinding, assembly, contour following, teaching a robot to turn a crank, lock up a door, write on or wipe a table, etc. Self-improving principles are applicable in the operational phase in a very direct way. Even if only the paths and not the reaction forces-/torques are of interest, i.e. we disregard any sensors in the robot's wrist, the sensor-ball teach-technique has turned out to be a very powerful method for fast, immediate robot teach-in.

REFERENCES

[1] Hirzinger, G.
Robot-Teaching via Force-Torque-Sensors, Sixth European Meeting on Cybernetics and Systems Research, EMCSR'82, Vienna, April 13-16, 1982.

[2] Plank, G. and Hirzinger G.
Controlling a Robot's Motion Speed by a Force-Torque-Sensor for Deburring Problems, 4th IFAC-IFIP Symposium on Information Control Problems in Manufacturing Technology, Gaithersburg, Maryland USA, Oct 26-28, 1982.

[3] Craig, J.J., Raibert, M.H.
A Systematic Method of Hybrid Position/Force Control of a Manipulator. IEEE Computer Software Applications Conference, Chicago, Nov. 1979.

[4] Mason, M.T.
Compliance and Control for Computer Controlled Manipulators. IEEE Trans. on Systems, Man and Cybernetics, SMC-11 (1981), No. 6, S. 418-432

[5] Whitney, D.E.
Quasi-Static Assembly of Compliantly Supported Rigid Parts. Journal of Dynamic Systems, Measurement, and Control, March 1982, Vol 104/65.

[6] Bejczy, A.K.
Smart sensors for smart hands. AIAA paper N. 78-1714, AIAA/NASA Conf. "Smart" Remote Sensors, Hampton, Virginia, 14-16 Nov. 1978.

[7] Schmieder, L., Mettin, F., Vilgertshofer, G.
Kraft-Drehmoment-Fühler. Patent E01L1/22 applied 6.6.1979.

[8] Heindl, J., Hirzinger, G.
Kraft-Momenten-Sensorgriff und Verfahren zum kombinierten Programmieren von Roboterbewegungen und Bearbeitungskraeften bzw. -momenten. Patent P3240251.1 applied 30.10.1982.

[9] Bejczy, A.K.
Advanced teleoperators. Astronautics and Aeronautics 17, 20-31 (1979).

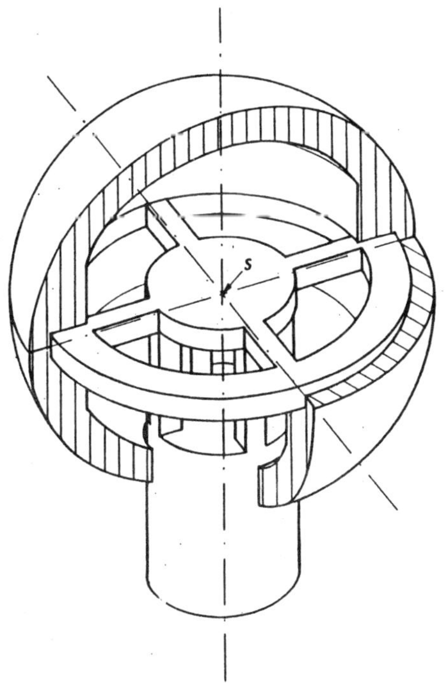

fig. 1 Sensor ball with DFVLR-force-torque-sensor integrated

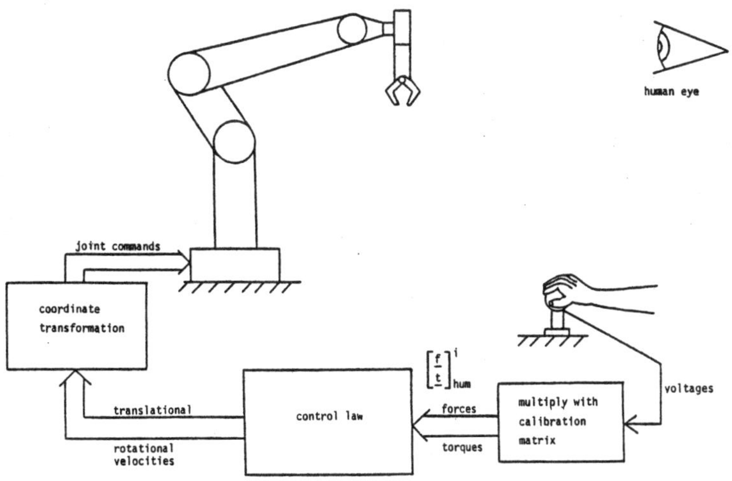

fig. 2 Path-only-teach-in with fixed sensor ball

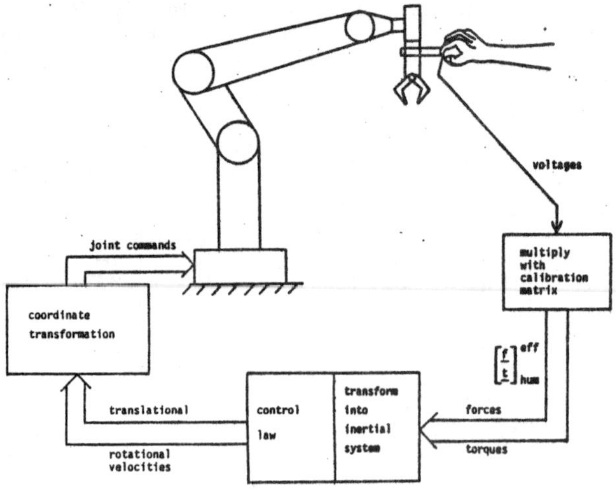

fig. 3 Path-only-teach-in with robot mounted sensor ball

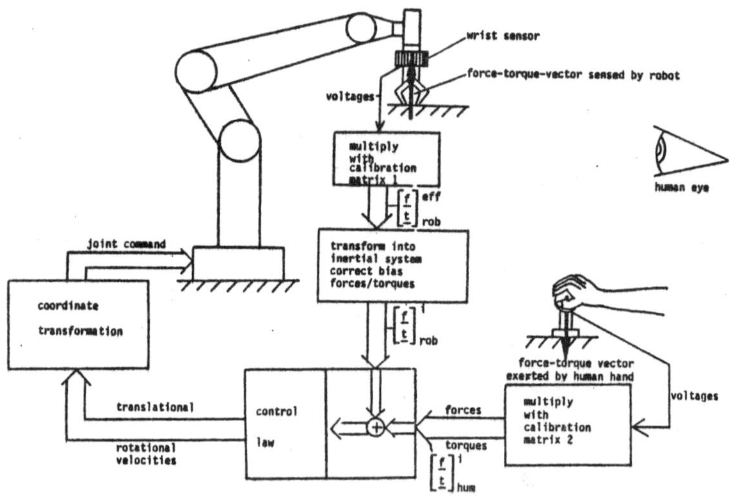

fig. 4 Simultaneous path-force-torque-teach in with fixed sensor ball

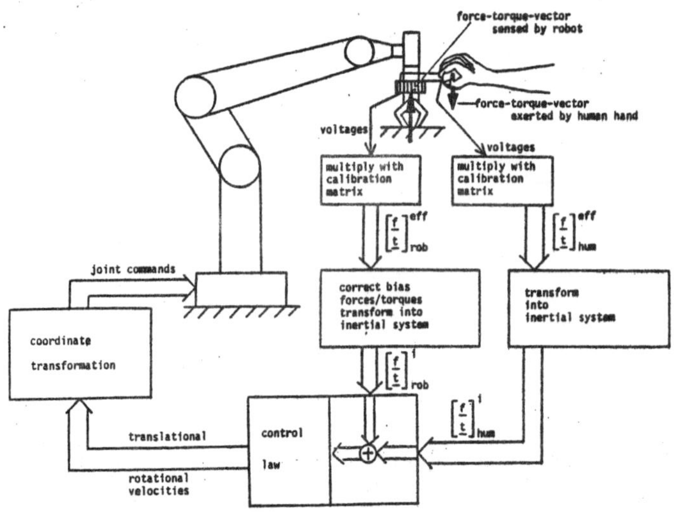

fig. 5 Simultaneous path-force-torque-teach-in with robot mounted sensor ball

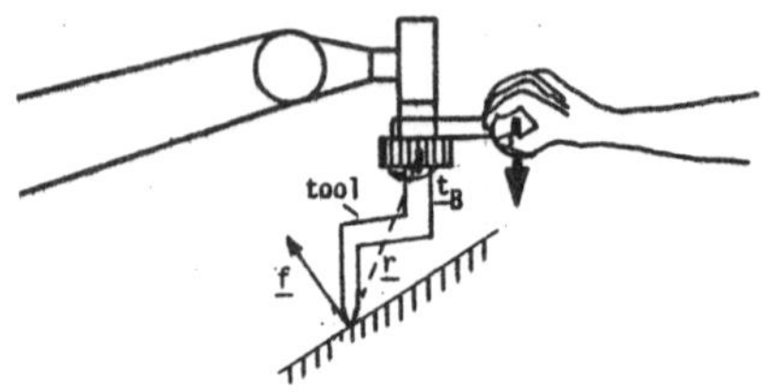

$\underline{t}_B = \underline{f} \times \underline{r}$

fig. 6 Bias torques generated by forces

TRACKING OF AN UNDERWATER INSPECTION MODULE

SUIVI D'UN MODULE SOUS-MARIN D'INSPECTION

A. FAURE*, P. LEROLLAND**, et P. POUPEL**

* U.E.R.S.T. - LE HAVRE - Laboratoire d'Automatique
B.P. 4006 - 76077 - LE HAVRE Cédex France

** SOTRAPLEX - 553, Bld J. Durand - BP 1176
76064 - LE HAVRE Cédex France

Introduction :

La plupart des expertises de structures immergées sont actuellement réalisées en cale sèche. Cette procédure oblige à l'arrêt momentané de l'exploitation du navire ou de l'engin flottant considéré. Afin d'éviter un passage systématique en forme de radoub et d'examiner fréquemment des structures opérant loin des ports (usines flottantes, barges de travaux,...) il est proposé l'utilisation d'un module d'inspection automatique, convenablement instrumenté, et guidé à distance [1] . Il est dès lors nécessaire de déterminer avec précision la position de ce module afin d'établir des comparaisons pertinentes d'une inspection à l'autre. D'un point de vue quantitatif, l'éloignement maximum est de 300 mètres, l'immersion maximum dans l'eau de mer est de l'ordre de 30 mètres et la vitesse de déplacement prévue pour le mobile suivi est au maximum de 1 m/s. La solution proposée fait appel à un mobile annexe suivant une trajectoire parfaitement connue. Ce dernier est l'origine d'un système de coordonnées permettant alors de repérer la position du module d'inspection. Compte tenu du milieu aquatique d'évolution, les informations échangées entre les deux mobiles ont un support ultrasonore par suite de la bonne propagation de ces ondes dans le milieu liquide en général. On considère que les deux mobiles sont mécaniquement découplés, que les trajectoires sont coplanaires dans un plan parallèle à la partie plane de la coque immergée, et que la densité linéaire des obstacles perturbant la vision entre les deux modules est, sinon nulle, très faible dans la plupart des applications envisagées. Une étude portant respectivement sur les conditions de propagation acoustique, sur le traitement de signaux et sur l'asservissement des trajectoires a été réalisée à échelle réduite afin de démontrer la faisabilité du procédé.

I - OBTENTION ET TRAITEMENT DES SIGNAUX

Dans son principe, le système comprend donc un module d'inspection libre de se déplacer dans un plan (P) parallèle au fond immergé de la coque du navire ; et un module suivant une trajectoire connue, matérialisée par exemple par le bordé de carène ou la quille de roulis (figure 1).

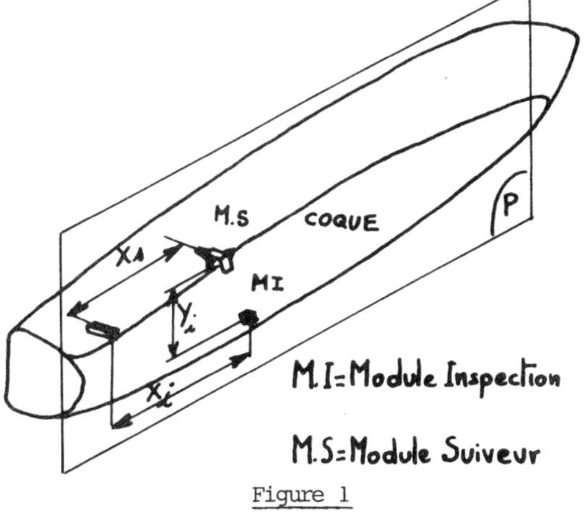

Figure 1

L'objectif est donc de déterminer les coordonnées X_i et Y_i, dans le plan P, du module d'inspection. Ce dernier porte un transducteur émetteur, tandis que le suiveur est muni d'un dipôle de réception. La mesure du temps de parcours entre ces deux éléments permet d'évaluer l'ordonnée Y_i. Un transducteur auxiliaire permet d'évaluer la seconde coordonnée X_i. On assigne au transducteur émetteur d'être constamment sur l'axe du dipôle suiveur. Pour cela le propulseur du suiveur est commandé de façon à maintenir l'écart x entre les abscisses égal à zéro. Cette donnée, obtenue par traitement des signaux ultrasonores, est tributaire des contraintes d'émission et de propagation des ondes employées. L'emploi d'ultrasons émis en permanence est exclu par suite des innombrables perturbations prévisibles. On considère alors des émissions organisées sous forme discrète afin de minimiser au moins l'influence des interférences.

Toutefois, cette procédure ne met pas à l'abri de la détection intempestive d'échos multiples faisant suite à une seule émission ; ce qui rend nécessaire la sélection de l'écho utile. Pour cela, on peut utiliser les deux méthodes principales, basées sur le filtrage tant spatial que temporel.

Par filtrage spatial, on entend le résultat de l'utilisation de transducteurs directifs. Si la directivité augmente le filtrage, donc l'élimination d'échos parasites, elle exige une grande précision dans le maintien de l'alignement. Par filtrage temporel, on entend le résultat de la mise en oeuvre d'une sélection électronique des échos. Le chronogramme des signaux mis en oeuvre est représenté sur la figure 2. La porte temporelle commande le passage du signal vers la chaîne de réception.

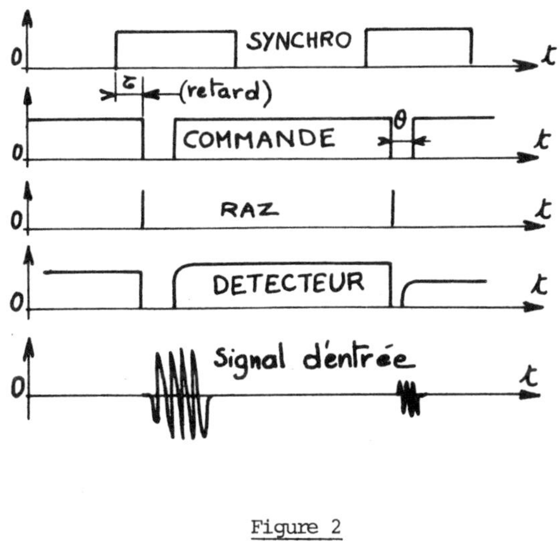

Figure 2

Cependant, ce seul dispositif ne permet pas de garantir la primeur de l'écho considéré. Soit θ l'intervalle de temps séparant deux émissions successives: E_n et E_{n+1} ; soit $t_{(En,\ Rn)}$ la durée du trajet direct émetteur-récepteur. Un réflecteur parasite renvoie un écho de durée de trajet à $t_{(En,\ Pn)}$. Cet écho parasite sera considéré comme le premier écho utile si :

$$t_{(E_n, R_n)} + \theta > t_{(E_{n-1}, P_{n-1})}$$

Les solutions pour éviter cette confusion consistent soit à ajuster le seuil de détection sur l'écho d'amplitude la plus grande, soit à augmenter l'intervalle θ . C'est la dernière solution que nous avons choisie. L'influence sur les performances du dispositif restant à examiner.

Les essais sont réalisés à l'aide d'une maquette à échelle réduite. Afin de respecter la similitude des conditions de propagation, on utilise un montage conservant la directivité. Cette grandeur est conditionnée par le facteur angulaire u défini par l'expression :

$$u = (\pi.d.f/c) . \sin \beta$$

d est le diamètre du transducteur, f la fréquence, c la célérité, β l'angle de site. Le système de réception est constitué par un doublet d'entre axe égal à ℓ . Soit h l'ordonnée d'un point situé sur l'axe de symétrie du doublet. La conservation du facteur angulaire permet de trouver les paramètres du système construit à échelle différente. Nous avons en effet :

$$\frac{H}{L} \# \left(\frac{D.F}{d.f}\right) . \frac{h}{\ell}$$

où les majuscules sont relatives au dispositif semblable.

Les modules évoluent à faible distance de la coque du navire. De ce fait, l'angle d'incidence des rayons émis est voisin de π/2, de sorte que la réflexion est pratiquement totale. Ces deux conditions liées peuvent conduire à la formation d'interférences dont les franges noires seraient autant de zônes de réception nulle (effet L loyd). Si r est l'éloignement entre l'émetteur et le récepteur, et h la distance les séparant de la coque. Si la source est ponctuelle, la pression p observée au niveau du récepteur est donnée par :

$$p = (2\ p_o/r) . \sin\ (2\pi\ h^2\ f/rc)$$

toutes choses égales par ailleurs, la réception est théoriquement **nulle** pour les distances r suivantes :

$$r = (2\ h^2\ f/\alpha c) \qquad \alpha \in \mathcal{N}$$

Sur la figure 3 nous avons reporté respectivement les résultats du calcul théorique et le relevé expérimental correspondant aux mêmes conditions. Il apparaît qu'en pratique, l'effet L. Loyd n'est heureusement pas marqué ; en raison sans doute de l'effet d'intégration réalisée par les transducteurs réels utilisés. Ces derniers présentent en effet une force parlante d'une certaine étendue qu'il n'est pas envisageable d'assimiler à un seul point rayonnant. Toutefois, on peut noter une décroissance en $1/r^2$ et non pas en 1/r, comme dans le cas d'une onde sphérique en espace libre.

II - ETUDE DE L'ASSERVISSEMENT

La fonction de l'asservissement considéré est de maintenir constamment égales l'abscisse X_s du module suiveur, et celle X_i du module d'inspection (fig.1). La distance y_i est évaluée à l'aide d'ultra-sons par la mesure du temps de parcours entre les deux modules. La célérité du son est obtenue grâce à un étalonnage effectué sur le site. Le schéma de l'asservissement est représenté sur la figure 4. L'écart x_i est obtenu à l'aide d'une paire de transducteurs délivrant un signal électrique

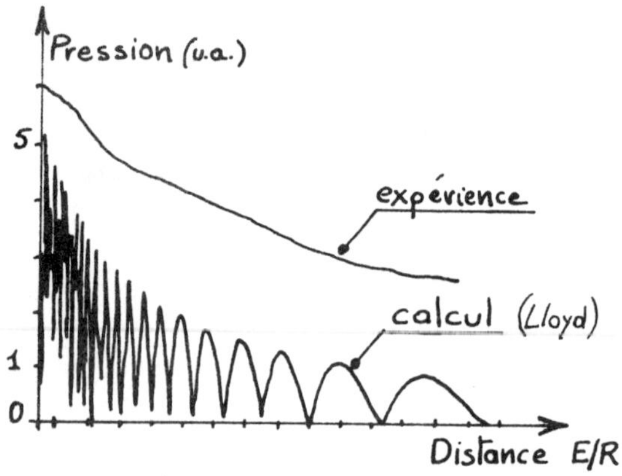

Figure 3

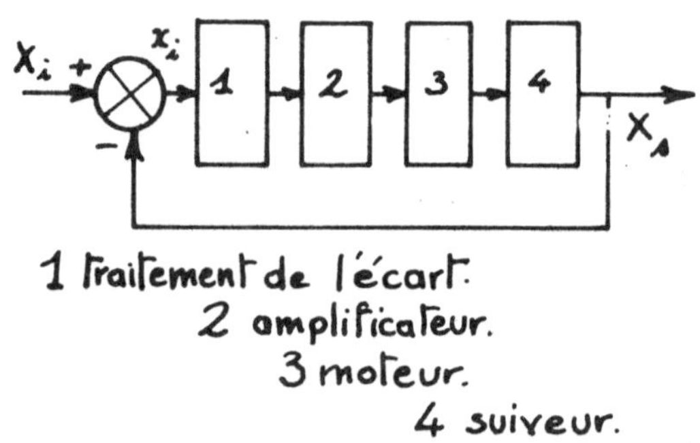

Figure 4

différentiel. Après amplification, la tension de commande est appliquée avec le sens convenable sur le moteur de propulsion.

2.1. Etude du capteur

Il est souhaitable d'obtenir une tension électrique directement proportionnelle à l'écart x_i de position. Comme les transducteurs utilisés sont du type piston circulaire, le rayonnement à l'émission (ou la propriété de réception) n'est pas uniforme, et la fonction de directivité présente des lobes secondaires (donc des zéro) pouvant affecter cette linéarité. Afin d'étudier la réception, on évalue la différence ε entre les signaux reçus par la récepteur, en fonction de l'écart x_i. Si v_1 et v_2 sont les signaux, proportionnels à la pression, recueillis par le récepteur, on a :

$$\varepsilon\ (x_i,\ y_i) = v_1 - v_2$$

Si b( ) est la fonction de directivité des transducteurs, l'écart normalisé entre les signaux est :

$$\varepsilon\ (x_i,\ y_i) = \{\ b^2\ (\beta_1) - b^2\ (\beta_2)\ \}$$

car les angles de site, β_1 et β_2 sont différents pour les deux récepteurs. Pour un piston circulaire 2

$$b(\beta) = 4(J_1(u)/u)^2$$

$$\text{avec} \quad u = (\pi d \sin(\beta / \lambda))$$

d étant le diamètre, λ la longueur d'onde, J_1 la fonction de Bessel. Les résultats théoriques et expérimentaux sont représentés sur la figure 5. Il apparaît un excellent accord entre les deux. Si l'influence des lobes secondaires semble non négligeables, il apparaît que cet écart présente des extrema, de sorte qu'il est nécessaire de mémoriser le point de fonctionnement s'il atteint un de ces extrema.

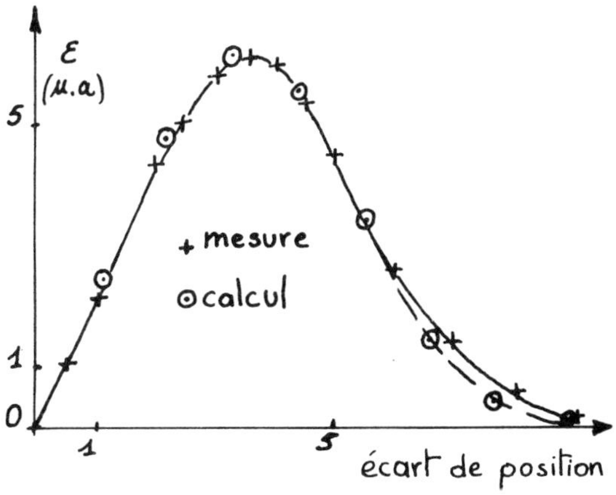

Figure 5

2.2. Performances du système

Afin d'étudier les performances du système, nous avons abordé en premier lieu l'identification de la partie dynamique comprenant le chariot muni des transducteurs, le moteur d'entraînement et le dispositif de recopie de la position (potentiomètre démultiplié). Pour cette étape, le capteur n'est pas utilisé pour générer le signal de commande. Ce dernier est délivré par un générateur, ce qui permet de faire varier les signaux d'écart.

Des méthodes classiques d'identification ont été utilisées. La fonction de transfert G(p) est alors de la forme suivante :

$$G(p) = k.\frac{a}{p\ (p+a)}$$

avec $a = 13{,}22$ et $k = 3{,}54.10^{-1}$; p est l'opérateur de Laplace. On règle alors le gain de l'amplificateur pour obtenir, en boucle fermée et en fonctionnement continu, une fonction de transfert du second ordre d'amortissement égal à 0,7.

Toutefois, dans l'application envisagée, le fonctionnement est différent d'une part en raison de l'échantillonnage du signal d'erreur, et d'autre part en raison du caractère non linéaire de son évolution en fonction de l'écart géométrique. La première influence est réduite par l'utilisation d'un bloqueur d'ordre zéro alimentant l'amplificateur de puissance. L'influence de la non linéarité persiste tout en étant ramené à une saturation. Un dispositif à seuil délivre une tension électrique ε répondant à la définition suivante :

si $|x_i| < x_M$ $\qquad \varepsilon = \alpha.\ x_i$

α étant constant.

si $|x_i| \geqslant x_M$ $\qquad \varepsilon = \varepsilon_M \operatorname{sgn} x_i$

On obtient ainsi une mémorisation du déplacement du chariot d'inspection.

Afin d'étudier théoriquement le fonctionnement réel dans ces conditions, on adopte la méthode du gain variable. En passant dans l'espace d'état $< x_1, x_2 >$ où x_1 et x_2 sont respectivement la position et la vitesse échantillonnées, les équations du fonctionnement s'écrivent :

$$\begin{Bmatrix} x_1(n+1)\\ x_2(n+1)\end{Bmatrix} = \begin{bmatrix}(1-B) & C\\ (-kaC) & A\end{bmatrix}\cdot\begin{Bmatrix}x_1(n)\\ x_2(n)\end{Bmatrix} + x_i(n)\begin{Bmatrix}B\\ kaC\end{Bmatrix}$$

$$y(n) = \begin{bmatrix}1 & 0\end{bmatrix}\cdot\begin{Bmatrix}x_1(n)\\ x_2(n)\end{Bmatrix}$$

$B = (k/a)\ (aT - 1 + A)$ $\qquad A = \exp - aT$

$C = (1/a)\ (1 - A)$

T = période d'échantillonnage, c'est la durée séparant deux émissions successives.

Le gain k est considéré comme une variable définie par :

$k = k$ $\quad$ si $\quad |x_i| \leqslant \frac{\ell}{2}$

$k = k/x_i$ $\quad$ si $\quad |x_i| > \frac{\ell}{2}$

Sur la figure 6 est reportée la réponse indicielle dans ce cas. Le fonctionnement étant non linéaire, le dépassement maximum est fonction de l'amplitude de la sollicitation. Si on utilise deux gains, l'un pour la région linéaire, et l'autre pour la saturation, on obtient la réponse indicielle de la figure 7. On remarque alors que le dépassement maximum est bien diminué. Enfin,

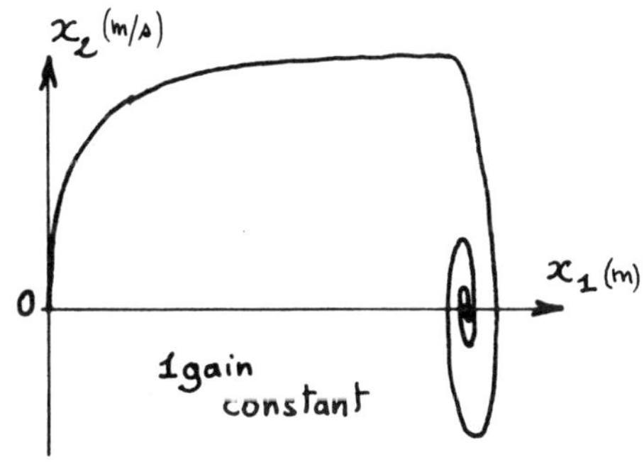

Figure 6

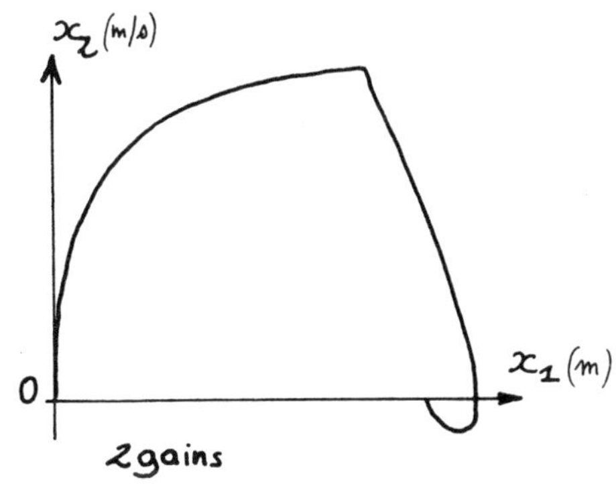

Figure 7

si le gain augmente de trop, le tracé dans le plan de phase devient un cercle limite (figure 8).

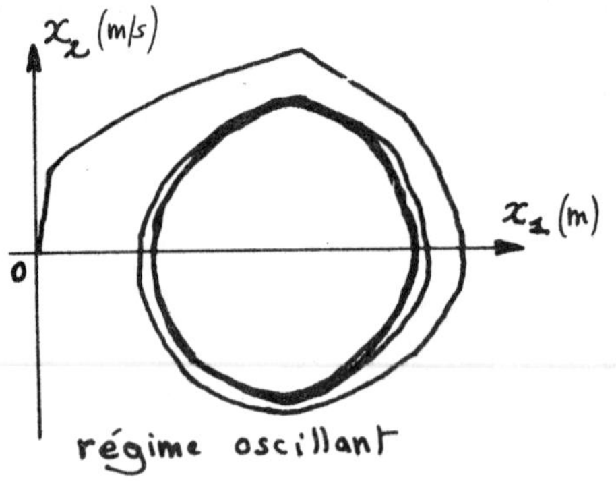

Figure 8

Un essai de suivi est effectué avec les règlages correspondants à ceux de la figure 6. On obtient le relevé expérimental de la figure suivante, qui montre un excellent asservissement du chariot suiveur sur le chariot d'inspection, et ce, malgré un déplacement très supérieur à la région de fonctionnement linéaire du capteur ultrasonore.

CONCLUSIONS

Les résultats obtenus au cours de cette expérimentation montrent sans équivoque la faisabilité du suivi par ultrason d'un module d'inspection se déplaçant dans l'eau. Cette opération peut être réalisée à proximité d'une grande surface plane (fond de la coque d'un navire interface mer/air) car, en égard aux techniques mises en oeuvre, l'effet LLOYD) ne perturbe pas sensiblement le déroulement du suivi. L'effet du filtrage spatial, opéré de fait par les transducteurs naturellement directifs est compensé par la mémorisation du dernier écart enregistré. Le système dynamique non linéaire correspondant se comporte quasiment comme un système linéaire compte tenu, principalement du faible gain statique utilisé. L'effet de perturbations inopinées n'a pas été envisagé, et ne devrait pas modifier sensiblement les bons résultats obtenus.

BIBLIOGRAPHIE

1 A. FAURE, P. LEROLLAND et P. POUPEL "Procédé et dispositif pour le repérage en position d'un engin autonome mobile sur une structure immergée" (Brevet Français N° 7832466).

2 R.S. URICK "Principles of underwater sound" Mc Graw Hill Editeru, 1975, p. 55

DESIGN OF HIGH-PERFORMANCE CONTROLLERS FOR REMOTE MANIPULATOR SYSTEMS WITH EXPLICIT ACTUATOR DYNAMICS

Professor B Porter

Centre for Instrumentation and Automation
and
Department of Aeronautical and Mechanical Engineering
University of Salford
Salford M5 4WT
England

ABSTRACT

It is shown that singular perturbation methods for the design of high-performance tracking systems can be directly deployed to yield simple controllers with high-accuracy non-interacting joint control for non-linear remote manipulator systems with explicit actuator dynamics. The effectiveness of this methodology is illustrated by designing a closed-loop manipulator system for the positioning of a payload in the cargo bay of the space shuttle.

1. INTRODUCTION

The design of high-performance controllers for remote manipulator systems, such as that used for the positioning of a payload in the cargo bay of the space shuttle, is complicated by the non-linearity of the mechanical manipulators and the presence of explicit electromechanical actuator dynamics in the joint drive systems. These difficulties have conventionally been addressed by the deployment of independent joint control with [1] or without [2] dynamic decoupling. However, the resulting controllers with dynamic decoupling [1] are unnecessarily complicated whilst the controllers without dynamic decoupling [2] are insufficiently accurate.

In this paper, it is therefore shown that singular perturbation methods for the design of high-performance tracking systems [3][4][5][6] can be directly deployed to yield simple controllers with high-accuracy non-interacting joint control for non-linear remote manipulator systems with explicit actuator dynamics. The effectiveness of this methodology is illustrated by designing a control system for the space-shuttle manipulator previously considered by Yuan [1] for which the results of computer simulation studies are presented.

2. MANIPULATOR SYSTEM

In the case of a manipulator arm consisting of a chain of rigid links connected at n joints, each with a single degree of freedom, it is convenient to fix a reference axis to each link and to define the joint angles by the vector

$$\gamma = [\gamma_1 , \gamma_2 , \dots , \gamma_n]^T \tag{1}$$

where γ_1 is the angle between the axis of the first link and an inertial reference direction, and γ_i is the angle between the axes of the ith and (i-1)th links. The equation of motion of such a manipulator arm can be expressed in the form

$$A(\gamma)\ddot{\gamma} = f(\gamma,\dot{\gamma}) + T \tag{2}$$

where the elements of the inertia matrix $A(\gamma) \in R^{n\times n}$ are continuously differentiable functions of the joint angles, the elements of the vector $f(\gamma,\dot{\gamma}) \in R^n$ are functions of joint angles and joint angular velocities corresponding to coriolis and centripetal accelerations such that $f(\gamma,0) = 0$, and the elements of the vector

$$T = [T_1 , T_2 , \dots T_n]^T \tag{3}$$

are the joint control torques. In order to design a controller for the manipulator system, it is convenient to linearise equation (2) about a nominal trajectory with angular velocities which are sufficiently small to permit the neglect of the non-linear vector $f(\gamma,\dot{\gamma})$. The resulting linearised equation of motion of the manipulator arm thus assumes the form

$$A(\gamma_o)\delta\ddot{\gamma} = \delta T \tag{4}$$

where γ_o is the quasistatic nominal angular trajectory, and $\delta\gamma$ and δT are respectively the vectors of perturbed joint angles and torques.

In order to apply singular perturbation methods for the design of high-performance tracking systems [3][4][5][6] to the design of a controller for the manipulator system, the state equation (4) of the open-loop manipulator arm can be conveniently expressed in the state-space form

$$\begin{bmatrix}\delta\dot{\gamma}\\ \delta\ddot{\gamma}\end{bmatrix} = \begin{bmatrix}0 , & I_n\\ 0 , & 0\end{bmatrix}\begin{bmatrix}\delta\gamma\\ \delta\dot{\gamma}\end{bmatrix} + \begin{bmatrix}0\\ A(\gamma_o)\end{bmatrix}\delta T \tag{5}$$

with an associated measurement equation [4][5] of the form

$$\delta\omega = [I_n , M]\begin{bmatrix}\delta\gamma\\ \delta\dot{\gamma}\end{bmatrix} \tag{6}$$

where $M \in R^{n\times n}$ is the measurement matrix. The error-actuated controller for the manipulator is governed by a control-law equation [4][5] of the form

$$\delta T = g(K_1\varepsilon + K_2\zeta) \tag{7}$$

where $K_1 \in R^{nxn}$ and $K_2 \in R^{nxn}$ are the controller matrices, $g \in R^+$ is the controller gain parameter, $\varepsilon = \delta\nu - \delta\omega \in R^n$ is the error vector, $\zeta = \zeta_0 + \int_0^t \varepsilon dt \in R^n$ is the integral of error vector, and $\delta\nu$ is the commanded perturbed joint angle vector. Then as $g \to \infty$, the eigenvalues of the 'slow' modes of the resulting closed-loop manipulator system correspond to the elements of the set $Z_s = Z_1 \cup Z_2$ where [4][5]

$$Z_1 = \{\lambda \in C : |\lambda I_n + K_1^{-1} K_2| = 0\} \quad (8)$$

and

$$Z_2 = \{\lambda \in C : |\lambda I_n + M^{-1}| = 0\} \quad , \quad (9)$$

and the eigenvalues of the 'fast' modes of the resulting closed-loop manipulator system correspond to the elements of the set $Z_f = Z_3$ where [4][5]

$$Z_3 = \{\lambda \in C : |\lambda I_n + gMA(\Upsilon_o)K_1| = 0\} \quad . \quad (10)$$

Therefore, if

$$MA(\Upsilon_o)K_1 = \Sigma \quad (11)$$

where $M \in R^{nxn}$ and $\Sigma \in R^{nxn}$ are positive diagonal matrices, and

$$K_2 = \rho K_1 \quad (12)$$

where $\rho \in R^+$ is a positive scalar, it is evident from equations (8), (9), and (10) that $Z_1 \subset C^-$, $Z_2 \subset C^-$, and $Z_3 \subset C^-$ where C^- is the open left half-plane.

The closed-loop manipulator system is therefore stable so that $\delta\Upsilon \to \delta\nu$ in the steady state and, moreover, is asymptotically non-interacting as $g \to \infty$ for such controller and measurement matrices [4][5]. Furthermore, in the presence of explicit electro-mechanical actuator dynamics in the joint drive systems, these stability and non-interaction characteristics are preserved provided only that such actuator dynamics are sufficiently 'fast' [6].

3. SIMULATION RESULTS

The effectiveness of this methodology can be conveniently illustrated by designing a control system for the space-shuttle manipulator with two rigid links for which equation (2) assumes the explicit form [1]

$$\begin{bmatrix} J_1+J_2(1+\alpha^2)+J_2\alpha\cos\Upsilon_2 , & J_2(1+\alpha\cos\Upsilon_2) \\ J_2(1+\alpha\cos\Upsilon_2) , & J_2 \end{bmatrix} \begin{bmatrix} \ddot{\Upsilon}_1 \\ \ddot{\Upsilon}_2 \end{bmatrix} = \begin{bmatrix} J_2\alpha\sin\Upsilon_2(2\dot{\Upsilon}_1+\dot{\Upsilon}_2)\dot{\Upsilon}_2 \\ -J_2\alpha\sin\Upsilon_2\dot{\Upsilon}_1^2 \end{bmatrix} + \begin{bmatrix} T_1 \\ T_2 \end{bmatrix} \quad (13)$$

where $J_1 = m_1\ell_1^2$, $J_2 = m_2\ell_2^2$, and $\alpha = \ell_1/\ell_2$. In this case, the arm parameters are $m_1 = 72.97$ kg, $m_2 = 14594$ kg, $\ell_1 = 6.096$ m, and $\ell_2 = 9.144$ m, and it is required to control the manipulator in the neighbourhood of $\Upsilon_o = [0^o \ , \ 45^o]^T$ such that [1]

$$|\dot{\Upsilon}_1| < 0.23 \ ^o/\text{sec} \quad (14)$$

and

$$|\dot{\Upsilon}_2| < 0.34 \ ^o/\text{sec} \quad . \quad (15)$$

The responses of the controlled non-linear manipulator when

$$M = \begin{bmatrix} 0.25 , & 0 \\ 0 , & 0.25 \end{bmatrix} \quad , \quad (16)$$

$$\Sigma = \begin{bmatrix} 1 , & 0 \\ 0 , & 1 \end{bmatrix} \quad , \quad (17)$$

and $\rho = 2$ in equations (11) and (12) are shown in Figs 1 and 2 when the controller is tuned such that $g = 500$ and the time constant of each of the servomotors generating the control torques is 0.001 sec [1]. The commanded perturbed joint angles are $\delta\nu = [3^o \ , \ 0^o]^T$ and $\delta\nu = [0^o \ , \ 3^o]^T$ in Figs 1 and 2, respectively, and it is evident that the constraints (14) and (15) are satisfied in each case with minimal transient interactions between the joint motions. Finally, the responses shown in Figs 1 and 2 are significantly faster than those obtained by Yuan [1] using a rather complicated independent joint control system incorporating dynamic decoupling.

4. CONCLUSION

In this paper, it has been shown that singular perturbation methods for the design of high-performance tracking systems [3][4][5][6] can be directly deployed to yield simple controllers with high-accuracy non-interacting joint control for non-linear remote manipulator systems with explicit actuator dynamics. The effectiveness of this methodology has been illustrated by designing a control system for the space-shuttle manipulator previously considered by Yuan [1].

REFERENCES

1. J S-C Yuan, "Dynamic decoupling of a remote manipulator system", Proc IEEE Trans Automatic Control, vol AC-25, pp 713-717, 1978.

2. D E Whitney, "Resolved motion rate control of manipulators and human postheses", Proc IEEE Trans Man-Machine Syst, vol MMS-10, pp 47-53, 1969.

3. B Porter and A Bradshaw, "Singular perturbation methods in the design of tracking systems incorporating high-gain error-actuated controllers", Int J Systems Sci, vol 12, pp 1169-1179, 1981.

4. B Porter and A Bradshaw, "Singular perturbation methods in the design of tracking systems incorporating inner-loop compensators and high-gain error-actuated controllers", Int J Systems Sci, vol 12, pp 1193-1205, 1981.

5. B Porter, "Design of high-performance tracking systems", Technical Report AFWAL-TR-82-3032, US Air Force Wright Aeronautical Laboratories, Wright-Patterson AFB, Ohio, July 1982.

6. B Porter, "Singular perturbation methods in the design of tracking systems incorporating error-actuated controllers for plants with explicit actuator dynamics", Int J Control, vol 35, pp 383-389, 1982.

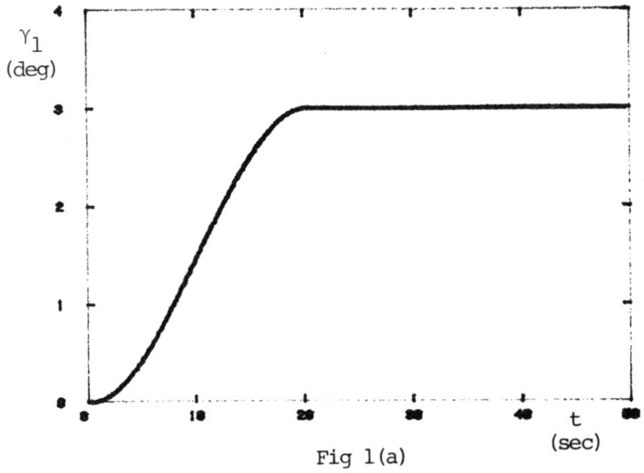

Fig 1(a)

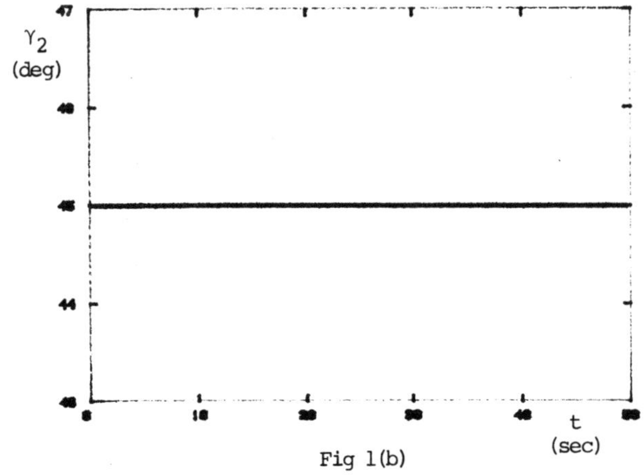

Fig 1(b)

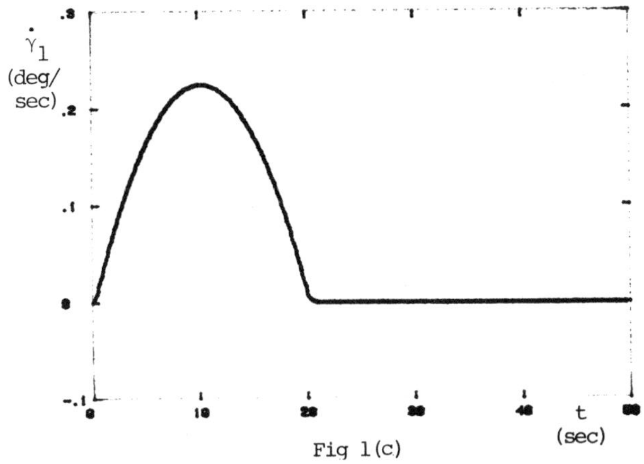

Fig 1(c)

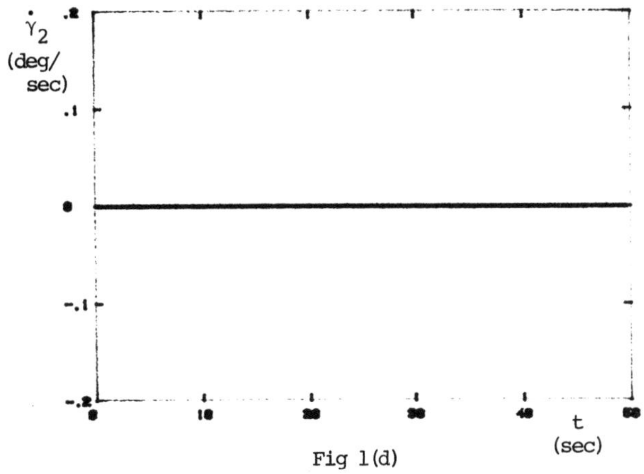

Fig 1(d)

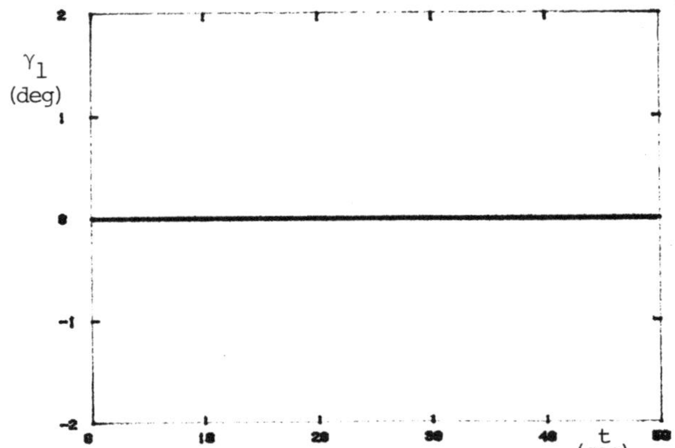

Fig 2(a)

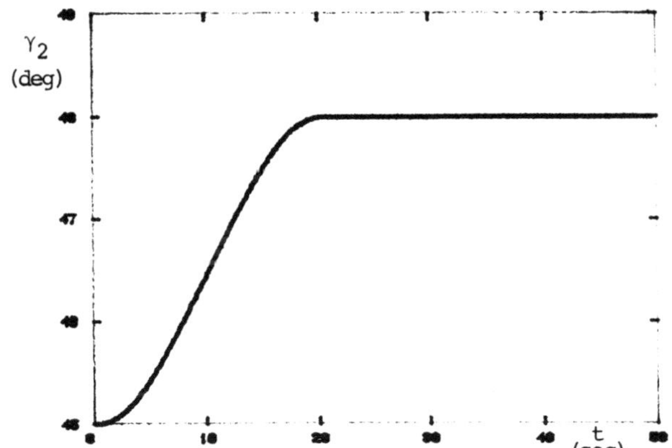

Fig 2(b)

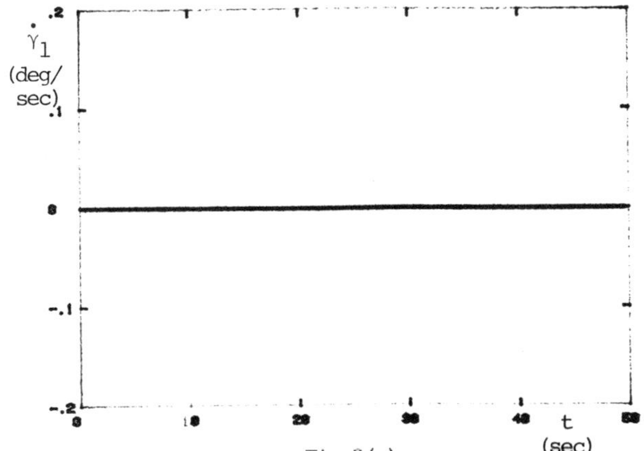

Fig 2(c)

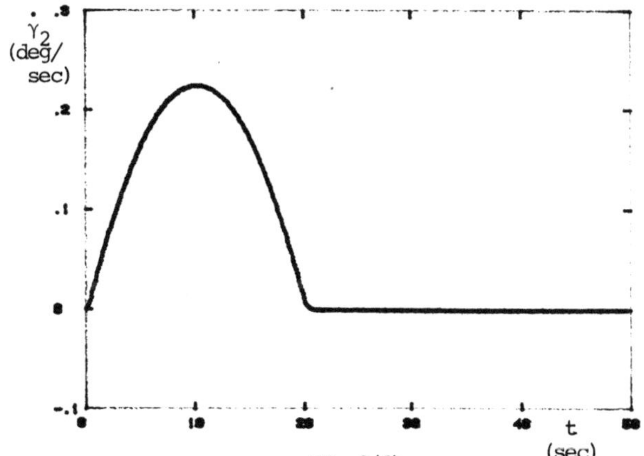

Fig 2(d)

ROBOT PROGRAMMING BY DEVELOPMENT TOOLS

DE CARLINI Ugo, ESPOSITO Aldo

Istituto Centro di Calcolo Facoltà di INGEGNERIA
Piazzale Tecchio (80125) NAPOLI (ITALIA)

The software cost for robot programming increases with continuity, but, in a short time, an improvement can be obtained by using computer science to present robots more sistematically. At present, every robot is normally individually programmed, without program optimization, with lack in documentation and with long testing and debugging referred to problem complexity. That implies an individual software cost depending on the specific robot application, the difficult implementation of libraries and parametric procedures and, above all, subsequent problems for maintenance and improvement. These drawbacks can be reduced at low cost, by applying development tools already available in other fields and by reorganizing software activity. For such reasons, in this paper some supports provided by software engineering (automatic documentation, off-line programming in host computer, tools for restructuring programs,...) are described; their possible application to robotics is also evaluated.

1. INTRODUCTION

Nowadays robotics seems to be a scientific and technological science very complex: in recent years many studies have been devoted to the improvement of this subject, which is, at present, closely connected to many other engineering fields. Recent results in computer science research will contribute widely to the development and proliferation of robots. The most important of these results are in the following areas:

- artificial intelligence /1/,
- high level programming languages /2/,
- operating systems,
- computer graphics /3/,
- data-base management systems,
- local networks.

At present these researches (carried out both at Universities and in industries) are at a very advanced stage. However, there is still a long way before we reach a completely integrated system. In fact, the various fields cannot be developed in such a way as to be totally indipendent among them or of other technological sectors. In particular, the artificial intelligence will supply the software for visual pattern recognition or vocal communication and also the situation recognition for failure detection, diagnosis and recovering strategies; consequently it must be closely connected to the local resources of the robot (hardware and basic software). Very high level programming languages (such as PASCAL and ADA) will make programming faster and more easily transferable,with remarkable advantages for software documentation and maintenance. New specialized operating systems directed to unskilled operators, local networks and distributed databases will both help distribute and integrate CAD/CAM systems.

With the global contribute of all the above-mentioned supports, with the parallel development of mechanical and electronic hardware and with the internal reorganization of production, we will be able to tend to unmanned factory. Apart from some social and political problems, this will be likely to happen only in a long time, both because of the further and necessary investigations and especially because of the high costs of investments required. For a wider use of computer science into robotics considerable additional expenses for the management software (operating systems), the supporting software (artificial intelligence, data-bases management systems, high level programming languages) and the additional hardware (networks and computer graphics) are required, with heavy restrictions against the practical application of these knowledges.

Theory progresses, but traditional simple robots are in many cases still produced: where this happens, an intermediate solution, first step to a more advanced one, is to be preferred.A very important role in this direction will be played by software engineering, which is now used in robotics only indirectly, that is by the software experts who study new languages and operating systems. The robot building firms haven't so far felt the need of a greater and direct use of software engineering in robot programming for several reasons (lack of specific internal competences, not too complex operations, hardware resources limits in robot, lack of standards, etc.).

2. SOFTWARE ENGINEERING AND ROBOTICS

The increased complexity of robot electronic components, together with the reduction in prices of the hardware, has recently allowed an increase of investments in application software: the use of facilities for the development of software, already existing in other sectors, could allow a rapid amortization of the initial greater costs. The positive effects of a systematic use of software engineering are essentially the following:

- flexibility,
- easy implementation,
- transportability,
- readability,
- reduction in maintenance costs.

Actually solutions directed towards the simplification of program writing and testing phase in a robot have already been proposed: these solutions differ form conventional systems of artificial intelligence because they don't replace the user in his process of reasoning, but they provide means of support to increase the flexibility and the simplicity of implementation. The industrial robot is a technologically complex product: requirements for an industrial robot for general use are very sophisticated with a wide range of specifications. It is enough to have only the required functions for each application; each industrial robot is the best for its application and each

---

This work was supported by the Italian "Ministero Pubblica Istruzione"

installation is limited to one application. This drawback may be reduced by introducing the concept of flexibility. In industries, the aim of simplifying the robot re-programming for new tasks has been partially reached, while a greater effort must be made for the programming of more complex tasks /4/. This activity does not involve only software: for example, strategies of avoiding errors can be only realised by using additional hardware. At the present state of technology, this is obtained by connecting a control computer to the robot. At the same time, thorough studies of programming languages have been carried out, choosing among general languages fairly standardized (FORTRAN, BASIC, PASCAL) and specific languages (AL, LAMA, AUTOPASS,...) /5,6,7/.

The common aim is to express the processes intended to being carried out and it is justified by the increasing complexity of tasks assigned to robots. Very often, the compiler is developed on a host computer and it produces an intermediate code coinciding with the robot programming language /8/.

In all the above-mentioned cases, the use of hardware and software resources outside to the robot is becoming wider and wider /9,10,11/. However, there are still big problems concerning the transportability, readability, maintenance and improvement of the programs developed for a set of robot or installations. Therefore, the need arises of a wider organization with reference particularly to the software and to the organization of the development with a more systematic use of programming techniques and of the software life cycle.

3. SOFTWARE TOOLS

In the last 10 years, with the continuos reduction in hardware costs, the development of software becomes the predominant economic factor. This fact tends to increase, because it is required a more and more complex software which is at the same time easy to use. This trend is being then changed, by introducing new tools and new methodologies of work aiming at optimizing and automating the whole processing of software production. Software tools, besides fulfilling the economic needs above mentioned, provide other advantages, such as:

- better communication among programmers, managers and customers,
- better global knowledge of every single project,
- better training of the end-user,
- control of the whole development made easy,
- better quality of the product at any level,
- greater integration with the outside environment.

Software tools can be applied to a lot of fields and are divided into 3 categories /12/:

- cognitive (for programmers with limited experience);
- augmentative (for programmers with great experience);
- notational (languages).

Until a short time ago, advanced software tools were only available on big computer systems, while on minicomputers there existed only some simple utilities, put in the basic software. In particular, in present robots, the internal hardware is limited (the memory has a capacity dimensioned on the specific working program); consequently, the software tools are practically non-existent.

The reduction in hardware costs and the increased total performances of minicomputers allow nowadays the overcoming of these disadvantages directly in the robot or externally with the support of a minicomputer. It's therefore possible a greater use of these tools, so as to have the software work changed from "manual" to "automatic". In particular, software tools are placed in a proper model said "software development life cycle", which describes the various phases, according to which a software project must be implemented /13/.

4. SOFTWARE DEVELOPMENT LIFE CYCLE

The development cycle is simply a series of orderly interrelated activities, leading to the successful completion of programs: the periods of time during which these activities take place are called "phases". With this model it is possible to describe the development of the project, the overlap of the various activities and the type of human resources required. In fig. 1 a possible schema of the software development life cycle is shown.

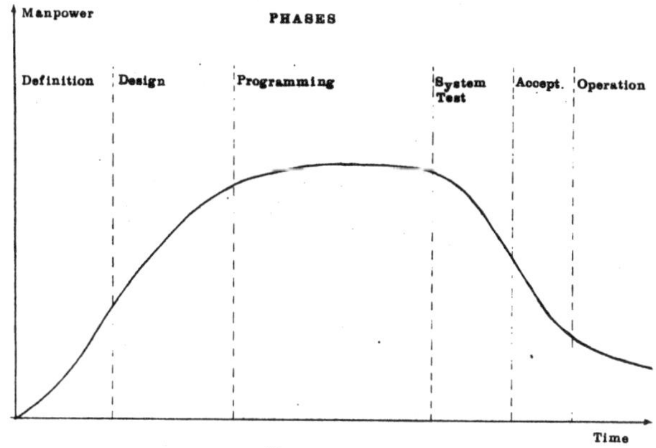

Fig. 1

It is divided into 6 phases (definition, design, programming, system test, acceptance, installation and operation), with the qualitative indication of the time and human resources required. The actual quantification depends on the type of problem, on the available supports and the organization of the work, but, in any case, the shape of the curve is confirmed; the included area represents the total work required (expressed in manyears, for example)W. The peak of the curve occurs approximately at the end of the programming phase (tM): in this period, in fact, a lot of activities are carried out at the same time (detail programming, fixing remaining bugs, writing documentation, system and acceptance test preparation) /14:.

The curve may be expressed by this kind of equation:

$$y = a\, t\, e^{-b\, t^2}$$

where a and b are values dipendent on W and tM. W and tM may of course have different values and the curve changes accordingly.

According to the specific sector, the shape and magnitude of the curve must be optimized; the typical objectives are essentially 3:

- reduction of W,
- reduction of tM,
- reduction of maintenance costs.

In order to reduce W, it is necessary to find fast and cheap automatic programming processes, with tools rapidly redeemable on few installations: a possible answer to that is the Automatic Program Generation (APG), which requires the development of a suitable software processor, interacting with the user, but which simplifies or even eliminates some activities from the software life cycle /15/. The necessity of reducing tM obviously depends on delivery times and it implies a greater concetration of manpower in the initial phases. The maintenance cost of a "manual" software product weighs heavily on the total cost. Even if the work is carried out very carefully, after the acceptance phase there generally still remain some problems which requires an additional manpower. According to various statistics (from simple programs to complex procedures), it comes out that installation and maintenance require a percentage usually included between 20 and 50% of the whole work. This range depends above all on the bugs remaining after the testing activity, on a wrong initial analysis and on changes which occur only when the software system is put in the actual working environment. It is quite impossible, in non-trivial cases, to find out all the mistakes until the program is actually used. Moreover, the necessity of increasing or modifying program performances often comes up after the acceptance and installation.

In the industrial robotics,times are not important: the robot production time is highly superior to the one necessary to the development of program. Besides, the weight of W on the total cost is not such as to justify on its own a change of the present execution methodologies. It is necessary, on the contrary, to reduce the maintenance and development work to a minimum, for the following reasons:

- the time necessary to single out and eliminate bugs during the development phase can be evaluated in a few hours, after the installation this time increases considerably;
- if the software development is not carried out methodologically, the programs (even if working) can be out of requirements, in terms of execution time or memory size, with an additional "a posteriori" refinement work;
- possible changes (even if of little importance) after the installation appear logistically difficult, because the present lack of suitable personnel, the great internal mobility and above all the considerable turnover.

A reduction of these drawbacks involves a better final product and also, subsequently, a reduction of manpower required for each different robot produced.

5. MAINTENANCE REDUCING

The above-mentioned problems are common to all the programming sector and so it is possible, in theory, to adopt the techniques provided by the software-houses for the development of application packages. In practice, this is not always possible and it must be verified case by case. The main obstacles, typical of robotics, are:

- lack of specific internal organization (the whole project is generally given to a mechanical engineer);
- limited hardware and software resources;
- lack of technological standards;
- difficulty of generalization of the product.

Taking all the considerations above said into account, there follow some low cost proposal with the evaluation of their effects.

A) DOCUMENTATION

In the software production process, documentation initially played a quite marginal and passive role, being it written only at the end of the process. It must be developed, on the contrary, during the life cycle phases, by assuming an active role as a means of communication inside the operating structure. It should also involve the end-user so as to simplify the definition phase of the specifications. The possible effects of documentation come up mainly during the maintenance, because it simplifies the interchanging of personnel. However, the use should not be exagerated, by making it become a simple bureaucratic tool. Documentation must be adapted to the specific environment (and so integrated with the documentation of mechanical and electronic projects) and possibly connected to other activities (for example, automatic documentation with the production of tables and paragraphs during the various activities of the software development. /16,17/.

B) PROGRAM RESTRUCTURING

Packages for the analysis and restructuring of programs can be very useful, because they allow an uniform programming style even if the programs are initially written by different people according to their personal optimization criteria. In such a way, the pre-existing libraries become more readable and useful for eventual subsequent changes. They perform, in particular, the following tasks:

- identification of unstructured segments and change with more efficient solutions;
- elimination of unnecessary GOTO;
- automatic flow-charting at different abstraction levels;
- insertion of comments to simplify the documentation.

So far, it's not been given great importance to software work because of its small effect on total cost: a program for assembly robot requires from 200 to 300 man-hours, that is about 10% of the total project. It must be considered, however, that it influences the whole work in terms of efficiency of the product. It is then useful that the programmers are led in their work in order to optimize the program. This is generally made when the product has already been obtained and by making subsequent attempts only when it is absolutely necessary. For example, a robot made an assembly in 6 secs, while the initial sepcifications had foreseen only 5 at the most. By calling software experts, the program has been rewritten, so reducing the time to 4.7 secs. It is better if this is done at first, eliminating all the unuseful operations with analysis and restructuring packages or with simulation models.

C) MODULAR PROGRAMMING

A lot of structured programming techniques are derived from the top-down approach to problem solving. Besides it is possible to build-up libraries of modules or segments to be utilized for different problems. This concept, used in some simple automatic generators /19/, is obstructed in robotics by lack of standards and by the limited power of present languages. It is, however, possible to single out some general macros, especially for world-modeling languages, directed to the routing control (for example, PAL) /20/. Moreover, by assigning to the programmers a structured technique, an uniform programming style is obtained with a better interchanging of people.

D) VALIDATION AND SIMULATION

Validation is a process to determine the level of adherence between the software system already realized and its specifications: it is a completion activity to testing and it is carried out during the design and programming phases. It can be performed at various levels, according to the available supports and it can represent a meaningful portion of the global cost of software. According to general statistics, validation shows only a little percentage of bugs which slipped the debugging, but they are usually serious errors (such as to compromise the final results or to stop the execution). For assembly robots programs the cost of validation varies between 30 and 60% of the development cost, but it reduces the maintenance activity considerably.

As an alternative to validation, simulation can be used, by developing programs in the simulated working environment. This can done on-line (some examples of this exist at present) or off-line (by using a mini host computer also available, for example, for robot programming). Off-line simulation is less efficient but simpler to be realized and it can be integrated with other available resources.

E) INTEGRATION WITH STANDARD SOFTWARE

Finally, if the robot processor is compatible with the supporting minicomputer one (for example LSI-11 for SIGMA Olivetti and PDP-11 as external host), the object programs produced on mini can be directly carried out on the robot, without the necessity of an intermediate code with subsequent translation. According to this hypothesis, it is possible to use the mini exclusively, by employing its standard products (general-purpose languages, text-editor, trace) and enriching it little by little with specific supports (modules' libraries, analysis and restructuring packages, simulation, documentation).

In fig. 2 the various phases of software development on an outside minicomputer are shown, with the use of the supports and the methodology above mentioned.

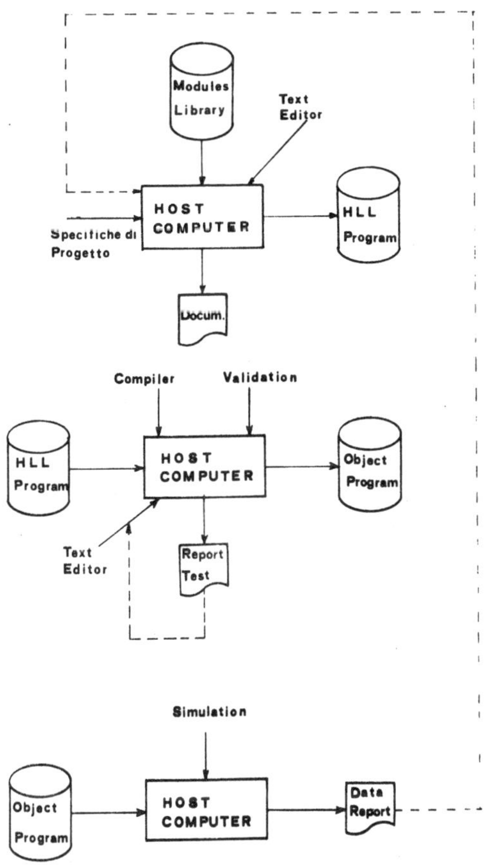

Fig. 2

In particular it:

a) shows the program writing in a high level language, starting from the project specifications, using libraries and producing the project documentation;
b) shows the subsequent phases of translation, standard diagnostics and validation; its outputs are the object program and the test reports;
c) shows the final simulation phase, with the production of data reports and with a possible return to the initial phase, in order to improve the operations of the programs, by correcting it, for example, with a text-editor.

The installation of the program on the robot can be made off-line, by means of intermediate supports (tape-cassettes, for example),or through direct connession. In the second case, the connection can be initially unidirectional (of the master-slave type) and afterwards it can be bidirectional, so as to reduce the program setting-up among the specifications and to obtain useful information on parameters and coordinates, according to a partial teaching-by-showing (fig. 3).

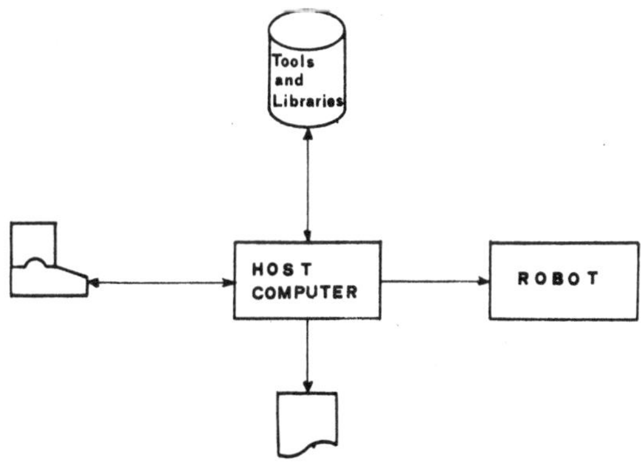

Fig 3

6. CONCLUSIONS

The techniques and the supports above mentioned are not meant to substitute more sophisticated solutions, but their purpose is the improvement of the programs output efficiency for standard robots still in production. The advantages of the proposed solutions are essentially the following:

- reduction in the testing and debugging activity,
- programs readability and modularity,
- reduction in the maintenance costs.

Besides, the use of the above-said supports is the starting point towards more advanced solutions, in that it is integrable in a wider environment.

REFERENCES

/1/ Somalvico, M. "Intelligenza artificiale e robotica industriale" Tecnica dell'Automazione e Robotica, v. XV, n. 12 (1981) and v. XVI, n. 1 (1982)

/2/ Will, P. "Very high level languages for robots" Proc. 9th Int.Symp.Ind.Robots, Washington (1979) 281-295

/3/ Dessimoz, J.D., Kunt, M., Zurcher, J.M. and Granlund, G.H. "Recognition and handling of overlapping industrial parts" Proc. 9th Int.Symp.Ind.Robots, Washington

(1979) 357-366

/4/ Gini, G. and Gini, M. "Sviluppo di un sistema di programmazione per l'automazione industriale" Riv. di Informatica, v. X, n. 3 (1980) 249-261

/5/ Carne, E.B., Chaney, J.E. and Zambuto, A. "An industrial robot with minicomputer control" Proc. 9th Int.Symp.Ind.Robots, Washington (1979) 215-229

/6/ Gini, G., Gini, M., Gini, R. and Giuse, D. "Introducing software system in industrial robot" Proc. 9th Int Symp.Ind.Robots, Washington (1979) 309-319

/7/ Popplestone, R.J., Ambler, A.P. and Bollos, I.M. "An interpreter for a language describing assemblies" Art. Intelligence, 14 (1980) 79-107

/8/ Salmon, M. "SIGLA- the Olivetti SIGMA robot programming language" 8th Ind.Symp.Int.Robots, Stuttgart (1978) 358-363

/9/ Perzley, W. " Robot path control by off-line computer" Proc.8th Ind.Symp.Int.Robots, Stuttgart (1978) 338-345

/10/Wech, M., Eversheim, W. and Zuhlke, D. "ROBEX: one off-line programming system for industrial robot" Proc. 11th Int.Symp.Ind.Robots, Tokio (1981) 297-307

/11/Ambler, A.P., Popplestone, R.J. and Kempf, K.G. "An experiment in the off-line programming of robot" Proc 12th Int.Symp.Ind.Robots, Paris (1982) 491-502

/12/Lauber, R.J. "Development support systems" Computer, v.15, n.5 (1982) 36-46

/13/Zelkowitz, M.V., Shaw, A.C. and Gannon, J.D. "Principles of software engineering and design" Prentice Hall (1979)

/14/Putnam, L.H. and Fitzsimmons, A. "Estimating software costs" Datamation, v.25, n.12 (1979) 137-140

/15/Priwes, M.S. "Automatic generation of computer programs" in "Advances in Computers" (Rubinoff, M; and Yovits, M.C. eds.) Academic Press (1977) 58-125

/16/Rosemberg, R. "Computer-aided documentation" AFIPS Proc Nat.Comp.Con., v.51 (1982) 351-358

/17/Benedusi, P. and De Carlini, U. "La strutturazione della documentazione del software prodotto con un generatore di applicazioni" accepted to AICA Congress, Naples (1983)

/18/Cantone, G. and Fadini, B. "An approach to a model for automatically generating a programs structured maintenance document" 9th IMACS Congress, Sorrento (1979) 1103-1111

/19/De Carlini, U., Esposito, A. and Fadini, B. "Un modello per la generazione automatica di programmi applicativi" accepted to AICA Congress, Naples (1983)

/20/Park, W.T. "Minicomputer software organization for control of industrial robots" 1977 Joint Aut.Cont.Con., S.Francisco (1977) 164-171

/21/ Adrion, W.R., Branstad, M.A. and Cherniavsky, J.C."Validation, verification and testing of computer software" ACM Comp.Surveys, v. 14, n.2 (1982) 159-192

A SIMPLE PATTERN RECOGNITION SCHEME BASED ON NEURAL NETWORK STRUCTURES

N. Dimopoulos and R.V. Patel
Department of Electrical Engineering
Concordia University
Montreal, Canada

ABSTRACT

In this paper, we develop a scheme for performing pattern recognition based on the neural network structures that can be found in the visual cortex. We first examine the problem of training a given network structure so that it will be able to distinguish between two classes of simple patterns. The patterns consist of line segments. The network is trained to respond to line segments of a given orientation while remaining silent for inputs representing any other orientation. This essentially corresponds to the behaviour of neurons in the visual cortex as observed by Hubel and Wiesel.

1. Introduction

Neural Networks are functional units, occupying con-connected regions within the central nervous system and consist of a multitude of elements called neurons. The neurons comprising a particular network form classes which are distinguished by their morphological and connectivity properties [1-3].

Neurons influence each other mostly through unidirectional functional junctions called synapses. As a consequence of Dale's Law [3], all the neurons of the same class exhibit identical behaviour in their influence, this being either excitatory or inhibitory. The neural classes, therefore, can be classified either as excitatory or inhibitory ones.

The above observations, which hold true for various neural networks (such as the cerebellum, the hippocampus, etc...) can be formulated as two generalized connectivity principles.

Microscopic Connectivity Principle Neurons influence, and are influenced by (i.e. synapse) neurons in their immediate vicinity.

Macroscopic Connectivity Principle Neurons are to be found in morphologically distinct classes - neural classes. Neurons from the same class synapse with neurons belonging to a genetically pre-determined group of neural classes.

Based on these two connectivity principles, it is possible to derive a mathematical model [1] that describes the behaviour of a neural network. Such a model will be presented in Section 2 and it has been used for the study of the various properties of neural networks attributed to their structure.

A fundamental property possessed by neural networks is that of adaptation and learning. This happens throughout the life-span of an organism and it is the result of environmental influence. Due to the two connectivity principles presented above, the gross structure of a given neural network, is genetically pre-determined and therefore, it cannot be altered during the life time of the organism. This of course limits the number of structures a network may achieve, but on the other hand, such genetically pre-determined structures ensure certain general properties for a given network. One such property, the asymptotic behaviour of some neural networks, was studied in [1].

In this work, we postulate that the alterable quantity that ultimately determines the exact function of a particular network, is the strength of the synapses of the neurons that compose the network. We will present a method of choosing the strength of the synapses so that a given structure will be "trained" to distinguish between classes of simple patterns.

2. The Mathematical Model

As derived in [1], a single neuron can be modelled by the following differential equation

$$\tau \frac{dx}{dt} + x = \sum_{i=1}^{k} w_i f_i(x_i) \tag{1}$$

where x is the membrane potential of a particular neuron. This neuron is influenced by other neurons and external inputs through k synapses. The strength of the ith synapse is described by the weight w_i while $f_i(x_i)$ is the frequency of firing of the ith neuron or external input which depends in a nonlinear fashion on the membrane potential x_i. This dependency is described by the functions f_i which belong to the class of neuromine functions defined as

$N = \{f: R \to R_+$ such that f is continuous, monotonically nondecreasing, satisfying a Lipschitz condition, and $\exists\ \theta \in R$ such that $f(\theta) = 0\}$.

As an example, the following function is a neuromine function.

$$f(x) = \begin{cases} x & \text{if } x > 0 \\ 0 & \text{if } x \le 0 \end{cases} \tag{2}$$

Having the model for a single neuron, (equation (1)) ,it is easy to derive the model for a neural network consisting of k neural classes each one of which has n_i neurons for which a total of $N = \sum_{i=1}^{k} n_i$ neurons in the network. The model is given as [1].

$$\underline{T}\dot{\underline{x}} + \underline{x} = W \cdot \underline{f}(\underline{x}) + \underline{u} \tag{3}$$

where $\underline{x} = [\ \underline{x}_1^T, \underline{x}_2^T, \ldots \underline{x}_k^T]^T =$

$$= [x_1, x_2, \ldots, x_{n_1}, x_{n_1+1}, \ldots, x_{n_1+n_2}, \ldots, x_N]^T$$

is the vector of the membrane potentials of the neurons in the network;

$T = \text{diag}\ [\tau_i]\ i = 1,2,\ldots,N$ is the membrane time constant matrix; vector $\underline{u}$ with non-negative elements is the input to the network;

$\underline{f}(\underline{x}) = [f_1(x_1)\ f_2(x_2)\ldots f_N(x_N)]^T$ is a vector of neuromine functions and it describes the frequency of firing of the neurons in the network while

$$W = \begin{bmatrix} W_{11} & W_{12} & \cdots & W_{1k} \\ \vdots & & & \\ W_{k1} & W_{k2} & \cdots & W_{kk} \end{bmatrix}$$

is the connectivity matrix describing the strength of the synapses and hence the structure of the network. The entries W_{ij} themselves, are matrices of dimensions $n_i \times n_j$ and describe the influence of the neurons of class j on the neurons of class i. All the entries of each one of the matrices W_{ij} have either the same sign or are identical to zero.

Thus, if $W_{ij} > 0$ then class j excites class i, if $W_{ij} < 0$ then class j inhibits class i and if $W_{ij} = 0$ then class j does not synapse with class i. We denote this condition on the signs of the elements of W_{ij} by

$$S_{ij} W_{ij} \geq 0$$

where

$$S_{ij} = \begin{cases} -1 & \text{if } W_{ij} < 0 \\ 0 & \text{if } W_{ij} = 0 \\ 1 & \text{if } W_{ij} > 0 \end{cases}$$

The aim of this work is to choose the elements of the non-zero matrices W_{ij} so that the resulting network will act as a pattern recognizer.

3. The Problem of Pattern Recognition

We are interested in the pattern recognition capabilities of neural networks modelled according to equation (3) which we repeat at this point.

$$T\dot{\underline{x}} + \underline{x} = W\,\underline{f}(\underline{x}) + \underline{u} \tag{4}$$

Note that $\underline{x} \in R^N$, $\underline{u} \in \mathcal{U}$ and $W \in R^{NxN}$. Usually the output, or the observable quantity, for such a system is the state of a limited number of neurons. This has also been substantiated by the experiments of Hubel and Wiesel [4] on the receptive fields in the cat's visual cortex. We therefore describe the output of such a network as

$$\underline{y} = V \cdot \underline{f}(\underline{x}) \tag{5}$$

where

$$V = [0 \quad I_{n_i} \quad 0] \tag{6}$$

with I_{n_i} the $n_i x\, n_i$ identity matrix.

Also, for such networks, since the function f is postulated as a Lipschitian one, there exists a unique solution and moreover the solution exhibits asymptotic behaviour provided that all the positive elements of the matrix W are located above the main diagonal [1].

Let the equilibrium point of (4) be denoted by $\underline{x}^e$. Then, equation (4) yields

$$\underline{x}^e = W\underline{f}(\underline{x}^e) + \underline{u} \tag{7}$$

The problem of a pattern recognizing network reduces to that of choosing the elements of the matrix W in such a way that

$$\forall u \in \mathcal{U}$$

$$\underline{x}^e = W\underline{f}(\underline{x}^e) + \underline{u} \tag{8}$$

$$\underline{y} = V \cdot \underline{f}(\underline{x}^e) > \underline{0} \quad \text{and} \quad S_{ij} W_{ij} > 0$$

In order to reduce the computational complexity associated with choosing the elements of W due to the presence of the nonlinearity $\underline{f}(\underline{x}^e)$, we first reduce (7) to a linear equation in $\underline{x}^e$. We do this by assuming that the elements of $\underline{x}^e$ are non-negative, so that using (2), we can write

$$\underline{x}^e = W\underline{x}^e + \underline{u}$$

$$\text{i.e. } M\underline{x}^e = \underline{u} \tag{9}$$

where $M = I_N - W$ and $\underline{u} \in \mathcal{U}$. The elements of M (and consequently of W) are chosen via a "training" scheme whereby a set of inputs $\underline{u}^i$, $i = 1, \ldots, p$ are chosen from the class $\mathcal{U}$, and M and $\underline{x}_i^e$ $i = 1, \ldots, p$ are determined such that they satisfy (9). The constraints on the outputs are modified to the form

$$\underline{y} = V\underline{x}^e \geq \underline{\alpha} \tag{10}$$

where the elements of $\underline{\alpha} \in R^{n_i}$ have some prespecified positive values.

By letting the vectors $\underline{u}_i$ and $\underline{x}_i^e$ be the columns of matrices U and X^e we can write equation (9) as

$$MX^e = U \tag{11}$$

where $X^e = [\underline{x}_1^e \,|\, \underline{x}_2^e \,|\, \ldots \,|\, \underline{x}^e]$ and $U = [\underline{u}_1 \,|\, \underline{u}_2 \,|\, \ldots \,|\, \underline{u}_p]$

The problem of choosing the elements of W then reduces to that of factoring $U \in R^{Nxp}$ into the product of $M \in R^{NxN}$ and $X^e \in R^{Nxp}$ subject to certain constraints on the elements of M and X^e. These constraints correspond to the constraints on the signs of W_{ij} and on the magnitudes and signs of $\underline{x}_i^e$. Since neither M nor X^e are known a priori, the problem of factoring U according to (11) with constraints on M and X^e is not a linear one. Here, we propose an iterative scheme to achieve this factorization.

Step 1: (1) Data: $U = [\underline{u}_1 \,|\, \underline{u}_2 \,|\, \ldots \,|\, \underline{u}_p]$, $\underline{\alpha}$ (>0), M (satisfying the constraints on W), V, C_{min} (largest acceptable value of cost function).

(2) For $i = 1, \ldots, p$.

$$\text{Minimize } c_i = \sum_{j=1}^{N} a_{ji} \tag{12}$$

subject to

$$[M \,|\, I_N] \begin{bmatrix} \underline{x}_i^e \\ \hline \underline{a}_i \end{bmatrix} = \underline{u}_i \tag{13}$$

$$\underline{x}_i^e \geq 0, \; \underline{a}_i \geq 0 \tag{14}$$

$$\text{and } V\underline{x}_i^e \geq \underline{\alpha} \tag{15}$$

(3) Let

$$c = \sum_{i=1}^{p} c_i \tag{16}$$

$$\text{and } X^e = [\underline{x}_1^e \,|\, \ldots \,|\, \underline{x}_p^e] \tag{17}$$

if $c \leq c_{min}$, stop; otherwise go to Step 2 with X^e given by (17).

Step 2: (1) Data: X^e(from Step 1.3), $U^T = [\hat{\underline{u}}_1 \,|\, \hat{\underline{u}}_2 \,|\, \ldots \,|\, \hat{\underline{u}}_N]$, Constraints on W, V, C_{min}

(2) For $i = 1, \ldots, N$

$$\text{Minimize } \hat{c}_i = \sum_{j=1}^{p} \hat{a}_{ji} \tag{18}$$

subject to

$$[X^{e^T} \mid I_p] \begin{bmatrix} \hat{\underline{m}}_i \\ \hat{\underline{a}}_i \end{bmatrix} = \hat{\underline{u}}_i \tag{19}$$

$$\hat{\underline{a}}_i \geq 0$$

$$-\hat{s}_{ji}\hat{m}_{ji} \geq 0 \quad i \neq j, \quad \hat{s}_{ii}(1-\hat{m}_{ii}) \geq 0 \tag{20}$$

where $\hat{s}_{ji}$ is +1, 0 or -1 depending on whether the corresponding element of W is positive, zero or negative.

(3) Let $\hat{c} = \sum_{i=1}^{N} \hat{c}_i$ (21)

and $M^T = [\hat{\underline{m}}_1 \mid \hat{\underline{m}}_2 \mid \ldots \mid \hat{\underline{m}}_N]$ (22)

if $\hat{c} \leq c_{min}$ stop;

otherwise go to step 1 with M given by (22).

Explanation of the Algorithm

In each step of the algorithm, we solve a linear programming problem. In the first step, we find a vectors $\underline{x}_i^e$ and $\underline{a}_i$ with non-negative elements such that

$$M\underline{x}_i^e + \underline{a}_i = \underline{u}_i \tag{23}$$

$$V\underline{x}_i^e \geq \underline{\alpha}$$

Here the vector $\underline{a}_i$ is some dummy vector with non-negative elements. The linear programming problem that we solve in this step minimizes the sum of all the elements of $\underline{a}_i$. This implies that if $c_i = 0$, then we have a solution for $M\underline{x}_i^e = \underline{u}_i$, which also satisfies all the constraints. It should be noted that the solution X^e given by (17) satisfies all the constraints including

$$MX^e + A = U \tag{24}$$

where $A = [\underline{a}_1 \mid \underline{a}_2 \mid \ldots \mid \underline{a}_p]$. Hence the given pair of matrices M and X^e has associated with it a cost c which is the sum of all the elements of A. Note that from the way the linear programming problem is formulated, the matrix X^e is determined to give the least possible value of c for the <u>given</u> value of M.

In Step 2, we find vectors $\hat{\underline{m}}_i$ and $\hat{\underline{a}}_i$ (with constraints on their elements) such that

$$X^{e^T} \hat{\underline{m}}_i + \hat{\underline{a}}_i = \hat{\underline{u}}_i$$

Here, it should be observed that $\hat{\underline{m}}_i$, $\hat{\underline{a}}_i$ and $\hat{\underline{u}}_i$ are columns of M^T, A^T and U^T respectively, so that at the end of Step 2 we have matrices M^T and A^T satisfying

$$X^{e^T} M^T + A^T = U^T$$

The pair of matrices X^{e^T} and M^T now has a cost $\hat{c}$ associated with it, the cost being the sum of all the elements of A^T. In this case M is determined such that the cost $\hat{c}$ is minimum for the given value of X^e. This property together with the corresponding one for Step 1, ensures that the algorithm converges. The parameter C_{min} determines when the cost is small enough to terminate the iterations.

The constraints on $\hat{m}_{ji}$ in (20) are obtained using the constraints on W and the fact that W = I-M. As described here, the algorithm solves p and N linear programming problems in Steps 1 and 2 respectively at every iteration. Each solution of the linear programming problem yields one column of X^e or M^T. It can be easily shown that using Kronecker products we can formulate the same problem so that X^e and M^T are obtained by solving only one linear programming problem at each step. This formulation, however, requires considerably more storage

4. Example

As an example we considered the problem of recognizing straight lines of a particular orientation. We assumed that the lines are drawn on a 4x4 raster by rows so that the vertical line in Fig. 1 is represented by 16x1 vector $\underline{u}_1 = [1\,0\,0\,0 \mid 1\,0\,0\,0 \mid 1\,0\,0\,0 \mid 1\,0\,0\,0]^T$.

For this example N = 16, $\underline{x}_i^e \in R^{16}$ and $W \in R^{16 \times 16}$. We assumed $n_i = 4$, $i = 1,\ldots,4$, so that $W_{ij} \in R^{4x4}$. Furthermore we assumed the following signs for the elements of W_{ij}:

$$[W_{ij}] = \begin{bmatrix} W_{11} & W_{12} & W_{13} & W_{14} \\ W_{21} & W_{22} & W_{23} & W_{24} \\ W_{31} & W_{32} & W_{33} & W_{34} \\ W_{41} & W_{42} & W_{43} & W_{44} \end{bmatrix}, \quad [s_{ij}] = \begin{bmatrix} - & + & - & + \\ - & 0 & - & + \\ - & 0 & - & + \\ - & 0 & - & + \end{bmatrix}$$

The matrix V was taken to be $V = [0_{4x4} \; I_4 \; 0]$ and the vector α was specified as $\underline{\alpha} = [10\;10\;10\;10]^T$.

In order to choose the elements of W, we first carried out a "training" process whereby we provided a set of inputs $\underline{u}_i \in \mathcal{U}$ where the class $\mathcal{U}$ consists of all inputs corresponding to vertical lines drawn on the raster. As a result of the learning process, we obtained the matrix X given in Table 1. To see if the elements of W had been chosen correctly, we performed several tests by taking inputs $\underline{u}_j$ from the class $\mathcal{U}$ other than those used in the training process as well as others not belonging to $\mathcal{U}$, and checking whether the corresponding outputs $\underline{y}_j$ satisfied the constraints i.e. $\underline{y}_j \geq \underline{\alpha}$. The results are shown in Table 2.

5. Conclusion

A pattern recognition scheme based on a mathematical model of neural network structures has been developed. The scheme consists of choosing the magnitudes of the elements of the the connectivity matrix W. An algorithm for determining these magnitudes has been proposed. The algorithm is iterative in nature and involves the solution of linear programming problems at each step.

6. References

1. Dimopoulos, N., "Organization and Stability of a Neural Network Class and the Structure of a Multiprocessor System", Ph.D. thesis, University of Maryland (1980).

2. Eccles, J.C., "The understanding of the Brain", McGraw Hill (1973).

3. Shepherd, G.M., "The Synaptic Organization of the Brain", Oxford University Press (1974).

4. Hubel, D.H. and Wiesel T.N., "Receptive fields, Binocular Interaction and Functional Architecture in the Cat's Visual Cortex". J. Physiol. (1962) 160, pp. 106-154.

$\underline{u} = [1\,0\,0\,0\,1\,0\,0\,0\,1\,0\,0\,0\,1\,0\,0\,0]^T$ $\quad\quad$ $\underline{u} = [0\,0\,1\,0\,0\,0\,1\,0\,0\,0\,1\,0\,0\,0\,0\,0]^T$

Figure 1 : Generation of input signals

$$
W = \left[\begin{array}{cccc|cccc|cccc|cccc}
-0.8 & -0.8 & -0.8 & -0.8 & 0.0 & 0.0 & 0.0 & 0.0 & -1.0 & -1.0 & -1.0 & -1.0 & 1.0 & 0.0 & 0.0 & 0.0 \\
-0.8 & -0.8 & -0.8 & -0.8 & 0.0 & 0.0 & 0.0 & 0.0 & -1.0 & -1.0 & -1.0 & -1.0 & 1.0 & 0.0 & 0.0 & 0.0 \\
-0.8 & -0.8 & -0.8 & -0.8 & 0.0 & 0.0 & 0.0 & 0.0 & -1.0 & -1.0 & -1.0 & -1.0 & 1.0 & 0.0 & 0.0 & 0.0 \\
-0.8 & -0.8 & -0.8 & -0.8 & 0.0 & 0.0 & 0.0 & 0.0 & -1.0 & -1.0 & -1.0 & -1.0 & 1.0 & 0.0 & 0.0 & 0.0 \\
\hline
-1.0 & -1.0 & -1.0 & -1.0 & 0.0 & 0.0 & 0.0 & 0.0 & -1.0 & -1.0 & -1.0 & -1.0 & 1.0 & 0.0 & 0.0 & 0.0 \\
-1.0 & -1.0 & -1.0 & -1.0 & 0.0 & 0.0 & 0.0 & 0.0 & -1.0 & -1.0 & -1.0 & -1.0 & 1.0 & 0.0 & 0.0 & 0.0 \\
-1.0 & -1.0 & -1.0 & -1.0 & 0.0 & 0.0 & 0.0 & 0.0 & -1.0 & -1.0 & -1.0 & -1.0 & 1.0 & 0.0 & 0.0 & 0.0 \\
-1.0 & -1.0 & -1.0 & -1.0 & 0.0 & 0.0 & 0.0 & 0.0 & -1.0 & -1.0 & -1.0 & -1.0 & 1.0 & 0.0 & 0.0 & 0.0 \\
\hline
-1.0 & -1.0 & -1.0 & -1.0 & 0.0 & 0.0 & 0.0 & 0.0 & -1.0 & -1.0 & -1.0 & -1.0 & 1.0 & 0.0 & 0.0 & 0.0 \\
-1.0 & -1.0 & -1.0 & -1.0 & 0.0 & 0.0 & 0.0 & 0.0 & -1.0 & -1.0 & -1.0 & -1.0 & 1.0 & 0.0 & 0.0 & 0.0 \\
-1.0 & -1.0 & -1.0 & -1.0 & 0.0 & 0.0 & 0.0 & 0.0 & -1.0 & -1.0 & -1.0 & -1.0 & 1.0 & 0.0 & 0.0 & 0.0 \\
-1.0 & -1.0 & -1.0 & -1.0 & 0.0 & 0.0 & 0.0 & 0.0 & -1.0 & -1.0 & -1.0 & -1.0 & 1.0 & 0.0 & 0.0 & 0.0 \\
\hline
0.0 & 0.0 & 0.0 & 0.0 & 0.0 & 0.0 & 0.0 & 0.0 & 0.0 & 0.0 & 0.0 & 0.0 & 1.0 & 0.0 & 0.0 & 0.0 \\
0.0 & 0.0 & 0.0 & 0.0 & 0.0 & 0.0 & 0.0 & 0.0 & 0.0 & 0.0 & 0.0 & 0.0 & 0.0 & 1.0 & 0.0 & 0.0 \\
0.0 & 0.0 & 0.0 & 0.0 & 0.0 & 0.0 & 0.0 & 0.0 & 0.0 & 0.0 & 0.0 & 0.0 & 0.0 & 0.0 & 1.0 & 0.0 \\
0.0 & 0.0 & 0.0 & 0.0 & 0.0 & 0.0 & 0.0 & 0.0 & 0.0 & 0.0 & 0.0 & 0.0 & 0.0 & 0.0 & 0.0 & 1.0
\end{array}\right]
$$

Table 1: Matrix W obtained after training the network

i =	1	2	3	4	5	6	7	8	9	10	11
$\underline{u}_i$	1	0	0	0	1	0	0	0	1	1	0
	0	1	0	0	0	1	0	0	1	1	1
	0	0	1	0	0	0	1	0	1	1	1
	0	0	0	1	0	0	0	1	1	0	1
	1	0	0	0	1	0	0	0	0	0	0
	0	1	0	0	0	1	0	0	0	0	0
	0	0	1	0	0	0	1	0	0	0	0
	0	0	0	1	0	0	0	1	0	0	0
	1	0	0	0	0	0	0	0	0	0	0
	0	1	0	0	0	0	0	0	0	0	0
	0	0	1	0	0	0	0	0	0	0	0
	0	0	0	1	0	0	0	0	0	0	0
	0	0	0	0	0	0	0	0	0	0	0
	0	0	0	0	0	0	0	0	0	0	0
	0	0	0	0	0	0	0	0	0	0	0
	0	0	0	0	0	0	0	0	0	0	0
Vertical line?	Y	Y	Y	Y	Y	Y	Y	Y	N	N	N
$y_i \geq \alpha$?	Y	Y	Y	Y	Y	Y	Y	Y	N	N	N

Y = Yes, N = No,

TABLE 2: Outputs y_i for some inputs u_i

AN OUTLINE OF A NEW APPROACH TO PATTERN RECOGNITION

Lev Goldfarb

School of Computer Science
University of New Brunswick
Fredericton, N.B.
Canada, E3B 5A3

ABSTRACT. The paper is an outline of the new approach to pattern recognition and data analysis in general. A more detailed description of the approach will appear soon in a monograph form [1]. Within the proposed framework the two principal approaches to pattern recognition - vector and syntactic - are unified.

I. INTRODUCTION

A century long tradition has generated a bias toward representing data, and patterns in particular, directly in an n-dimensional Euclidean vector space. The viability of this tradition in all applied fields can easily be explained by a well known fact that all the classical mathematical tools (such as those of analysis and linear algebra) are developed for data represented in a vector space.

During the last several decades, however, many researchers in different areas of pattern recognition began to move away from this "convenient" representation space to other forms of representation. As always, these transitions were possible, because the appropriate representation structures (graphs, grammars, etc.) had already been formally developed within mathematics. Thus, the "structural", or "syntactic" approach to pattern recognition appeared [2], [3], where a pattern is represented not as a vector but as a graph, a string from a formal language, etc. As a result, one gains a considerably greater representation freedom, but loses the convenience of the vector space. "The convenience" should be understood here in the practical sense: the analytical tools available in a vector space are not available in the other "spaces", and therefore, decision-theoretic (learning) algorithms in such "spaces" are more "primitive" and more costly (in computer terms). Indeed, the emergence of a new field - computational geometry [4] - is another fact supporting the last statement. Nevertheless, the structural approach is justified on the following ground: in many concrete cases the gains resulting from the possibility of reflecting in the pattern representation different "structural" pattern features seem to outweigh the advantages of the vector space representation.

Is it possible to develop a systematic approach to pattern recognition which would combine the advantages of the two basic approaches? Although the question has been raised repeatedly since the emergence of the structural approach, no satisfactory answer has been proposed. The author's answer is outlined below.

II. PSEUDOMETRIC SPACES AS A TOOL FOR INFORMATION PROCESSING AND ANALYSIS

One can rephrase the above problem as follows: Find a general approach to pattern recognition which 1) imposes no restrictions on the form of pattern representation, and 2) allows the utilization of the analytical tools available in a vector space. The last formulation implies that a satisfactory solution to the problem should find a way of transplanting the original sample to a vector space, preserving some discriminatory information contained in the original sample representation. An immediate question in connection with this is: What should constitute a basis of the transition from any pattern representation to a vector representation? I suggest that the answer lies in the notion of a dissimilarity measure.

DEFINITION 1. A pseudometric space is a set P together with a non-negative real-valued mapping

$$\Pi : P \times P \longrightarrow \mathbb{R}_1$$

satisfying the following two natural conditions

a) $\forall p_1, p_2 \in P \qquad \Pi(p_1,p_2) = \Pi(p_2,p_1)$

b) $\forall p \in P \qquad \Pi(p,p) = 0$

Both notations (P,Π) or P (for short) will be used, and the mapping Π will be called a pseudometric (distance) function.

It is important to note here that the distance function Π defined on P becomes of the prime importance. In other words, it is this distance function which we should try to preserve in any satisfactory vector representation of P, and, therefore, the choice of is crucial. Obviously, there are infinite number of distance functions on a set P (if the number of elements of P - $|P|$ - is greater than 1). The choice of Π, however, should be dictated by the chosen representation, as well as by the information in the representation which would allow to partition the data optimally. A simple example is in order.

Example 1. Suppose that the problem at hand is to classify (partition) a group of social scientists on the basis of the results of a preliminary questionnaire in which each scientist was asked to assign to a phenomenon p_i (from a set P) a reason c_j (from a set C)

that he(she) considers to be the main cause of p_i. Thus, each scientist can be identified with a mapping from the set P to the set C. Denoting by Map (P, C) the set of mappings from P to C, we can define on the set Map(P,C) the following distance function

$$(1) \qquad \Gamma(\mu_1, \mu_2) = \sum_{c \in C} \mu_1^{-1}(c) \Delta \mu_2^{-1}(c) \quad ,$$

where $\mu_1, \mu_2 \in$ Map(P,C) and $A \Delta B \overset{def}{=} (A \cup B) - (A \cap B)$ is a symmetric difference of the two sets. In other words, for every cause c C we count the number of phenomenons that one of the two scientists attributed to c and the other did not (Fig. 1).

This example can be supplemented by a large number of well known examples of metric spaces [5], as well as by others which are scattered throughout numerous papers in mathematics, taxonomy, pattern recognition, etc. [6].

However, the main reason for adapting a dissimilarity measure as a basis of pattern recognition is

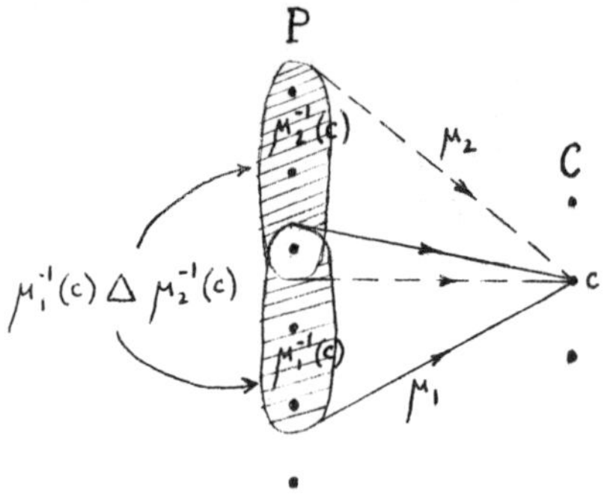

Figure 1

the uniformity it affords. Thus, for example, the importance of distance functions in structural pattern recognition has become quite apparent [3].

III. VECTOR REPRESENTATION OF FINITE PSEUDOMETRIC SPACES

The framework adopted in part II implies that the distance matrix

$$D = (d_{ij})_{1\le i,j\le k} ,$$

where $d_{ij} = \Pi(p_i,p_j)$ and $|P| = k$ (see Def. 1), is the only information which is to be preserved when constructing a vector representation of a finite pseudometic space (P,Π). In other words, by a vector representation of (P,Π) we shall mean a distance preserving isometric mapping of (P,Π) into a vector space. But it is well known that in order to be able to measure distances in a vector space the vector space should have an inner (scalar) product defined on it. It turns out (see corollary of Theorem 1 below) that the class of Euclidean inner product spaces is not large enough to accommodate any finite pseudometric (or even metric) space, necessitating consideration of a larger class of vector spaces - pseudoeuclidean spaces.

DEFINITION 2. A pseudoeuclidean vector space is vector space V together with an inner product $\langle .,. \rangle$ satisfying the following axioms

$$\forall x_1,x_2,y\in V \quad \forall c\in R$$

1) $\langle x_1+x_2,y\rangle = \langle x_1,y\rangle + \langle x_2,y\rangle$

2) $\langle cx_1,y\rangle = \langle c\ x_1,y\rangle$

3) $\langle x_1,y\rangle = \langle y,x_1\rangle$

4) $\langle x_1,y\rangle = 0\ \forall x_1\in V \Rightarrow y=0.$

It is easy to show that a pseudoeuclidean vector space may have vectors $x \neq 0$ such that $\langle x,x\rangle = 0$ (isotropic vectors) and vectors y such that $\langle y,y\rangle < 0$. The most known example of such a space is a 4-dimensional Minkowski space of special relativity theory. In general, a pseudoeuclidean space $\mathbb{R}^{(p,q)}$ of dimension n=p+q is "constructed" from two non-commensurable Euclidean spaces $\mathbb{R}^p$ and $\mathbb{R}^q$, as the Minkowski space is formed by the 3-dimensional space of the space-like vectors and 1-dimensional space of time-like vectors [1], Ch. 3; [7], p. 281. One can also show [1] that a pseudoeuclidean space $\mathbb{R}^{(p,q)}$ is a vector space with an inner product $\langle .,. \rangle$ which can be computed (in an orthonormal basis $(e_i)_{1\le i\le n}$) by the formula

$$\langle x,y\rangle = x^1y^1+x^2y^2+\ldots+x^py^p-x^{p+1}y^{p+1}-\ldots-x^{p+q}y^{p+q} ,$$

where x^i and y^i are the coordinates of x and y with respect to the basis $(e_i)_{1\le i\le n}$.

DEFINITION 3. A pair of non-negative numbers (n^+,n^-) will be called the vector signature of a finite pseudometric space (P,Π) if there exists a distance preserving mapping

$$\alpha : (P,\Pi) \rightarrow \mathbb{R}^{(n+,n-)} ,$$

such that for any other similar mapping of (P,Π) into $\mathbb{R}^{(n_1,n_2)}$ we have $n_1 \ge n^+$, $n_2 \ge n^-$. The above α will be called a vector representation of (P,Π).

In other words, (n^+,n^-) is the vector signature of (P,Π) if $\mathbb{R}^{(n+,n-)}$ is a "minimal" pseudoeuclidean vector space where (P,Π) can be isometrically represented (Fig. 2).

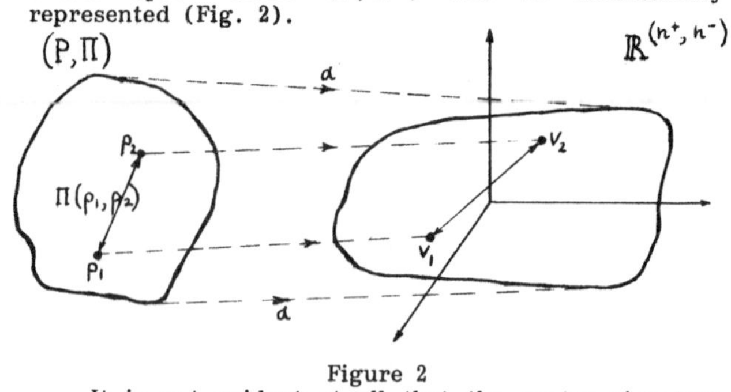

Figure 2

It is not evident at all that the vector signature exists for every finite pseudometric space, but this follows from the next theorem (the proof can be found in [1]).

Theorem 1. A finite pseudometric space (P,Π), where $P = \{p_i\}_{0\le i\le k}$, has the vector signature (n^+,n^-) if and only if the following quadratic form

$$(2) \qquad \Psi(x) = \sum_{i,j=1}^{k} \tfrac{1}{2}(d_{0i}^2+d_{0j}^2-d_{ij}^2)x^ix^j,\ x=(x^1,\ldots,x^k),$$

where $d_{ij} = \Pi(p_i,p_j)$, has the signature (n^+,n^-).

Corollary. A finite pseudometric space (P,Π) can be isometrically represented in the Euclidean n-dimensional space if and only if the corresponding quadratic form is positive and of rank less than n.

Quite fortunately, there exists a constructive proof of the above theorem, i.e., given the matrix $D = (d_{ij})_{0\le i,j\le k}$ of the interdistances of (P,Π) one can construct a set of k+1 vectors in some pseudoeuclidean vector space having D as their matrix of interdistances. Perhaps, a simple example can demonstrate the main idea of the embedding algorithm.

Example 2. Going back to Example 1 let us assume that we are given the results of the questionnaire answered by 4 social scientists. The results can be given by the following matrix in which each row corresponds to the answers of one scientist

	p_1	p_2	p_3	p_4	p_5	p_6
s_0	c_2	c_3	c_1	c_2	c_2	c_1
s_1	c_1	c_2	c_2	c_3	c_1	c_2
s_2	c_1	c_2	c_1	c_3	c_1	c_1
s_3	c_2	c_3	c_2	c_2	c_2	c_1

As one can see, $P = \{p_1, p_2,\ldots, p_6\}$, $C = \{c_1, c_2, c_3\}$. Using formula (1) it is easy to compute the interdistance matrix:

$$D = \begin{pmatrix} 0 & 12 & 8 & 2 \\ 12 & 0 & 4 & 10 \\ 8 & 4 & 0 & 10 \\ 2 & 10 & 10 & 0 \end{pmatrix}.$$

By formula (2) the quadratic form of (S,Γ), where $S = \{s_0, s_1, s_2, s_3\}$, is

$$\begin{aligned}\Psi(x) &= d_{01}^2 (x^1)^2 + (d_{01}^2 + d_{02}^2 - d_{12}^2)\, x^1 x^2 + \\ &+ (d_{01}^2 + d_{03}^2 - d_{13}^2)\, x^1 x^3 + d_{02}^2 (x^2)^2 + \\ &+ (d_{02}^2 + d_{03}^2 - d_{23}^2)\, x^2 x^3 + d_{03}^2 (x^3)^2 \\ &= 144(x^1)^2 + 204x^2x^3 + 48x^1x^3 + 64(x^2)^2 \\ &\quad -32x^2x^3 + 4(x^3)^2.\end{aligned}$$

Using Lagrange's algorithm ([8], p. 194) we can reduce Ψ to the following cononical form

$$\Psi(y) = (y^1)^2 - (y^2)^2 + (y^3)^2, \tag{3}$$

where $y = T \cdot x$ and

$$T = \begin{pmatrix} 12 & 8.5 & 2 \\ 0 & 2.8722813 & -11.489125 \\ 0 & 0 & 11.489125 \end{pmatrix}.$$

So the signature of the quadratic form Ψ is (2,1), and by Theorem 1 pseudoeuclidean vector space $\mathbb{R}^{(2,1)}$ is the minimal one where (S,Γ) can be isometrically represented. In fact, $\alpha: (S, \Gamma) \to \mathbb{R}^{(2,1)}$ given by (0-vector and columns of T)

$$\alpha(s_0) = v_0 = {}^t(0,\ 0,\ 0)$$

$$\alpha(s_1) = v_1 = {}^t(12,\ 0,\ 0)$$

$$\alpha(s_2) = v_2 = {}^t(8.5,\ -2.8722813,\ 0)$$

$$\alpha(s_3) = v_3 = {}^t(2,\ -11.489125,\ 11.489125).$$

The distance in $\mathbb{R}^{(2,1)}$ is computed by the following formula corresponding to (3) above

$$d(v,w) = [(v^1 - w^1)^2 - (v^2 - w^2)^2 + (v^3 - w^3)^2]^{\frac{1}{2}}$$

where $v = (v^1, v^2, v^3)$ and $w = (w^1, w^2, w^3)$. Using the last formula it is easy to check that, indeed, the distances between v_i's in $\mathbb{R}^{(2,1)}$ equal to the corresponding distances between s_i's.

A most efficient way to construct a vector representation is by adapting the so-called QR-algorithm well known in numerical analysis [1], Ch. 4.

It is important to note that the vector representation of a finite pseudometric space is essentially unique [1], Ch. 4.

IV. PATTERN RECOGNITION IN PSEUDOEUCLIDEAN SPACES

Having constructed a vector representation of the original sample in a pseudoeuclidean vector space one is confronted with the question: Can the classical vector space algorithms be used in this situation? The answer is - positively yes. Of course, in order to proceed one needs to develope the generalization of the appropriate classical concepts.

One of the most important concepts widely used in pattern recognition is that of the covariance matrix. Thus, it is only natural that one is concerned first of all with the generalization of this notion to pseudoeuclidean spaces.

Assume that, as above, $\alpha: (P,\Pi) \to \mathbb{R}^{(n+,n-)}$ is a vector representation of (P,Π) and $\forall p_i \in P \quad v_i = \alpha(p_i)$.

DEFINITION 4. Let $(v_i - \bar{v})$, $1 \le i \le k$, be the nx1 matrix of coordinates of the vector $v_i - \bar{v}$ in some fixed $\langle .,. \rangle$ -orthonormal basis of $\mathbb{R}^{(n+,n-)}$, $n^+ + n^- = n$, $\bar{v} = \frac{1}{k}\sum_{i=1}^{k} v_i$. Let us denote by A the n x k matrix whose ith column is $(v_i - \bar{v})$. The n x n matrix

$$M = M(P) = M(P,\alpha) \overset{\text{def}}{=\!=} A \cdot {}^tA \cdot J,$$

where

$$J = \begin{pmatrix} I_{n+} & 0 \\ 0 & -I_{n-} \end{pmatrix},$$

I_m is the identity m x m matrix, will be called <u>the covariance matrix of (P,Π) with respect to the vector representation α</u>.

It is easy to see that if (P,Π) is representable in a Euclidean space, then $J = I_{n+}$ and the above definition gives the classical one. The choice of the vector representation is not essential at all, since one can show that the change of corresponds to the motion (in the pseudoeuclidean space) of the original image.

Obviously, the covariance matrix can be represented as

$$M = \begin{pmatrix} M_1 & -M_{12} \\ {}^tM_{12} & -M_2 \end{pmatrix},$$

where M_1 and M_2 are symmetric matrices, which are the classical covariance matrices of the orthogonal projections of $\alpha(P)$ onto the two sub-spaces: $\mathbb{R}^{n+}$ and $\mathbb{R}^{n-}$. When $n^- = 0$ (Euclidean case), $M = M_1$. The opposite signs on (off) the main diagonal may be interpreted as a result of non-commensurability of the two groups of variables.

The covariance matrix is not a symmetric nor a non-negative matrix, but is a <u>symmetric and a non-negative matrix with respect to the inner product $\langle .,. \rangle$</u> of the pseudoeuclidean space $\mathbb{R}^{(n+,n-)}$ [1], Ch. 5. Moreover, we have the following important result.

<u>Theorem 2</u>. The covariance matrix $M(P,\alpha)$ has exactly n^+ positive and n^- negative characteristic values. Furthermore, there exists a $\langle .,. \rangle$ -orthonormal basis of $\mathbb{R}^{(n+,n-)}$ consisting of the characteristic vectors of $M(P,\Pi)$. The characteristic values are invariant under any vector representation of the finite pseudometric space (P,Π).

Thus, the characteristic values of the generalized covariance matrix are the true parameters of the finite pseudometric space (P,Π) representing the original data. To fully realize the significance of this result one should understand that the information provided by the characteristic values (which is of fundamental importance in many areas of data analysis) <u>cannot be obtained by any other means</u>; i.e. without <u>producing a vector representation of the</u> data one cannot measure these as well as many other important characteristics of the data. And without such numerical invariants the most popular data analysis techniques are not available. For example, the best known techniques for dimensionality reduction rely on the notion of the covariance matrix [9], so that within the proposed framework these techniques can now be applied to any data representation, in particular to syntactic patterns.

To illustrate the main thesis of this section - that all vector space algorithms of pattern recognition are transferable to pseudoeuclidean spaces - a single example is presented.

Example 3. We continue the line of thought started in the previous examples. This example demonstrates the application of Fisher's linear discriminant method for dimensionality reduction ([9], p. 114) to the structural patterns described in Example 1. So, the problem is to find a one-dimensional subspace of $\mathbb{R}^{(2,1)}$ (see Example 2) the projection onto which will minimize the total within-class scatter and maximize the between -class scatter. The classes are $C_1 = \{v_0, v_3\}$ and $C_2 = \{v_1, v_2, v_4\}$, where vector $v_4 = {}^t(10.1, 0, -0.9)$ is added to C_2 in order to avoid degeneracy. As in the classical case the orthogonal projection

$$\pi : \mathbb{R}^{(2,1)} \longrightarrow \mathbb{R}$$

can be given by the formula

$$\pi(v) = \langle v, x \rangle ,$$

where x is a normalized non-zero vector. Thus,

$$\pi(v) = {}^t x \cdot J \cdot v \quad ,$$

where for simplicity the coordinate columns of vectors v and x are denoted by the same letters as the vectors themselves, and

$$J = \begin{pmatrix} I_{n+} & 0 \\ 0 & -I_{n-} \end{pmatrix} .$$

It is also easy to show ([1], Ch. 7), that the within and the between scatter matrices of the projected samples are

$$\overline{M}_w = {}^t x \cdot J \cdot M_w \cdot x \cdot J ,$$

$$\overline{M}_b = {}^t x \cdot J \cdot M_b \cdot x \cdot J ,$$

where M_w and M_b are the within and the between scatter matrices. As in the Euclidean case, we take the criterion function to be maximized

$$\phi(x) = \frac{\det(\overline{M}_b)}{\det(\overline{M}_w)} ,$$

which (taking the volume as a measure of the scatter) gives the ratio of the between-class scatter to the within-class scatter. Next, recalling that

$$M_b = M_b^0 \cdot J \ , \ M_w = M_w^0 \cdot J ,$$

where M_b^0 and M_w^0 are the classical between and within scatter matrices, we can simplify the expression for ϕ

$$\phi(y) = \frac{\det({}^t y \cdot M_b^0 \cdot y)}{\det({}^t y \cdot M_w^0 \cdot y)} , \qquad y = J \cdot x .$$

The problem is reduced now to the classical one, for which we know the solution ([9], p. 117):

$$y_0 = (M_w^0)^{-1} \cdot (\bar{v}_2 - \bar{v}_1) .$$

Setting $\bar{y}_0 = \frac{1}{\|y_0\|} \cdot y_0$ ($\|\cdot\|$ in $\mathbb{R}^{(2,1)}$) we have

$$\pi(v) = {}^t x \cdot J \cdot v = {}^t\bar{y}_0 \cdot J \cdot J \cdot v = {}^t\bar{y}_0 \cdot v .$$

Substituting the data into the formulas we obtain

$$\bar{v}_1 = {}^t(1, -5.7445625, 5.7445625)$$

$$\bar{v}_2 = {}^t(10.2, -0.9574271, -0.3)$$

$$(M_w^0)^{-1} \doteq \begin{pmatrix} 3.782847 & -4.408575 & -5.088170 \\ -4.408575 & 5.369547 & 6.162675 \\ -5.088170 & 6.162675 & 7.092927 \end{pmatrix}$$

$${}^t y_0 = (44.453509, -52.104816, -60.182979)$$

$${}^t\bar{y}_0 = (0.8278833, -0.970378, -1.1208223) .$$

Finally, the two classes projected onto the $\bar{y}_0$ are

$$C_1 = \{\pi(v_0), \pi(v_3)\} = \{0, -0.0727067\}$$

$$C_2 = \{\pi(v_1), \pi(v_2), \pi(v_4)\}$$

$$= \{9.9345996, 9.8242066, 9.3703614\}.$$

As one can clearly see, the separation of the reduced patterns is indeed remarkable.

The last example should give the reader some idea about tremendous possibilities offered by the proposed approach: originally structural patterns are represented in a one dimensional space.

V. CONCLUSION

The paper outlines a new approach to pattern recognition (and to data analysis in general). Since no restrictions on the form of pattern representation are imposed all the advantages of the "structural" approach are automatically incorporated in the proposed approach. The difference between the two approaches is that the latter, unlike the former, allows one to use the analytical machinery available in a vector space for making the decisions. The practical consequences of this fact are obvious: for example, linear and various non-linear decision surfaces can be used for supervised classification of "syntactic" patterns, or different scattering criterions can be used for clustering of "structural" patterns.

Once the original sample has been represented in a vector space,a new pattern can be represented in the space by the method of "abstract orthogonal projection" [1], § 7.2.

The proposed approach reduced to the classical vector apporach if the given patterns are represented as a finite set of Euclidean vectors with the Euclidean distance. Thus, one can clearly see that within the discussed framework the two existing approaches to pattern recognition are unified.

REFERENCES

1. L. Goldfarb, A unified approach to pattern recognition, 1981. (submitted for publication to "Progress in Pattern Recognition" ed. L. Kanal and A. Rosenfeld).
2. T. Pavlidis, Stuctural Pattern Recognition, Springer-Verlag, 1977.
3. K.S. Fu, Syntactic Pattern Recognition and Applications, Prentice-Hall, 1982.
4. M.I. Shamos, Computational geometry, Ph.D. thesis, Yale University, 1978.
5. J. Dieudonne, Foundations of Modern Analysis, Academic Press, 1960.
6. E. Diday and J.C. Simon, Clustering Analysis, Communications and Cyberatics 10, Springer-Verlag, 1976.
7. W. Greub, Linear Algebra, Springer, 1974.
8. A. Mal'cev, Foundations of Linear Algebra, Freeman, 1963.
9. R.O. Duda and P.E. Hart, Pattern Classification and Scene Analysis, J. Wiley & Sons, 1973.

GENERATION OF STRAIGHT LINE MOTION FOR ROBOTS

W.KHALIL
Faculty of Engineering, Mansoura, Egypt
Current Address: E·N·S· M ,Nantes ,France

ABSTRACT

This paper presents an algorithm for the generation of trajectory in the cartesian coordinates of robots.The trajectory passes close the specified points,it is composed of straight line segments connected by circular arcs. The maximum position error on the arc of the actual path from the corner of two adjacent straight line segments can be specified . The trajectory is continuous on position and velocity. The constraints of maximum velocity and acceleration are taken into account. The trajectory can be calculated on-line,only two points ahead should be known

I - INTRODUCTION

Control languages of robots specify tasks as sequences of vectors giving the orientation and location of the hand through which the control frame is to pass [1,2] . Conventional controls are designed in such a way that the robot stops at each point,consequently the travelling time is not efficiently utilised.

Trajectory calculation can be carried out either in the joint space of robots or in the cartesian space. In [3,4,5,6] different algorithms to generate continuous motion in the joint space of the robots are presented.This type of motion needs less computational cost,it is limited only by the maximum acceleration and velocity of the joints.The motion in the space is not along straight lines or any well defined path.Degenerancies of the robot do not cause any problem in this type of motion,it is used when the robot is moving clear of any objects.

In [3,5,7,8] algorithms which generate motion along straight lines are presented.In this type of motion constraints specified in cartesian coordinates such as avoiding obstacles can be easily taken into account.Cartesian motion has the disadvantages that: it needs the transformation of the path points into joint coordinates,and that cartesian motion breaks down when passing by singular points. In the given references ,either the path should be preplaned or the position error from the straight line path is not predicted.

This paper presents an algorithm in which the path is composed of straight lines connected by circular arcs.The maximum position error on the arc of the actual path from the corner of two adjacent straight line segments can be controlled.The trajectory can be calculated on line.The constraints on the velocity and acceleration in the space are taken into account,The motion is continuous on the position and velocity.

II-PROBLEM DEFINITION

The problem is to generate a trajectory which passes by a set of m points,defined in the cartesian coordinates by the vectors

$$\underline{X}^1 , \underline{X}^2, \ldots , \underline{X}^m$$

where

$$\underline{X}^i = | x_1^i \quad x_2^i \quad x_3^i |^T \tag{1}$$

The components of each vector represent the coordinates in the direction of the X_o ,Y_o ,Z_o of the fixed frame.

The maximum velocity and acceleration are given by Kv and Ka respectively.

The departure and arrival velocities are zero,the trajectory is continuous on the position and velocity,

The maximum position error of the actual path from point i is equal to e_i .

III- PATH CALCULATION

Let the segment between point i and point i+1 be called segment i , The length of segment i will be defined by d_i .The transition from segment i-1 to segment i will follow an arc of circle C_i, this circle is tangent to segment i-1 and segment i at the points $\underline{P}^i$ and $\underline{S}^i$ respectively,Figure(1) depicts the situation.Let the distance between $\underline{P}^i$ and $\underline{X}^i$ be represented by ℓ_i ,and the radious of C_i by r_i. The angle between segment i-1 and segment i will be θ_i. where

$$\theta_i = \cos^{-1} (\underline{X}^i - \underline{X}^{i-1}).(\underline{X}^{i+1} - \underline{X}^i) /(d_i)(d_{i-1}) \tag{2}$$

$$0 \leq \theta_i \leq \pi$$

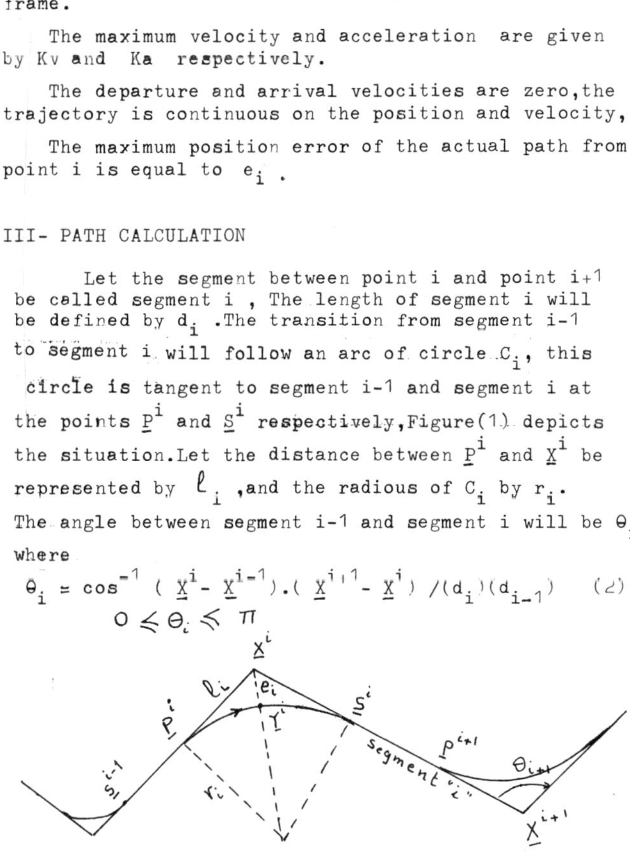

Figure 1

III-1 Calculation of The Cicles Radious

From Fig.1 ,it can be seen that the maximum error position between the circle C_i and the point $\underline{X}^i$ is given by the distance $|\underline{X}^i - \underline{Y}^i|$.It can be proved that:

$$r_i = (r_i + e_i) \sin \frac{\theta_i}{2} \tag{3}$$

Then the maximum permitted radious of the circle C_i , in order to satisfy the position error condition is given by

$$r_i = e_i \sin \frac{\theta_i}{2} / (1 - \sin \frac{\theta_i}{2}) \quad (4)$$

We choose that the motion on the circle C_i will be carried out by a constant linear velocity V_i, thus the linear acceleration will be defined by α_i, then

$$\alpha_i = V_i^2 / r_i \quad (5)$$

Putting α_i = Ka the linear velocity of the circular path can be obtained.
It is to be noted that:

1- If V_i is found to be greater than Kv, r_i must be modified, by the use of Eq.(5), but in this case V_i will be put equal to Kv and α_i = Ka.

2- The distance ℓ_i is equal to:

$$\ell_i = r_i \cot \frac{\theta_i}{2} \quad (6)$$

3- The distance between the points $\underline{S}^{i-1}$ and $\underline{P}^i$, will be denoted by f_{i-1}, where

$$f_{i-1} = d_{i-1} - \ell_{i-1} - \ell_i \quad (7)$$

4- the distance f_{i-1} must be sufficient to attain V_i with respecting the acceleration condition, i.e

$$|(V_i)^2 - (V_{i-1})^2| \text{ must be } \leqslant 2 \text{ Ka } f_{i-1} \quad (8)$$

If the condition 8 is not verified another value of r_i can be calculated by the use of Eqs(5-6-7-8).

5- ℓ_i will be chosen $\leqslant \frac{1}{2} d_i$, unless the velocity at the point $\underline{X}^{i+1}$ is known to be zero, in this case the condition will be that the distance $d_i - \ell_i$ must be sufficient to stop the robot.

6- The error at points 1 and m are zero, and the circles are just points.

7- If θ_i equals 180°, the points $\underline{X}^{i-1}$, $\underline{X}^i$, and $\underline{X}^{i+1}$ are lying on the same straight line, then the path will continue along straight line with $e_i = 0$;

8- If e_i is specified to be zero while $\theta_i \neq 180°$, then the robot must stop at that point.

9- If $\theta_i = 0$, then r_i will be equal to zero and the robot must stop at that point.

III-2 Calculation of the Motion On the Straight Line:

The motion on the straight line between the point $\underline{S}^i$ and the point $\underline{P}^{i+1}$ will be represented by polynomials with constant acceleration, with the distance from $\underline{X}^i$ as the dependent variable and the time as the independent variable, the motion will be represented by:

$$\ddot{q}_k^i(t) = A_k^i$$
$$\dot{q}_k^i(t) = A_k^i t + \dot{q}_k^i(0) \quad (9)$$
$$q_k^i(t) = A_k^i t^2/2 + \dot{q}_k^i(0)\, t + q_k^i(0)$$

where i represents the segment number, and k the polynomial number on segment i(k$\leqslant$3). Each polynomial is valid during an interval T_k^i.

It can be proved that:

$$q_k^i(T_k^i) - q_k^i(0) = T_k^i (\dot{q}(T_k^i) + \dot{q}(0))/2 \quad (10)$$

This relation will be used to determine the time of each polynomial if the terminal velocities and the distance are known.

$T_i = \sum_k T_k^i$ is the time of travelling f_i. (11)
The coordinates along the axis will be determined from the relation:

$$x_j(t) = \{(x_j^{i+1} - x_j^i)\, q_k^i(t)/d_i\} + x_j^i \quad (12)$$
$$j=1,2,3$$

The velocity components can be calculated from:

$$\dot{x}_j(t) = (x_j^{i+1} - x_j^i)\, \dot{q}_k^i(t) / d_i \quad (13)$$

The accelerations components can be calculated from Eq.(13), by replacing $\dot{x}_j$ by $\ddot{x}_j$ and $\dot{q}_k^i$ by $\ddot{q}_k^i$.

III-3 Calculation of the Polynamials representing The Motion on the Straight Line

Suppose that the motion will be carried out by two polynomials. The first by an acceleration equal to Ka, and during a distance ℓ_1^i, while the second by an acceleration equal to -Ka, Fig. 2. and during a distance equal to ℓ_2^i. It can be proved that

$$(V_i)^2 + 2\ell_1^i \text{ Ka} = (V_{i+1})^2 + 2\ell_2^i \text{ Ka} = (V_{im})^2 \quad (14)$$

There are two cases:
1- If $V_{im} \leqslant$ Kv
then it can be proved that

$$\ell_2^i - \ell_1^i = (V_{i+1})^2 - (V_i)^2 / 2 \text{ Ka} \quad (15)$$

and the coefficients of the polynomials can be calculated.

2- If $V_{im} >$ Kv

The motion will be represented by three polynomials the first during a distance equal to

$$\ell_1^i = (\text{Kv})^2 - (V_i)^2 / 2 \text{ Ka} \quad (16)$$

the third during a distance ℓ_3^i, where

$$\ell_3^i = (\text{Kv})^2 - (V_{i+1})^2 / 2 \text{ Ka} \quad (17)$$

the second by a constant velocity equal to Kv and during a distance

$$\ell_2^i = f_i - \ell_1^i - \ell_3^i \quad (18)$$

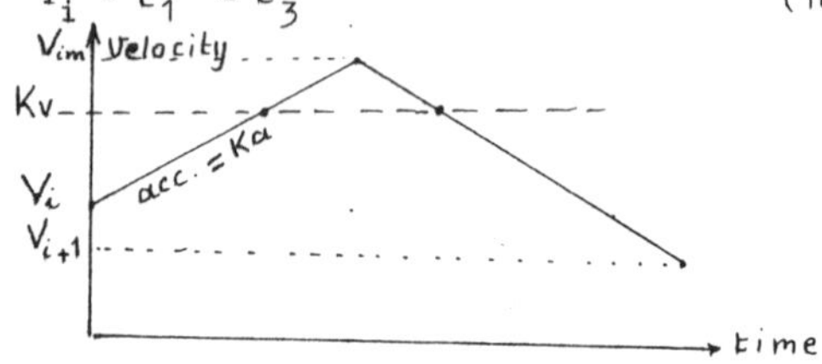

Figure 2

III-4 Calculation of the motion on the circular Arc C_i

To get the point $\underline{S}^i$ starting from point $\underline{P}^i$ along the circular arc C_i, an angle equal to $\pi - \theta_i$ must be travelled.(Fig.3). Let the angle travelled at time t be $\Phi(t)$, where t=0 is defined at the point $\underline{P}^i$. A system of coordinate R_i is defined as shown in Fig.3 such that the circle C_i will be in the $X_i^! Y_i^!$ plan. It can be proved that the components of the position, velocity, and acceleration referred to frame $R_i^!$ will be :

$$x'(t) = r_i \sin \Phi(t) \quad , \quad y'(t) = r_i \cos \Phi(t) \quad (19)$$

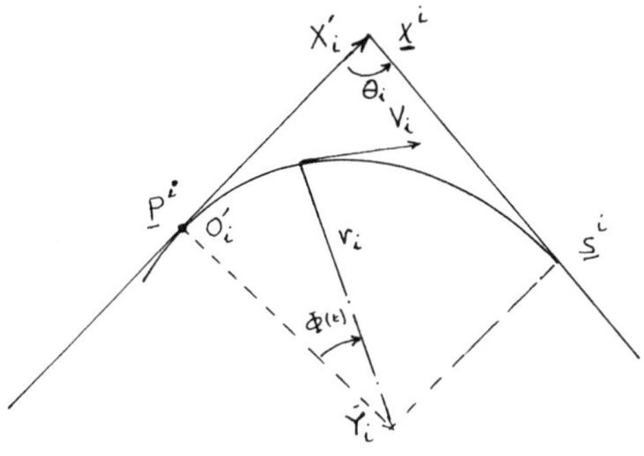

Figure 3

$$\dot{x}'(t) = V_i \cos \Phi(t) \quad , \dot{y}'(t) = -V_i \sin \Phi(t) \qquad (20)$$

$$\ddot{x}'(t) = -\alpha_i \sin \Phi(t) \quad , \ddot{y}'(t) = -\alpha_i \cos \Phi(t) \qquad (21)$$

The position,velocity and acceleration can be calculated with respect to the fixed frame by the use of the homogenous transformation matrix:

$$T'_0 = \begin{vmatrix} \underline{i}_i & \underline{j}_i & \underline{k}_i & P^i \\ 0 & 0 & 0 & 1 \end{vmatrix} \qquad (22)$$

where $\underline{i}_i$ is the unit vector in the direction of X'_i

$$\underline{i}_i = \begin{vmatrix} (x_1^i - x_1^{i-1}) / d_{i-1} \\ (x_2^i - x^{i-1}) / d_{i-1} \\ (x_3^i - x^{i-1}) / d_{i-1} \end{vmatrix} \qquad (23)$$

$$\underline{k}_i = \underline{i}_i \wedge \underline{i}_{i+1} \qquad (24)$$

$$\underline{j}_i = \underline{k}_i \wedge \underline{i}_i \qquad (25)$$

The time required to travel the circular arc C_i can be determined from the following equation

$$T_{ci} = (V_i / r_i) / (\pi - \theta_i) \qquad (26)$$

IV-APPLICATION

Consider the trajectory treated in[7],which consists of the following points:

$$\begin{aligned}
\underline{X}^1 &= | 20 \quad -20 \quad 10 |^T \text{ cm} \\
\underline{X}^2 &= | 30 \quad 0 \quad 2 |^T \text{ cm} \\
\underline{X}^3 &= | 30 \quad 20 \quad 2 |^T \text{ cm} \\
\underline{X}^4 &= | 30 \quad 20 \quad 10 |^T \text{ cm} \\
\underline{X}^5 &= | 10 \quad 20 \quad 10 |^T \text{ cm} \\
\underline{X}^6 &= | 10 \quad 20 \quad 2 |^T \text{ cm}
\end{aligned}$$

Ka = 10 cm/sec^2 , Kv = 40 cm/sec

The maximum error at point i is defined by:

$$e_{i,m} = \frac{1}{4} \min (d_{i-1} , d_i)$$

Table I summarises the result of the solution by the proposed algorithm.The total time of the trajectory is equal to 9.565 sec ,while in[7]the time is equal to 8.024 sec when using the linear programming and a preplanned trajectory and the time was equal to 10.019 when using an implicit method with an undetermined error.

V - CONCLUSION

This paper presents an algorithm for the generation of the trajectory between some specified points. The generated trajectory is calculated in the cartesian space.The calculation can be made on line ,two points ahead should be known.The trajectory is composed of straight lines connected by circular arcs,it passes close the given points with a selected error position. The time of the trajectory is seen to have a value between that of the method uses the linear programming and that of the method with implicit solution and relaxed error position.

REFERENCES

[1] "User's Guide to VAL,a Robot Programming and Control System",Unimate Inc.,1979.

[2] Latombe.J.C,Mazer.E,"Definition d'un Language de Programmation Pour La Robotique(LM),Analyse Fonctionnelle de L'interpreteur",Rapport de Recherche n° 197,I.M.A.G.,Grenoble,Mars 1980.

[3] Paul.R. "Robot Manipulators,Mathematics,Programming and Control",The MIT Press Series in Artificial Intelligence 1981.

[4] Binford.T.O,et al."Exploratory Study of Computer-Integrated Assembly Systems",Stanford A.I.Memo n° 2854.June 1977.

[5] Taylor.R.H.,"Planning and Execution of Straight Line Manipulator Trajectories",I.B.M,J.Res.Develop. vol.23,n° 4, july 1979,p.424 - 436.

[6] Khalil.W.,"Trajectories Calculation in the Joint Space of Robots",Advanced Software in Robotics,May 1983, Liege ,Belgium.

[7] Paul.R.,et al.,"Advanced Industrial Robot Control Systems",Purdue Univ.,TR-EE 78-25,1978.

[8] Luh.J.Y.S.,Lin.C.S,"Optimum Path Planning For Mechanical Manipulators",Trans.of the A.S.M.E.,J.of Dyn.Syst.Meas.and Control,vol.102,june 1981,p. 142-151.

i	1	2	3	4	5
d_i	23.74	20	8	20	8
θ_i		32.632°	90°	90°	90°
$e_{i,m}$	0	5	2	2	2
r_i	0	1.96	4.8	4.8	4.8
r'_i	0	1.96	4.0	4.0	4.8
V_i	0	4.145	6.3	6.3	6.95
T_{ci}	0	1.135	0.99	0.99	1.08
f_i	17.07	9.33	0	11.17	3.17
T_i	2.26	1.14	0	1.169	0.795

Table I

THE DEVELOPMENT OF MODULAR SIMULATION PROCEDURES FOR THE DESIGN OF TASK DEPENDENT INDUSTRIAL ROBOTS

R.C. Michelini, R. Molfino, G. Acaccia

Istituto di Meccanica Applicata

Università di Genova - (Italy)

INTRODUCTION

The introduction of the flexible mechanisation to improve the productivity for batch manufacturing, needs a complete redesign of the fabbrication systems (machines and procedures).

The traditional scheduling may engage the machine tools for times as lows as the 5%, so that the fabbrication cycle is mostly spent for operations not related with the conventional technological transformations.

Temptative solutions are sought analysing the parts, materials and tools flows, to organise fully or partially automated facturies with production cells endowed with special purpose machining centers and servoed robots, so to increase the production operations (machining, inspection, etc) as compared to the auxiliary ones (parts positioning, tools changes, etc), simultaneously rising the operative phases of the fabbrications processes, as compared to the selection, fatching, transport, storage, movimentations, etc phases.

The worthiness of integrated manufacturing systems depends on the plants costs and on the operation costs. Such costs have not, normally, to be compared with the related costs in the actual context (products and lay-outs); sensible improvements, indeed, could be expected with a parts redesign, that matches with the possibilities offered by alternative technological cycles. The simulation procedures for manufacturing industries deserves great attention, since they are a necessary performances evaluation tool, for setting the economical limits to the complexity of a production cell and to its mechanisation level. At the moment, in fact, the added costs of the digital hardware devices have a comparatively low incidence as comapred to the development costs of the functionally oriented machines, that have to be appointed, in order to make economical the automated factury solutions. Such situation hinders the feasability of well taylored integrated manufacturing systems, until sufficient experience will be available, for selecting the sophistication level of the machining, inspection, etc units and of the related servoed robots.

In the last decade these problems have started to deserve increased attention, within, both, industrial and academic institutions as it is exemplified by the studies [1] , to [5] more recently organised as "Nucleo di Progettazione meccanica applicata alla robotica industriale", of the Italian National Council of the Researches for developing special purpose control strategies for industrial robots [6] , to [10] .

In the present paper the main lines of the researches are recalled and few example results are presented specifically related to the investigation of the robots controlling strategies.

DEFINITION OF THE DESIGN MODULES

Industrial robots are functionally oriented machines having, to a convenient level, "intelligence" of the outside world they interact with, to perform, within given limits of autonomy, the tasks they have been assigned. The design is based on a set of simulation procedures, ranging into hierarchic subsets, that distinguish fixing the goals of each subsystem. The individual goals are the constrains emploied to condition the modules of different level, so that, as the project progresses, the amount of informations embedded as "system hypotheses" expands. The modularity of the design procedure is exploited, distinguishing the information representation and processing for each subsystem.

In the design procedure we distinguish:

- a parametrisation loop through which the selected (a priori) informations, embedded as system hypotheses for the modelling of the machine or of the task, can be reset and modified according to some performances criterion, as the simulation work progresses;
- an identification loop, through which the selection of the system hypotheses is reprocessed to match with the (a posteriori) informations, gathered as experimental data, while the machine is interacting with the environment to perform its prescribed task.

The existance of the identification loop shows that the design phases overlap the running phases of the robots. The fact characterises the functionally oriented machines, whose setting is fixed by the operative environment they have to match with; it is, indeed, the reason why the simulation procedures are greately exploited to face the CAD-CAM requirements of a partially or fully automated factory.

The modules of the parametrisation loop are grouped and organised in relation with the job they performs, for instance:

- selection of the robot configuration for given manipulating requirements and operating cycles;
- simulation of the robot dynamics for testing the operation functional and structural performances;
- setting of the robot control strategies, for given information loops on the tasks progression.

The same modules are further specialised by means of suitable application indicaters, such as:

- assembling robots, that corresponds to a rather sophisticated task, where dexterity, precision and speed have, mainly, to be compromised with price;
- heavy duty robots, that corresponds to low sophistication tasks, where manipulator compliance (sometimes stiction and backlash) and price have to be matched.

Once the application indicators are preset, the development of the selected operation robot moves through simulation sequences, that encompass the job indicators abovementioned. Then the design phases are continued, considering

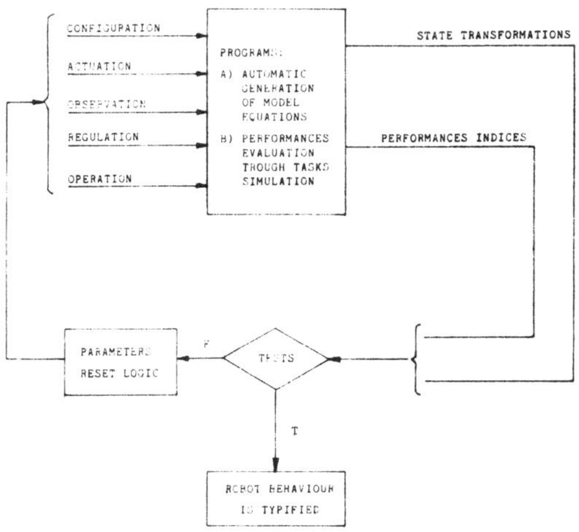

FIG 1 - FLOW CHART OF THE INTERACTIVE MODULE.

the requirements outcoming from embedding the robots into properly selected production lines.

The modules of the identification loop are developed to obtain the reconfiguration of the individual robot when inserted into a manufacturing line as a special purpose peripheral of the automated factory. In fact the task assignament to the individual robot has to be optimised, with criteria that depend on the overall parts planning and process scheduling and on the level of integration required for the manufacturing of the considered final product.

In correspondance a different series of computer simulation programs are developed, following the design job they perform, namely:

. selection of the process planning for given production cells and operating scheduling;
. simulation of the robots interaction for testing the task distribution and investigate alternative process plannings;
. setting the manufacturing process integration, for given product part planning.

The design modules of the first kind consider the generation in real time, of the steering requirements for an automated factory with assigned operating codes and operating times. The information system in the second kind design modules, needs cuncurrent programming capabilities for the robot (peripheral) computers, among them, and with the central supervising unities. With the third kind design modules, a general purpose simulation language is necessary, where both discrete state and continuous state blocks are present and their structure has to be identified in line as the oepration progresses and the tasks are rescheduled between the peripheral and the supervising units.

In the following section few results are summarised to shows the effectiviness of the procedure.

The explanatory example ranges at the level of selecting the control strategies once the robot configuration and the robot dynamics sophistication have been preset; then the corresponding design modules of the parametrisation loop, mainly concerned in the project of a system, having a given level of sophistication, to match with comparable performances fucntional requirements.

The presentation offers an heuristic explanation of the distinction introduced between the "parametrisation" and the "identification" loops.

The robots are assumed as task-dependent machines, simulated by models whose "parameters" need a setting and a trimming; the actual mechanisation is, thus, replaced by its corresponding task assignement and, accordingly, we move to the "identification" of suitable models that materialise the fucntional description, in keeping as central point, the robots operation performances.

PROGRAM ORGANISATION AND LAY-OUT

The computer aided procedure, ranging at the level of parametrisation loop, is sketched in the Fig 1. Five sets of parameters are considered, namely:

. configuration parameters, specifying the geometrical and the structural requirements
. actuation parameters, specifying the task performances and constraints
. observation parameters, specifying the sensing (direct and undirect) capabilities
. regulation parameters, specifying the steering and control strategies
. operation parameters, specifying the govern and the supervisory rules.

The performances indices depend on the tasks requirements; they are referred in terms of (static and dynamic) accuracy,

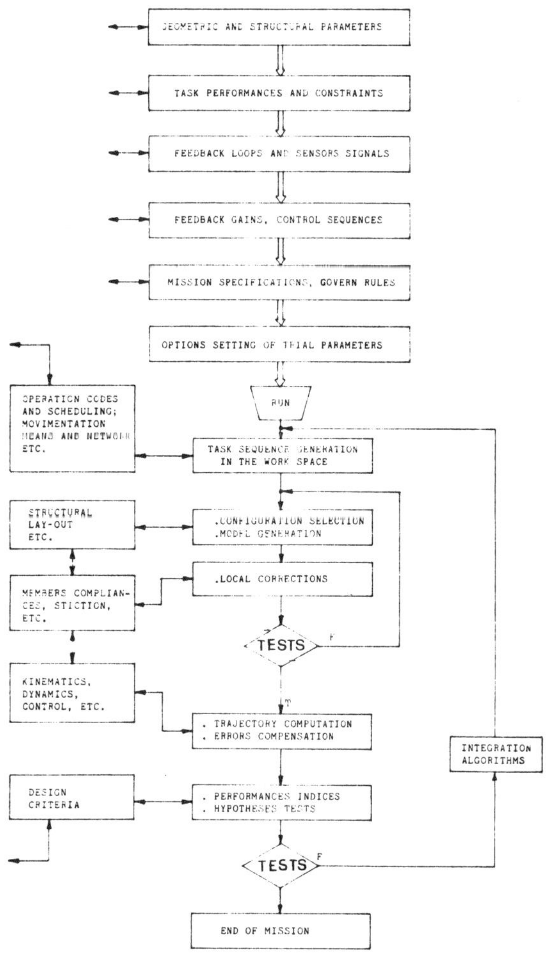

FIG 2 - BLOCK DIAGRAM OF AN EXAMPLE PROGRAM MODULE.

with additional figures concerned with the operation cycle time, the energy consumption, etc. The robot dynamics equations are automatically generated, once, given the configuration (say the types of members and joints), a work trajectory is selected. Among the options provided by the program, we recall:

. the step by step generation of the nonlinear dynamical couplings, etc
. the computation of the members compliance, the evaluation of the transmission stifness and backlash effects, etc.

The Fig 2 presents a simplified block diagram of the program module emploied for the setting of the robot control strategies. One should note the existance of a number of "service blocks": few of these provide an operational linking with the design modules of higher order (through, e.g., the fabbrication scheduling, movimentation network, etc requirements, or through performance criteria, and the like); few others close internal information loops (for updating the simulation computations with the optionals embedded by the CAD procedure). Typical options, at the regulation strategy level, are:

. position optimal dynamical-follower
. adaptive-control trajectory pursuer

and the feedback loops gains are reckoned optionally through:

. matrix inversion of local linearisations
. nonlinear uncoupling with dynamical compensation.

The availability of a rather sophisticated model is particularly important for the many applications that uses robots having none or very limited sensing capabilities. For these the investigation on the accuracy of the prescribed tasks is mainly intrusted on the simulations consistency, since the local errors are evaluated in terms of simulated observations.

EXAMPLE RESULTS AND CONCLUSIONS

The theoretical background used to develop the robot simulation program is presented in the references already quoted, or can be desumed from the current technical literature on the subject say e.g. [11] to [19] . Moreover the procedure for the automatic generation of the dynamical equations has been specifically developed [20] , due to the flexibility requirements of the CAD modules. New options and alternative service modules are considered to

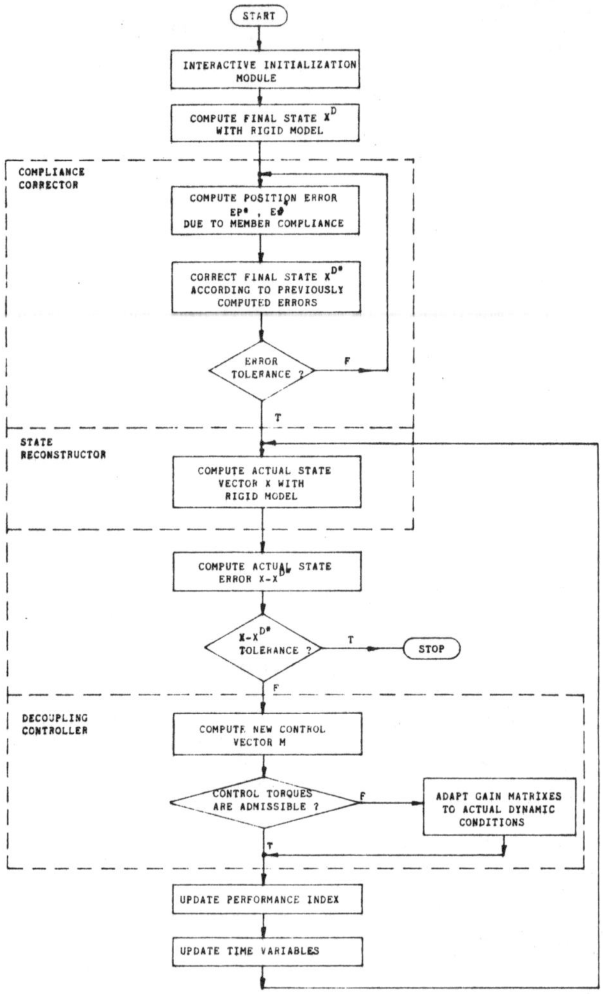

FIG 4 - BLOCK DIAGRAM OF THE SIMULATION PROGRAM.

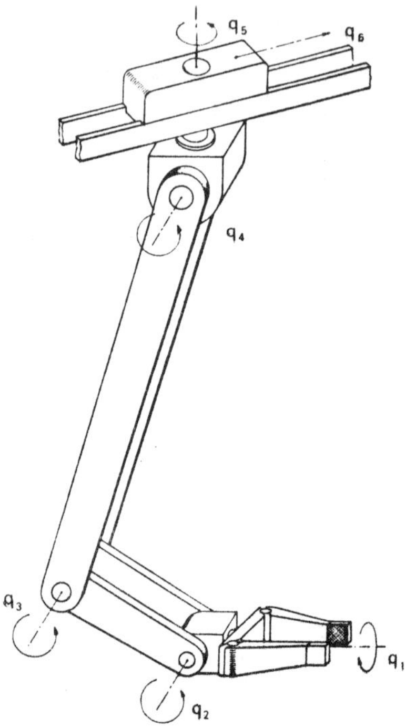

Fig. 3 - ROBOT ARM CONFIGURATION

expand the program and cover the applications of practical interest as they are prospected. This expansions reach the CAM requirements where special purposes design interactive procedures have been developed [21] to [23].

The industrial robots are essentially operational machines of the automated factury. However the universal and perfect machine (say: a robot endowed with expanded intelligence) may yield to little benefit, if the production flow is not properly reorganised. The factory reorganisation is the central point to be investigated; the tasks assignement and performances specification to the robots ought be taylored to the individual flexibility needs.

We may consider as example application the service manipulator, sketched in the Fig 3. The configuration corresponds to a six degrees of freedom robot, with the additional mobility of the work tool or grip. The Fig 4 presents the essential block diagram of the program modules involved in the design procedure. Terminal sensors are, whenever possible, avoided for heavy duty robots, due to reliability constraints (and costs limitations). Then the control strategies have in charge the position steering accuracy. The simulations results make straightley comparable different solutions, that distinguish for numbers of design possibilities as the following:

- structural data (say: geometrical parameters; centers of gravity position; members masses: members mass second moments; members densities, compliance computation procedures; etc);
- actuation data (actuators types: step motors, d.c. motors, hydraulic motors, etc; actuation specifications: max torques, transmission ratios;
- observation data (sensors options: ideal technical behavior, noise figures, replacements with indirect measurements, non existance; sensor specifications, if any, etc);
- regulation data (controllers options: observers with nonlinearities total compensation, with partial decoupling, adaptive control via local linearisations etc; controllers specifications: gains, stability margin, etc);

- operation data (mission type: absolute positioning, threshold positioning, trajectory follower, etc; loading requirements: tools weight, work forces etc; accuracy performances: initial and final positioning, time constant, settling time, etc; error weighting factors and performances indices for each particular operation, etc).

The design procedures obtained through simulations apply for the ingenuity of the engineer and, as a rule of thumb, experience is better help than a theoretical criterion. The kind of the outputs that may be obtained in the present application, are exemplified by the Fig 5 and 6. The first gives the manipulation tip coordinates versus time for a given positioning task, and different controling or correction condition. The operation strategies have been splitted previously setting the local station and orientation (say: two degrees of freedom out of six), then controling in line the tip position and orientation (say: the left four degrees of freedom). The Fig 6 gives the corresponding relative angular displacements versus time for a set of conditions; say: different gain matrix, phase margin (mantained or not with torque motor saturation, etc). The compliance corrections are particularly noteworthy in the example application, so that the simulation studies have proved to be an effective help in the design phase of such service robots.

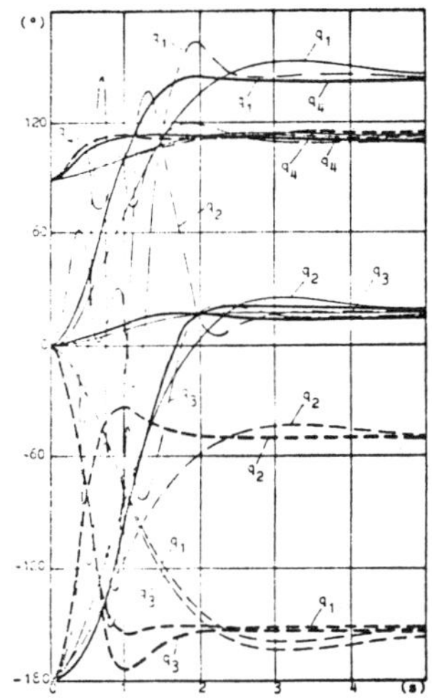

FIG 5 - CONTROL TYPICAL OUTPUTS.

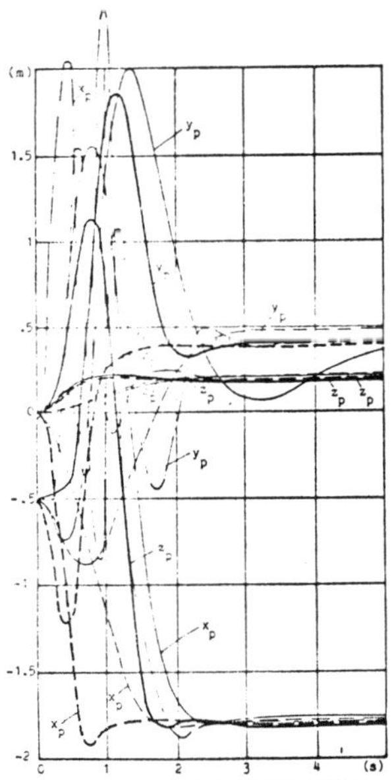

FIG 6 - TASK VARIABLES TYPICAL OUTPUTS.

REFERENCES

[1] Michelini R.C. "Meccanizzazione dei processi di assiemaggio con l'impiego di robot", Convegno sulle Ricerche di Robotica Industriale, CISM, Udine, 18-19 Aprile 1974.

[2] Michelini R.C. "Linee di sviluppo di robot atti ad assiemare gruppi meccanici", Automazione e Strumentazione, Vol. 23, n°2, 1975.

[3] Michelini R.C. "La meccanizzazione flessibile nell'evoluzione dei sistemi di produzione", Ingegneria Meccanica, n°5, Maggio 1978.

[4] R.C. Michelini; P.L. Polledro; C. Marcantoni-Taddei, "Position steering of industrial robots by statistical controllers", VIII International Symposium on Industrial Robots, May 1978, Stuttgart.

[5] Michelini R.C. "I sistemi di manipolazione: classificazione e linee di sviluppo", in, Stato attuale ed evoluzione della Robotica industriale, AIRI, Roma, 1979.

[6] C.F. Marcantoni-Taddei; R.M. Molfino; G.M. Acaccia "On the design of high performances industrial manipulator by on line modeling and simulation of the system dynamics", I Symposium IASTED on Modelling, Identification and Control, Davos, February 1981.

[7] R. Molfino Cepolina, G.M. Acaccia, C. Marcantoni-Taddei "Development and experimentation of a high accuracy position control for service manipulators", IV CISM-IFToMM Symposium on theory and practice of robots and manipulators, Warsaw, September 1981.

[8] R.M. Molfino, G.B. Rossi, G.M. Acaccia "Controllo di robot industriali con compensazione delle nonlinearità del modello dinamico", VI Congresso Nazionale AIMETA, Genova, Ottobre 1982.

[9] R.M. Molfino, G.B Rossi, G.M. Acaccia "Controllo di robot industriali con linearizzazione locale e calcolo in linea dei guadagni ottimali", VI Congresso Nazionale AIMETA, Genova, Ottobre 1982.

[10] R.M. Molfino, C. Marcantoni-Taddei, G.M. Acaccia "Progettazione di robot industriali per impieghi gravosi" Progettare n°33, Febbraio 1983.

[11] Dubowsky, D.T. Forges "The application of model-referenced adaptive control to robotic manipulators" ASME, Journal of Dynamic Systems, Measurement and Control, Spet. 1979.

[12] Luh J.Y.S., Walker M.W., Paul R.P.C. "On-line computational scheme for mechanical manipulators" Journal of Dynamics Systems, Measurements, and Control, June 1980.

[13] Vukobratovic M. Cvetkovic V. "Concept of manipulation robots simulator" IV CISM IFToMM Symposium on theory and practice of robots and manipulators, Warsaw, September 1981.

[14] Vukobratovic M., Potkonjak V., Hristic D. "Contribution to the computer-aided design of industrial manipulators" IV CISM IFToMM Symposium on theory and practice of robots and manipulators, Warsaw, September 1981.

[15] Paul R.P. "Robot manipulators: mathematics, programming, and control" MIT Press, 1981.

[16] Vukobratovic M., Kircanski Nevrad "Computer-oriented Method for linearisation of Dynamic Models of Active Spatial Mechanisms" Mechanism and Machine Theory, Vol.17, n°1, 1982.

[17] Vukobratovic M., Stokic D. "Dynamics of Manipulation Robots" Theory and Application, Springer Verlag (1982).

[18] Vukobratovic M., Stokic D. "Control of Manipulation Robots" Theory and Application, Springer Verlag (1982).

[19] The Japan industrial robot association "The robotics industry of Japan, today and tomorrow" Fuji Corporation 1982.

[20] R.M. Molfino, G.M. Acaccia, R.C. Michelini "Simulatore di robot industriali per l'assemblaggio Quaderno IMAGE, SSP.05/R.628, Marzo 1983.

[21] R.M. Cepolina Molfino, L. Torri "Digital simulation for a computed controlled production center", Simulation 80, International Symposium, Interlaken, 25-27 June 1980.

[22] R.M. Cepolina Molfino, G. Acaccia "Progettazione interattiva di impianti flessibili di produzione" VII Convegno Nazionale di Impiantistica, S.Margherita Ligure, 1980.

[23] G.M. Acaccia, R. Cepolina Molfino, C. Marcantoni-Taddei "Tecniche di simulazione di sistemi flessibili di produzione" Pixel n°2, 1981.

PERFORMANCE EVALUATION OF AN AUTOMATION SYSTEM I - METHODOLOGY

ANTTI AIMO
Valmet Corp., Instrument Works
Tampere, Finland

HEIKKI N. KOIVO
Tampere University of Technology
Tampere, Finland

ANTERO HÄLIKKÄ
Valmet Corp., Instrument Works
Tampere, Finland

Abstract. The paper presents a new method for a performance evaluation of an automation system. First general properties of digital automation systems are defined to limit the study. The main body of the paper discusses new performance requirements of an automation system. The requirements form multi-level hierarchical structure that is presented with block diagrams. Altogether 95 items of concern to evaluate the performance of an automation system are determined.

1. INTRODUCTION

The number of digital automation systems in process industry has increased strongly during the last few years (Borslei and Pavlik, 1980; Büsing, 1980; Dewaele, 1982; Enslow, 1978; Weitzman, 1980). Comparative evaluation of the respective performance characteristics of different automation systems has become an important factor in the planning of process automation and in the integration of the automation system as part of the production plant.

Existing methods of evaluating the performance of an automation system are quite scarce (Dobrowolski, 1981; Golemanov, et.al., 1981; Heidlich and Prütz, 1982). These methods do not treat automation systems with sufficient totality. Performance evaluation tends to center around evaluation of minor technical details. In this paper a new method is developed to eliminate this flaw. This paper is supported by a second one (Aimo, Koivo, and Hälikkä, 1983 b) where a real industrial process representing an average of given data of automated processes is determined. The process can be used as a "benchmark" process. In the third paper (Aimo, Koivo, and Hälikkä, 1983 c) the developed method is applied to the determined average process.

Here general properties of automation systems are first defined in order to set limits for the study. Then a survey of existing methods for performance evaluation of automation systems is given. The main body of the paper discusses new performance requirements of an automation system. A block diagram representation of many hierarchical levels is chosen. A performance evaluation method is then an easy corollary.

2. PROPERTIES OF AUTOMATION SYSTEMS

Several excellent treatments of automation systems exist in the literature. These include (Borsi and Pavlik, 1980; Büsing , 1980; Dewaele, 1982; Früh, 1980; Weitzman, 1980) discussing the general structure of process automation systems and the advantages of distributed systems. Bux (1981) and Wittie (1981) consider the distributed system structure from the information transmission point of view and Bux also estimates the performance of different network structures. Schmidt and Sendler (1980) treat the factors effecting the system reliability. Here only the main properties of automation systems are given in order to set limits for the study.

Digital automation system in this study satisfy the following six requirements:

1. An automation system must be suitable for management of different industrial process.
2. The operation with an automation system is performed in a separate control room where the process operation is controlled with the aid of a videomonitor and associated control device.
3. An automation system measures and monitors the process variables, and regulates the process.
4. An automation system is distributed. It satisfies the following requirements (Enslow, 1978):
   - It contains both physical and logical recources that can be combined to different tasks dynamically.
   - The resources are distributed physically and they communicate with each other via a communication network.
   - It contains a high level control system, which combines and coordinates the operation of the distributed components.
   - The system is transparent so that different tasks can only be called by their names.
   - Members are autonomous so that they can act independently if communication is lost.
5. An applicable system is created for an automation system according to the needs of the application.

6. Sensors and actuators need not be products of the automation system manufacturer.

The function required by an automation system can be divided into four categories: process (1) supervision, (2) control, (3) regulation, and (4) measurement functions.

3. REVIEW OF PERFORMANCE EVALUATION METHODS FOR AN AUTOMATION SYSTEM

Methods for performance evaluation of automation systems used in process industry are based on estimating somehow the internal structure of an automation system. The methods use a weighing scheme, which is useful but the things weighed have been partly indeterminate and nonunique.

In Golemanov et.al. (1981) a uniform method for comparing automation systems has been developed. In this method an automation system is divided into a set of operation units describing physical structure characteristics of the system. An automation system is divided into operation units as follows:

- control station
- redundant control station
- process interface unit
- programmable logic unit
- information transmission controller
- control room (3 different models)

Each operation unit is given a number of capacity parameters.

The listed parameters are determined as absolute values after which they are normalized so that the largest value gets value 1 and others are scaled with the largest value. The proportional values are then added for each system and the automation system having the best total is given number one ranking.

The comparison method has been used in practice to test characteristics of ten automation systems well-known in Europe. The end result of the test can be considered in many ways inconclusive, since the basis of the test method is not particularly successful. Information transmission and physical characteristics of automation systems are chosen as starting points. The automation system is treated as a separate entity when it should be considered as an integral part of the control hierarchy of a production plant, and as a tool for a person managing the process. Many parameters are taken into account that do not have much practical significance as far as the user or the controlled process is concerned. An advantage of this method is that no limitations are set on the characteristics of an automation system in contrast to those suggested in section 2 of this paper.

Both methods reviewed are partially based on performance values of separate operation units. An automation system is not viewed as entity, as a process management tool. This results in weighing technical, often less significant details too much in comparing automation system with each other.

Schmidt and Sendler (1980) consider selection of a distributed automation system. They give a detailed classification of automation system characteristics. This may be used to clarify the important characteristics for a purchaser. A mutual comparison of automation systems is also possible, but the end result is practically the same as that of the second method presented above. Dobrowolski's selection table (1981) gives a brief list of automation system characteristics. Comparison of automation system on the basis of this table is practically impossible.

4. REQUIREMENTS AND PERFORMANCE EVALUATION OF AN AUTOMATION SYSTEM

4.1 General

There are a great number of requirements for an automation system at different levels of operation. The starting point must be the requirements set by the production plant management about the most efficient and economical use of the production process. This basic philosophy is followed in the sequel.

Block diagrams are used at different hierarchical levels beginning from the management requirements and ending in the detailed study of an automation system. The end result forms a group of evaluation items for the functions and characteristics of an automation system. A performance evaluation method is based on this.

Plant management wants (A) the process to be controllable under all process situations and (B) the usability degree to be as high as possible. Fig. 1 shows the requirements to be satisfied by an automation system at this level.

Requirements (A) about the optimal process controllability can be further divided to the following categories for an automation system shown in Fig. 2: (A.1) production and controllability, (A.2) operational characteristics, (A.3) process controll unit, and (A.4) information transmission ability.

Definition 1. A process control unit is a unit that can act independently if needed. It measures, controls, regulates, and monitors process variables according to a programmed application task. It also takes into account in its operation instructions arriving from outside the unit.

Here only the (A.2) branch is followed to indicate the line of reasoning of the study. Further details of the complete structure may be found in (Aimo, Koivo, and Hälikkä, 1981 a).

4.2 Operational requirements of an automation system (A.2)

The operational requirements set by the operators for the usability of an automation system are presented in Fig. 3.

Definition 2. The extent of the process is determined by the limit which consists of either the maximum number of interface connections or the maximum size of application software of the automation system built for the process automation depending on which limit is reached first.

Ergonomic requirements (A.2.a) are not detailed since the system user can influence the ergonomic level with his own actions. An automation system manufacturer can only effect the monitor and control device ergonomy, and restraining the system noise level.

4.2.1 Control of disturbance situations (A.2.1)

A disturbance situation occurs when the product quality is in danger of becoming poorer. In the worst case the process must be stopped bacause of the disturbance. An automation system must be able to recognize the cause and the object of the disturbance. Here the cause of the disturbance is assumed to be in the process.

4.2.2 Efficiency of control room operation (A.2.2)

The requirements for efficient control room operation can be divided according to Fig. 4. Here a definition is needed to clarify monitoring and control functions.

Definition 4. Process monitoring occurs with the aid of a monitor and the control of the process operation with a control device.

Both A.2.2.1 (Fig.5) and A.2.2.2 can be divided to several sublevels. First define the process data and the process operation parameters.

Definition 5. Measurements, setpoints, and controls are process data.

Definition 6. Regulation, control, and alarm parameters are process operation parameters.

Definition 7. A monitor together with a control device belonging to it forms an operation point.

The second level requirements A.2.2.1.1-4 in Fig. 5 can further be divided as seen in e.g. Fig:s 6-7.

4.3 Performance evaluation

Once the overall requirements are determined, performance of automation systems can be evaluated at the points of interest. A weighing scheme can be used to rank the automation systems at different levels and then to give an overall ranking. There are altogether 95 items of concern so that the method requires good knowledge of the process and takes considerable amount of time. It can, however, be used at different hierarchical levels to achieve a quicker, but not as thorough, overall evaluation of automation systems.

A cursory example of the method is given in Aimo, Koivo, and Hälikkä (1983 c).

5. CONCLUSION

A new method for a performance evaluation of an automation system is developed. It is based on the requirements set by the plant management about the most efficient and economical use of the production process. The requirements form a multi-level hierarchical structure that is presented with blockdiagrams. Altogether 95 items to evaluate the performance of an automation system are determined.

REFERENCES

AIMO, A., KOIVO, H.N., and HÄLIKKÄ, A.: Performance evaluation of an automation system I - Methodology. Report 9-83, Control Engineering, Tampere University of Technology, Tampere, Finland (1983 a).

AIMO, A., KOIVO, H.N., and HÄLIKKÄ, A.: Performance evaluation of an automation system II - Determination of a benchmark process. Report 10-83, Control Engineering, Tampere University of Technology, Tampere, Finland (1983 b).

AIMO, A., KOIVO, H.N., and HÄLIKKÄ, A.: Performance evaluation of an automation system III - Application of methodology to a benchmark process, Report 11-83, Control Engineering, Tampere University of Technology, Tampere, Finland (1983 c).

BORSI, L., and PAVLIK, E.: The Concepts and Structures of Distributed Process Automation Systems. Process Automation 1980:2, s. 63-70.

BUX, W.: Local-Area Subnetworks: A Performance Comparison. IEEE Transactions on Communications COM-29 (1981) 10, s. 1465-1473.

BÜSING, W.: Distributed Process Automation Systems: Requirements and Interfacing. Process Automation 1980:2, s. 71-76

DEWAELE, W.: Structures of digital process automation systems. Journal A 23 (1982).

DOBROWOLSKI, M.: Guide to Selecting Distributed Digital Control Systems. InTech 1981, June, s. 45-56.

ENSLOW, P.H. Jr.: What is "distributed" Data processing system? Computer, Januar, 1978.

FRÜH, K.F.: "Optimal Structure" of Distributed Process Automation Systems. Process Automation 1980: 1, s. 1-2.

GOLEMANOV, L.A., ANGELOV, V., and NIKIFOROV, G. & BAHAROV, B.: A unified approach in the analysis and design of integrated production management systems. Sofia C, Bulgaria 1981. Ministry of the Chemical Industry. 12 s.

HEIDLICH, S. and PRUTZ, G.: Selection Criteria for the Use of Digital Distributed Control Systems. Process automation 1982:2, s. 66-74.

SCHMIDT, G. and SENDLER, W.: A Failure tolerant Multi-microcomputer Controller for Process Control Applications. Process Automation 1980:2, s. 77-84.

WEITZMAN, C.: Distributed Micro/Minicomputer Systems. New York, N.Y., USA, Prentice-Hall, Inc., 1980. 400 s.

WITTIE, L.D.: Communication Structures for Large Network of Microcomputers. IEEE Transactions on Computers C-30 (1981) 4, s. 264-273.

Figure titles.

Fig. 1. Requirements set by the plant manager.

Fig. 2. Requirements of optimal process controllability.

Fig. 3. Operational requirements and their solutions.

Fig. 4. Rough division of requirements for efficient control room operation and their solution.

Fig. 5. Requirements for monitoring and control of process operation and solutions suitable for these.

Fig. 6. Requirements of process data monitoring and their solutions. Graphical representation in A.2.2.f means for example a column display.

Fig. 7. Requirements of managing process operation with programmed control and their solution.

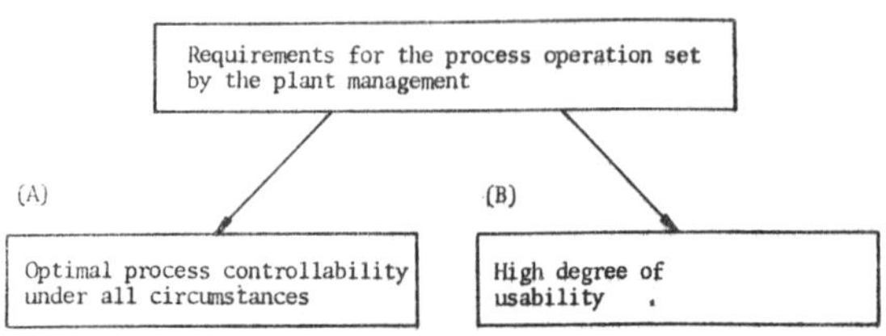

Fig. 1.

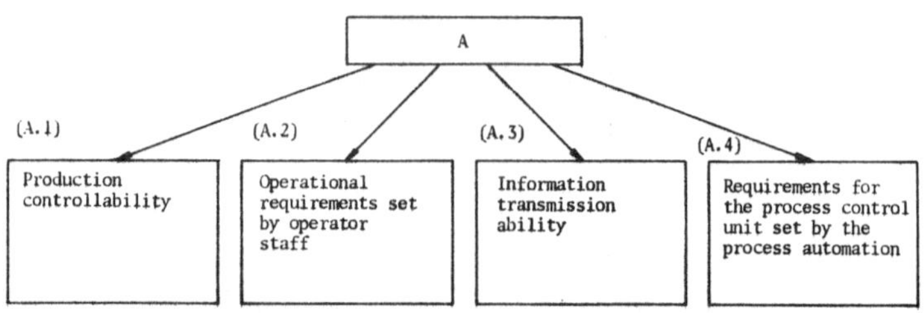

Fig. 2.

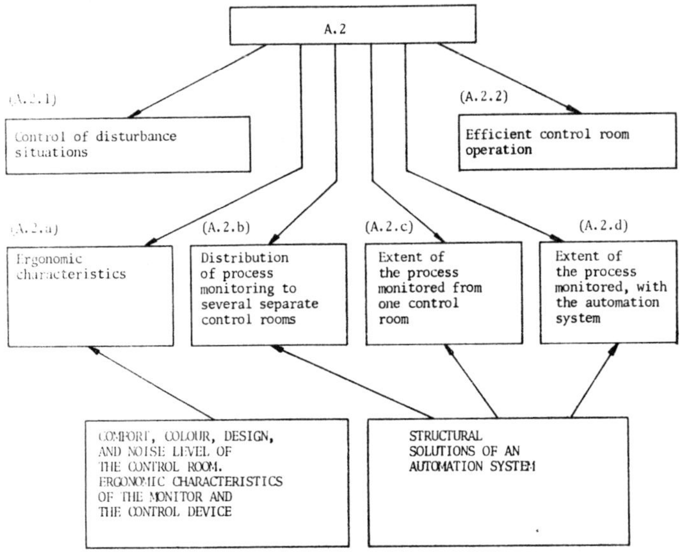

Fig. 3.

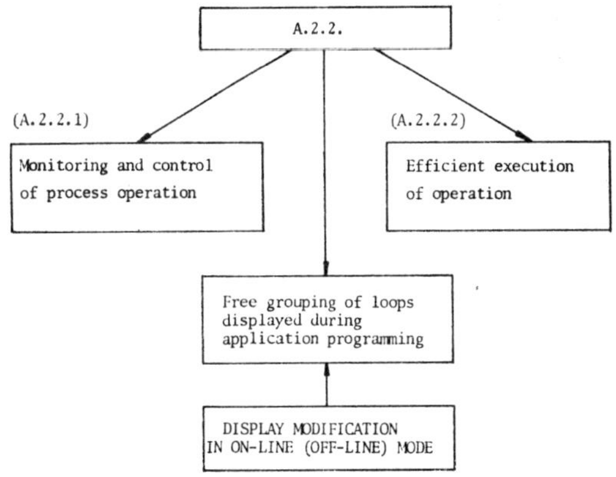

Fig. 4.

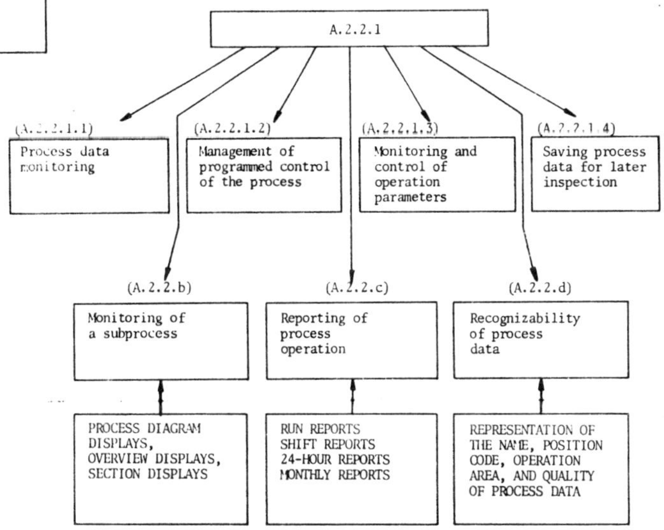

Fig. 5.

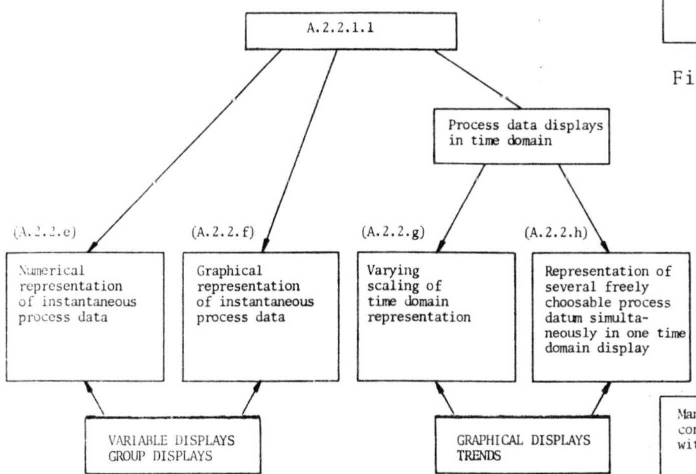

Fig. 6.

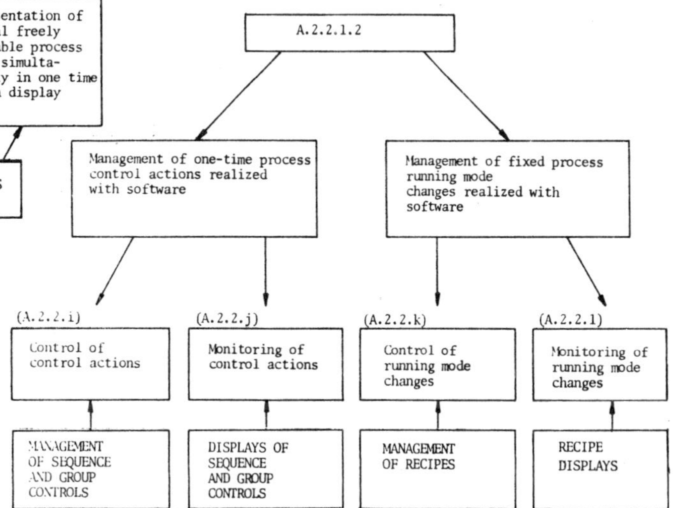

Fig. 7.

USE OF A MEASUREMENT AND CONTROL PROCESSOR IN GROUNDWOOD MILL CONTROL

Jarmo Mustapää
Kajaani Oy
Kajaani, Finland

Heikki N. Koivo
Tampere University of Technology
Tampere, Finland

Juha Kalliokoski
Kajaani Oy
Kajaani, Finland

Abstract. Computer control of processes has developed more and more into distributed and decentralized direction. The system functions are distributed to different levels. In this paper the possibility of using a commercial measurement and control processor at a lower level of a computer control system is investigated. Execution speed tests are performed with the built measurement and control system to determine its feasability in the groundwood mill control. The biggest test system includes all the functions needed to control ten grinders. The system then has to handle 100 analog inputs, 80 digital outputs, 20 computing algorithms, 30 control loops, 70 digital outputs, and 20 analog outputs.

On the basis of the test results the studied measurement and control processor can well be used as an efficient interface unit between the main process control computer and analog instrumentation.

1. INTRODUCTION

Computer process control has developed more and more into distributed and decentralized direction. This means that the main computer does not perform all the control and measurement functions by itself. The system functions are distributed to different levels. At a lower level a processor specializing in taking and analyzing measurements and processing them for control may be used. In this research the use of a commercial measurement and control processor, HP 2250, in process control is studied.

In particular, the possibility of using the HP 2250 as part of a computer control system for controlling a groundwood mill is investigated. The HP 2250 processor is mainly used in controlling grinders. The paper first gives a brief description of a groundwood mill and its control system. Based on this and taking into account the high level control system a task selection for the HP 2250 is specified. After the task definition the system software, including both the processor and the needed higher level computer software, is described. The paper concludes with execution speed tests to determine the feasibility of the built measurement and control processor HP 2250 in the groundwood mill control.

2. PROCESS AND ITS CONTROL SYSTEM

Groundwood is made using a rotating grinding stone fed with logs about one meter long. A groundwood line consists of machinery for grinding the wood and sorting the resulting groundwood. The grinders in a groundwood line can be divided into groups, each containing two or more machines. A grinder is shown in Fig. 1.

The logs are first cut to the right size (0.8...1.5 m) then they are peeled, i.e. their bark is removed. The pealed logs are then distributed for the grinders, in which wood is ground into pulp by pressing the logs against the wet grinding stowe.

All current groundwood grinder control strategies are realized by controlling the grinding pressure (Manner, 1978). Typical instrumentation and control circuits of a grinder are shown in Fig. 2.

A computer system for a groundwood mill is displayed in Fig. 3. The system includes an intelligent sensor for measuring the piston-speed. The computer system has an HP 100L computer with a 16 bit processor and 512 kword memory. It is at this level where the measurement and control processor is located if analog instrumentation is used.

In the lower level control system the following measurements are taken in each pocket and for each stone for the primary control loop:

- the speed of the piston (production) and its position
- power used by the electric motor

For the higher level control system the measurements are:

- the pulp temperature in the basin under the grinders
- the flow rate and the temperature of the shower waters
- the grinding pressure

The primary control strategy used for an individual grinder at the lower level control system is the speed control. This results in a uniform pulp quality, since a reliable speed measurement device exists.

3. TASK DEFINITION

The goal of the task definition is to build as universal and independent measurement and control system as possible. At the lower level there are a great number of measurements and the measurements have to be performed at short intervals. At the higher level both report and alarm formats should be taken into account.

The measurement and control processor is used only with analog instrumentation. An alternative to this is a totally digital instrumentation system. The processor must therefore carry out those tasks of the digital instrumentation system that cannot be realized with analog instrumentation.

Based on the execution speed tests FORTRAN and MCL were chosen as programming lanquages. FORTRAN is used for all signal analysis such as limit checks, filtering, computational algorithms, etc. In performing the measurement and control actions the language is MCL. Mostly integer arithmetic is used in the FORTRAN software.

The software consists of the following tasks:

- measurement task
- computing task
- control task
- output task
- digital signal task

Since the tasks are standard (e.g. Bibbero. 1977), no details are given. These can be found in Mustapää, Koivo, and Kalliokoski (1983).

4. SYSTEM SOFTWARE

The software can be divided into two parts: (1) the measurement and control software of the HP 2250 and (2) the higher level computer software for loading parameters and programs into the HP2250 processor.

In software design structural and modular programming together with TOP-DOWN method (McGowan and Kelly, 1975) are used. Different subtasks are realized as independent programs which are tested in small entities. Fixed parameters and changing information are completely separated from each other. The parameter tables related to different subtasks are also separated into separate buffers. The information transmission is limited so that only the minimum swich, computing, and control programs can fetch information from one location and take it to another location.

The software structure is shown in Fig. 4. Variables mean the MCL-language variables. All the information to be updated or needed in more than one program module is collected in this vector. In this way the higher level computer can read the desired number of variables starting from any chosen location of the vector. When buffers are read, the elements must always be read starting from the first element. The same holds when transmitting information into the processor variables and buffers.

The software consists of one continuously running program loop. The timing of different functions is performed at the end of the loop. The task execution always progresses without interruption from module 1 up to module 12.

5. EXECUTION SPEED TESTS

5.1. Test system signals

The purpose of the execution speed test is to determine how many grinding stones can be controlled and supervised with one HP 2250 processor. The tests were designed so that each grinding stone (two pockets/ grinding stone) the following signals are required:

1. Analog inputs (10 signals)
2. Analog outputs (2 signals)
3. Digital inputs (8 signals)
4. Digital outputs (7 signals)
5. Control signals (3 signals)
6. Computational algorithms (2 signals)
7. Minimum selectors (2 signals)

5.2. Test results

Altogether five different systems were generated for the execution speed tests. The rest results are collected in Table 1.

Program space in Table 1 means the memory space taken up by all programs. Memory used is obtained by adding the memory space required by the programs to that required by the buffer tables. The remaining memory will be slightly smaller than indicated in the table, since the device needs certain memory space for each interface card. During the test only three interface cards are connected to the processor.

The execution times in Table 1 cannot be considered quite exact, but they are close and give qualitatively right results. The exact execution times are impossible to determine, since the program proceeds along different paths during different execution cycles depending on the measurement and parameter values, for example, because the alarm limit is exceeded. The test results can, however, be treated as sufficiently reliable and exact for estimating the utility of the processor.

The analog measurements are carried out with a 16 channels ADC card in the test. In an industrial environment a separated extension card (relay card) must be used. Then a channel measurement takes one millisecond longer than with the card used in the test. This means that if e.g. 100 analog measurements are taken then 100 ms must be added to the test results. In addition the following execution times have to be taken into account when estimating the final execution speed: the execution of the exchange program (4 ms), an alarm acknowledgement (33 ms), and calibration of analog input cards (about 340 ms). So altogether about 400 ms must be added to the test results to assure that the basic cycle is not exeeded during calbration or an alarm acknowledgement.

The test results in Table 1 are divided according to the number of stones. Table 2 on the other hand summarizes the test results when the number of signals for each test system are displayed. This results in a better view of the system execution speed, since the realization possibilities of different systems can be computed.

The execution times in Table 2 are obtained so that all operations are carried out during every basic, two-second cycle. There are only five measurements for each stone that must be taken during the basic cycle:

- the piston speeds (2 measurements)
- the motor power
- the feedback signals of the analog controller setpoints (2 signals)

For all other measurements a longer execution interval (of 10 sec.) suffices.

5.3. System usability

A basic assumption in the study is that the grinding pressure control loop signals requiring great speed are realized with analog controllers. In this way the usability of the groundwood mill also improves since in the case of a computer fault the process can be driven with analoque controllers. The stated assumption implies that two seconds is sufficient for the basic system cycle. From the execution speed tests it is then clear that the processor can handle all such system configurations that fit in the processor.

The measurement and control processor can also handle the pressure control loops but in that case the basic cycle must be shortened to one second in order to make the pressure control loop fast enough. The number of the control loops for each grinding stone increases to five. The program size does not increase. The execution time for each control loop is about 5 ms.

If all pressure control loops are realized digitally with the HP 2250 and the basic system cycle is one sec., the system execution speed becomes the limiting factor as far as the system size is concerned. Based on the information in Table 1 and 2, the largest system size in this case is six stones.

If more than one HP 2250 processor are connected to the higher level computer, it too might encounter problems with the execution speed or memory size. Transferring the average calculations from the higher level computer to the processor, significant savings in execution time and memory space of the higher level computer can be achieved.

The main body of the work was carried out during 1981. At that time the pulse output card was only tentatively specified so it was decided to leave all the pulse output programs out of testing. The process control system often uses pulse outputs in the control of motors and valves, so that attachment of the pulse output card to the system is necessary. The additional memory needed lowers the number of the grinding stones to be controlled by the processor by one. The execution times for pulse output programs are somewhat longer than those for the analogue output programs, because the number of parameters increase. Thus the exact execution times given in Tables 1 and 2 have to be slightly modified in the case that analog controllers and actuators are controlled with pulse outputs.

6. CONCLUSION

The feasability of a measurement and control processor in a groundwood mill control system has been investigated. First an appropriate task selection for the processor is made. The system software is then designed. The chosen measurement and control strategies have been realized with the HP 2250 processor.

Execution speed tests have been performed with the built system using five different test system configurations of grinders. Both the memory size and the execution time are found to be sufficient in groundwood mill control, if the grinding pressure control loop is realized with analog controllers. If also the pressure control loops have to be realized then the processor can handle up to six stones.

Since the execution speed tests were carried out at a very early stage, the pulse output card was only tentatively specified. The pulse output programs were therefore left out. The process control system often uses pulse outputs in the control of motors and valves so the attachment of the pulse output card to the system will be necessary. The execution times in Tables 1 and 2 would be slightly modified in the case that analog controllers and actuators are controlled with pulse outputs.

The overall conclusion is that a measurement and control processor will be quite useful in a groundwood mill control system. It can be used as an efficient interface unit between the main process control computer and analogue instrumentation.

REFERENCES

BIBBERO, R.J. (1977). Microprocessors in Instruments and Control. Johan Wiley, New York, New York.

HP 2250 (1981 a). Hardware Reference Manual, Hewlett Packard Co., Roseville, Calif.

HP 2250 (1981 b). Programmer's Manual. Hewlett Packard co., Roseville, Calif.

MANNER, H. (1978). Selection of a control strategy for the grinding process. Paperi ja Puu 60, 633-650.

McGOWAN, C. and KELLY, J. (1975) Top-Down Structured Programming Techniques, Wiley, New York.

MUSTAPÄÄ, J., KOIVO, H.N., and KALLIOKOSKI, J. (1983). Use of a measurement and control processor in groundwood mill control. Report 12-83, Control Eng. Lab., Tampere University of Technology, Tampere, Finland.

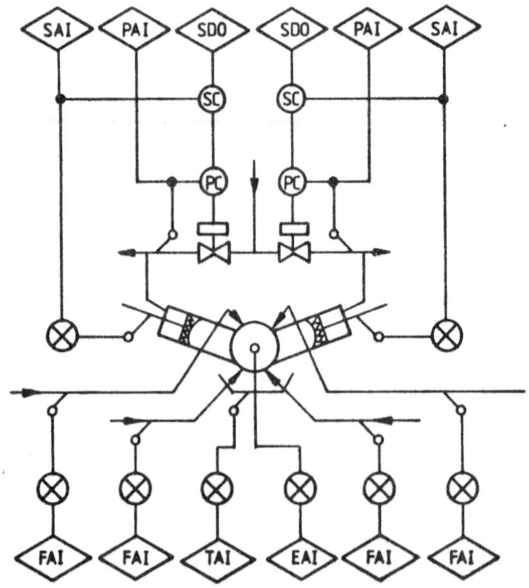

Fig. 2. Instrumentation and control loops of a grinder. Notation used: SAI = piston speed measurement, SC = piston speed control, SDO = piston speed setpoint, PAI = pressure measurement, PC = pressure control, FAI = flow measurement, TAI = temperature measurement, EAI = motor power measurement.

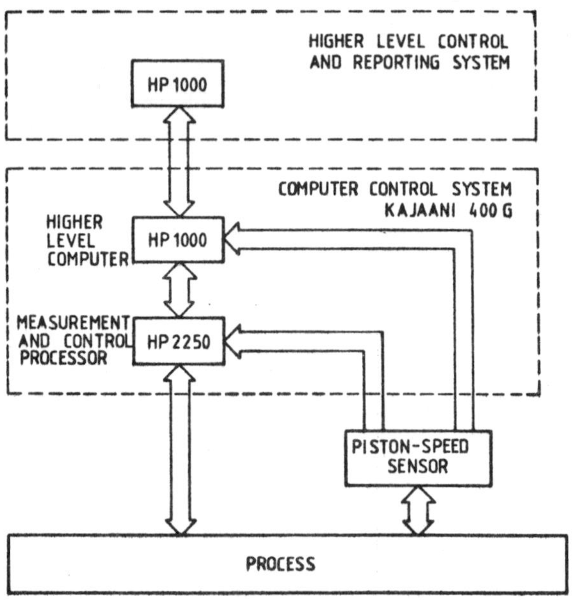

Fig. 3. Structure of a groundwood mill control system.

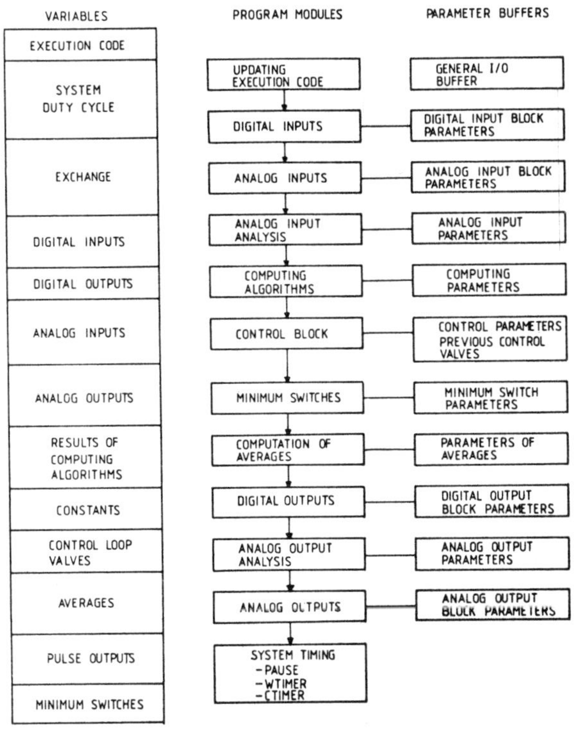

Fig. 4. Software structure.

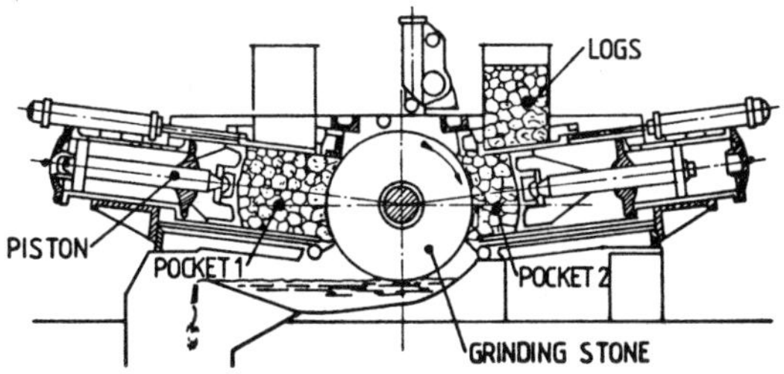

Fig. 1. A typical grinder with its shower water system.

Table 1. Results of execution speed tests.

No. of stones	Program space	Buffer tables	Memory used	Remaining memory*	Execution time (ms)
1	8136	968	9104	5858	125
2	8248	1365	9613	5349	180
5	8696	2488	11184	3778	420
10	9202	4333	13535	1427	780
10+avg's	9202	5001	14003	959	860

*Total memory = 14962 words

Table 2. Test system configurations.

signals \ system (no. of stones)	1	2	5	10	10+ avg's
Analog inputs	10	20	50	100	100
Digital inputs	8	16	40	80	80
Analog outputs	2	4	10	20	20
Digital outputs	7	14	35	70	70
Computing algorithms	2	4	10	20	20
Control loops	3	6	15	30	30
Minimum switches	2	4	10	20	20
Averages	1	1	1	1	40
Execution time (ms)	125	180	420	780	860

MAIN DE ROBOT A DEUX DEGRES DE LIBERTE ACTIONEE PAR MOTEUR A COURANT CONTINU GERE PAR MICROPROCESSEUR

J.C. DEMORISE / H. CUNHA JR. / J.P. L'HUILLIER

Centre de Recherche en Automatique de Nancy

L.E.E.A. - ENSEM - B.P. 54011 - Nancy Cedex - France

Resumé: Il s'agit d'une main capable de realiser un mouvement de pince et un mouvement de rotation du poignet, produits chacun par un micro-moteur à courant continu, suivi d'un réducteur.

L'etage de puissance est constitué par un hacheur. La commande est entierement faite par un microprocesseur 6802.

La fermeture et l'ouverture de la pince s'effectuent à vitesse maximum. On peut choisir l'angle d'ouverture et de fermeture par un octet que l'on écrit a partir du clavier dans la RAM du kit 6802 D5.

La main étant fermée sur un objet, il est possible de maintenir une force de serrage donnée.

Une boucle de vitesse rend la rotation du poignet indépendante du couple résistant. La vitesse échantillonnée est obtenue par mesure de la f.e.m, du moteur lorsque le courant s'annule dans le fonctionnement en hacheur. Cette mesure faite par un amplificateur différentiel, à trés grande impédance d'entrée, est transmise à un CAN. La référence de vitesse est donnée aussi par un octet de RAM. L'angle de rotation est obtenu en sommant dans un registre, les differents échantillons de vitesse. On peut en choisir la valeur. Cette rotation effectuée, on peut maintenir un couple constant, de valeur fixée par un ootet de RAM.

Les programmes commandant la pince et la rotation sont indépendants et peuvent être combinés pour obtenir des séquences telles que préhension - rotation - serrage.

1 - Description de la main

Cette main destinée à étre mise au bout d'un bras articulé est capable d'effectuer un mouvement de pince et un mouvement de rotation du poignet. Un micro-moteur à aimant permanent produit chacun deux mouvements par l'intermediaire d'un train de 4 pignons réducteurs à denture droite. Le dernier pignon de cette démultiplication à 4 étage attaque directement un secteur denté solidaire du pouce.

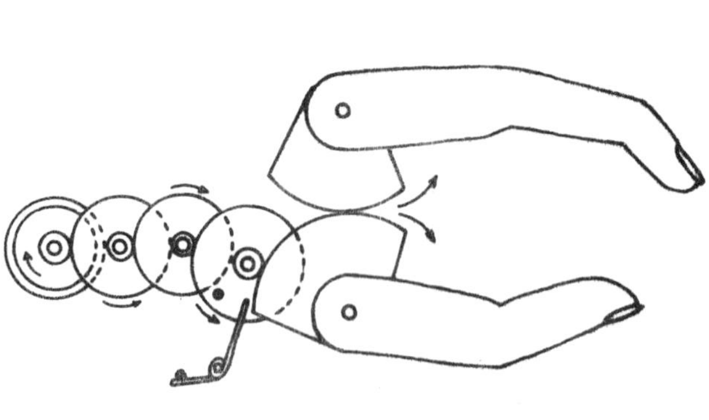

L'index est mû par un secteur denté estrainé par celui du pouce, et de même rapport. A un sens de rotation du moteur correspond l'ouverture de la pince, à l'autre sens correspond la fermeture. La rotation du poignet dans un sens et dans l'autre correspond de la même façon aux deux sens de rotation d'un deuxième moteur.

2 - Les Micro-Moteurs à Courant Continu à Aimant Permanent

Ils sont du type dit à induit en cloche. Ils ont pour équations éléctriques et mécaniques:

$$\begin{cases} U = RI + L\,dI/dt + kw \\ \Gamma m = kI \end{cases}$$

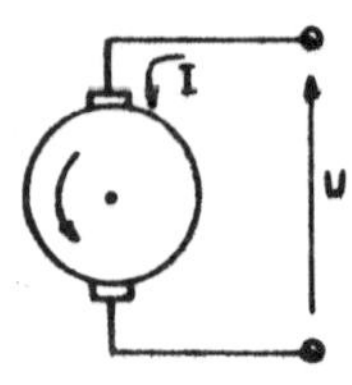

Le couple moteur Γm est equilibré de façon générale, par:

$$J \,.\, dw/dt + fw + Cp + Cu$$

où Cp est le couple de perte et Cu le couple utile.

Lorsqu'il y a peu de frottements f = o et Cp = o. Le regime permanent du moteur est alors fixé par les equations:

$$\begin{cases} U = RI + kw \\ kI = Cr \end{cases}$$ ou Cr est le couple résistant.

Soit $w = \dfrac{U - RCr/k}{k}$

Pour $Cr = \dfrac{kU}{R}$ le moteur se bloque.

Ces micro-moteurs ont la particularité d'avoir une resistance élevée. Une mesure de R a rotor bloqué nous a donné:

$$R = 12\,\Omega$$

La réponse de I à un echelon de tension de U, à rotor bloqué permet de determiner L = 0,37H

Pour déterminer K nous avons couplés mécaniquement deux micro-motors.

mini-banc d'essai

L'un alimenté par une tension réglable entraine l'autre qui fonctionne en géneratrice.

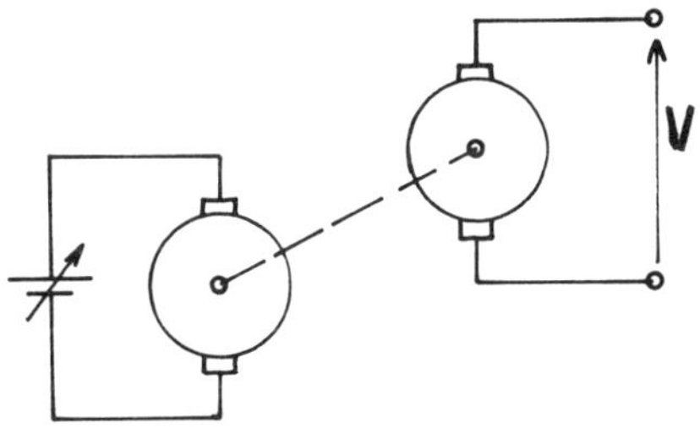

La tension mesurée aux bornes de la géneratrice à vide, vaut:

$V = k \cdot w$

En mesurant w au stroboscope, on en deduit $k = 82 \times 10^{-4}$ Vs/rd.

3 - Alimentation d'un Moteur

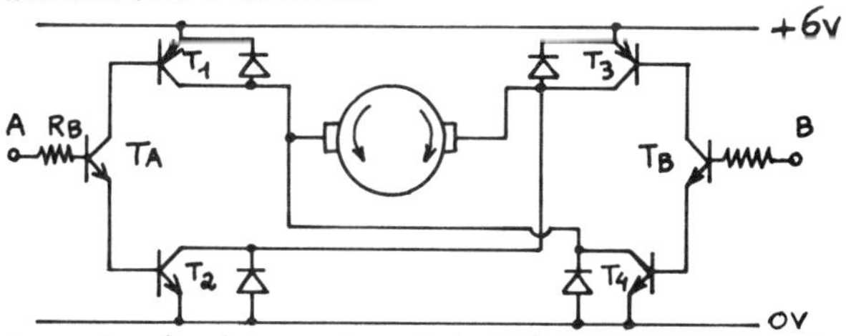

On peut grâce à ce montage à transistors, commander le sens de rotation du moteur à partir de niveaux logiques . Lorsque A = 0 et B = 1, le transistor T_A est bloqué, ce qui bloque T_1 et T_2 . Par contre T_B conduit, donc T_3 et T_4 aussi, ce qui fait circuler le courant du moteur dans un sens. Inversement la combinaison A = 1 B= 0 fait circuler le courant dans l'autre sens.

Le moteur consommant au maximum un courant

$I = 6V/R = 6V/12\,\Omega = 0,5$ A, il faut fixer le courant de base de T_A et T_B à $I_B = 0,5A/ 100 \times 200$. 100 et 200 étant les valeurs respectives du β de T_1 (ou T_2) et T_A.

Pour un niveau logique TTL on a $I_B = 5 - 2V_{BE} / R_B = 3,6/ R_B$. Ce qui donne $R_B \simeq 140$ K .

Aussi lorsque le moteur est bloqué, les transistors qui conduisent (T_1 et T_2 par exemple) sont saturés.

Pour un couple résistant Cr plus faible, appliqué au moteur, le courant de collecteur des transistors Ti qui conduisent, s'établit à la valeur Cr/k, de façon à ce que le couple moteur KI équilibre le couple résistant en régime permanent.

La vitesse se stabilise à une valeur telle que:

$R \cdot Cr/k + kw = 6$. C'est à dire que U prend sa plus grande valeur, correspondant à la saturation des transistors.

Ainsi les transistors Ti dés qu'ils conduisent, sont saturés et dissipent donc une puissance thermique trés faible. La vitesse est fixée par le couple résistant:

$$w = \frac{6 - R/k\,Cr}{K}$$

4 - Alimentation du Moteur en hacheur

Un moyen de commander la vitesse w, consiste a "hacher" le courant I du moteur, periodiquement en laissant conduire les transistors Ti correspondant à un sens de rotation, qu'a certains instants. Ceci sera obtenu en envoyant sur A ou B, suivant le sens de rotation, un signal carré de rapport cyclique variable.

Ainsi pour controler la vitesse dans un sens de rotation, on mettra B = 0 et sur A on enverra la commande suivante:

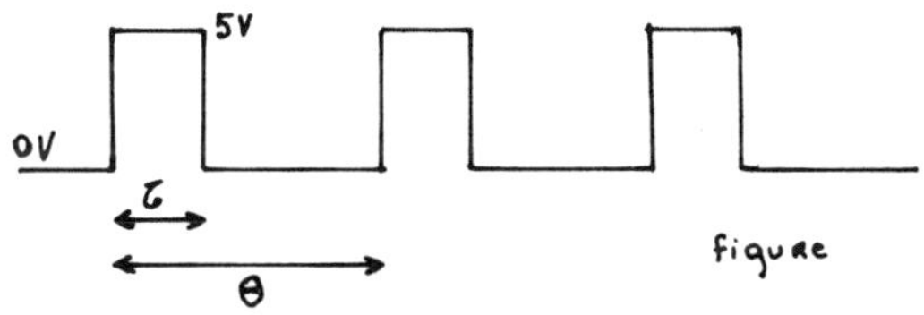

figure

Pendant l'intervalle de temps τ , T_1 et T_2 conduisent, et sont bloqués pendant l'intervalle de temps $\theta - \tau$ .On peut montrer que la vitesse ne dépend que du rapport $\frac{\tau}{\theta} = \alpha$ et varie linéarment avec celui-ci.

Lorsque $\alpha = 0$ on a $w = o$ et lorsque $\alpha = 1$ on a la vitesse maximale $w_M = \frac{6 - R/k\,Cr}{K}$ pour un couple Cr donné.

La vitesse ne dépend plus uniquement du couple resistant Cr, elle depend aussi de α , on a:

$$w = \alpha \frac{6 - R/k\,Cr}{k} \qquad \text{avec } 0 \leqslant \alpha \leqslant 1$$

5 - Commande en Boucle Ouverte, Commande en Boucle Fermée

Dans une application où Cr ne varie pas, pour maintenir w constante, il suffit de donner une valeur a α .

Cette commande en boucle ouverte suffit dans le cas de la fermeture et ouverture de la pince de la main, puisqu' aucune charge variable n'est entrainée par ce mouvement.

Par contre pour la rotation du poignet, il faut une commande en boucle fermée, puisqu'on peut faire tourner des objets de masses differentes et qu'un même objet au cours d'une rotation peut introduire un couple résistant variable.

Pour maintenir w constante pendant la rotation du poignet, il faut, au fure et à mesure que Cr varie, corriger α, grace à un signal de erreur.

Ce signal d'erreur est la difference entre la consigne de vitesse et la vitesse instantannée donnée par un retour.

Traditionnellement, la génération du signal carré de rapport α variable est réalisé par un systeme à amplificateur opérationnel et comparateur. Le retour est fait par une dynamo tachypétrique ou un codeur incrémental couplé mécaniquement à l'arbre du moteur.

6 - Mesure de la Vitesse

Nous utilisons pour mesurer la vitesse, une méthode qui a l'avantage de supprimer le couplage mécanique entre le moteur et le capteur et donc les oscillations engendrées par ce couplage.

Lorsqu'on alimente un moteur en hacheur, le courant oscille autour d'une valeur moyenne, fixé par α.

L'amplitude de cette oscillation dépend de θ. Il existe un mode de fonctionnement dit "mode discontinu" où le courant s'annule lorsqu'on bloque les transistors Ti.

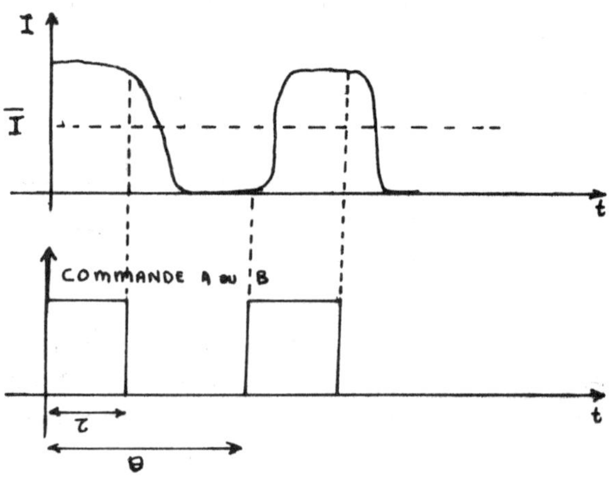

A un instant où le courant I est nul, la difference de potentiel U aux bornes du moteur, vaut:

U = kw , elle represente la vitesse.

La mesure de U est faite par un amplificateur differentiel à trés haute impedance d'entrée, de façon à ne pas pérturber le fonctionnement du moteur par la mesure.-

Cette mesure est declanché avec un certain retard, determiné experimentalement, par rapport au front descendant du crêneau de commande envoyé an A ou B.

Aprés conversion, l'echantillon de vitesse est transmis par un CAN à un μ P.

7 - Mesure de l'Angle de Rotation d'un Moteur

Le μ P fait la somme des échantillons de vitesse que lui envoie le CAN. Cette somme $\sum w(k\theta)$ peut être assimilé à l'integrale de la vitesse $\int w(t)dt$ et represente donc l'angle de rotation θ.

8 - Génération du Crenéau de Commande Envoyé en A ou B

Le carré o - 5V dont α varie, est généré par programme par une méthode classique. Chacun des temps θ et τ sont obtenus en chargeant un registre, qu'on decrémente ensuite jusqu'a zéro et on fait boucler ce programme.

Nous avons utilisé un kit 6802 D5 pour réaliser l'ensemble de la commande de la main.

Ce créneau de commande sort sur un fil du port d'entrée - sortie du PIA 6821 (programmé en sortie)

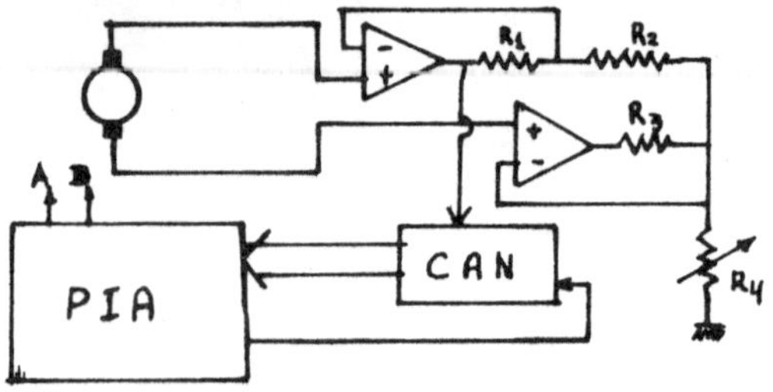

9 - La boucle Fermée de Vitesse

La référence de vitesse est un octet de RAM, entrée à partir du clavier.

Dans la boucle fermée de vitesse, le μ P compare l'octet transmis par le CAN, à cette référence et en fonction du résultat, corrige α de la façon suivante:

$$\tau_{k+1} = \tau_k - \varepsilon$$

$$\theta_{k+1} - \tau_{k+1} = (\theta_k - \tau_k) + \varepsilon \quad . \text{ Avec } \varepsilon > o \text{ ou } < o$$

θ reste ainsi constant au cours du temps.

ε est déterminé experimentalement pour que l'asservissement de vitesse soit stable.

10 - Réglage de la Force de Serrage

Lorsque le moteur est arrêté, en faisant varier α, on fait varier la valeur moyenne du courant I et donc la valeur moyenne du couple moteur. Il en résulte une force de serrage reglable par α. Cette réference de force est un octet de RAM entré à partir du clavier. Cette force est imposée en boucle ouverte, sans retour. Le système permet d'imposer dans un programme, à un instant determiné, la consigne α.

L'experience montre en effet que pour une valeur de α donnée, la force ne varie pas , car elle n'est perturbée par rien.

11 - Programmes Réalisés

Le programme qui suit, permet de réaliser la rotation du poignet à la vitesse constante, d'un angle donné.

Dans l'ordinogramme T_1^o et T_2^o representent les références respectives et τ et $\theta - \tau$.

IND est un indicateur qui lorsqu'il est a 1 indique que la mesure de U est faite.

w_r : reference de vitesse

θ : reference d'angle de rotation

Nous avons utilisé CA_2 et B_2 du PIA pour les commandes en A et B.

Il existe aussi un programme qui peut être justaposé

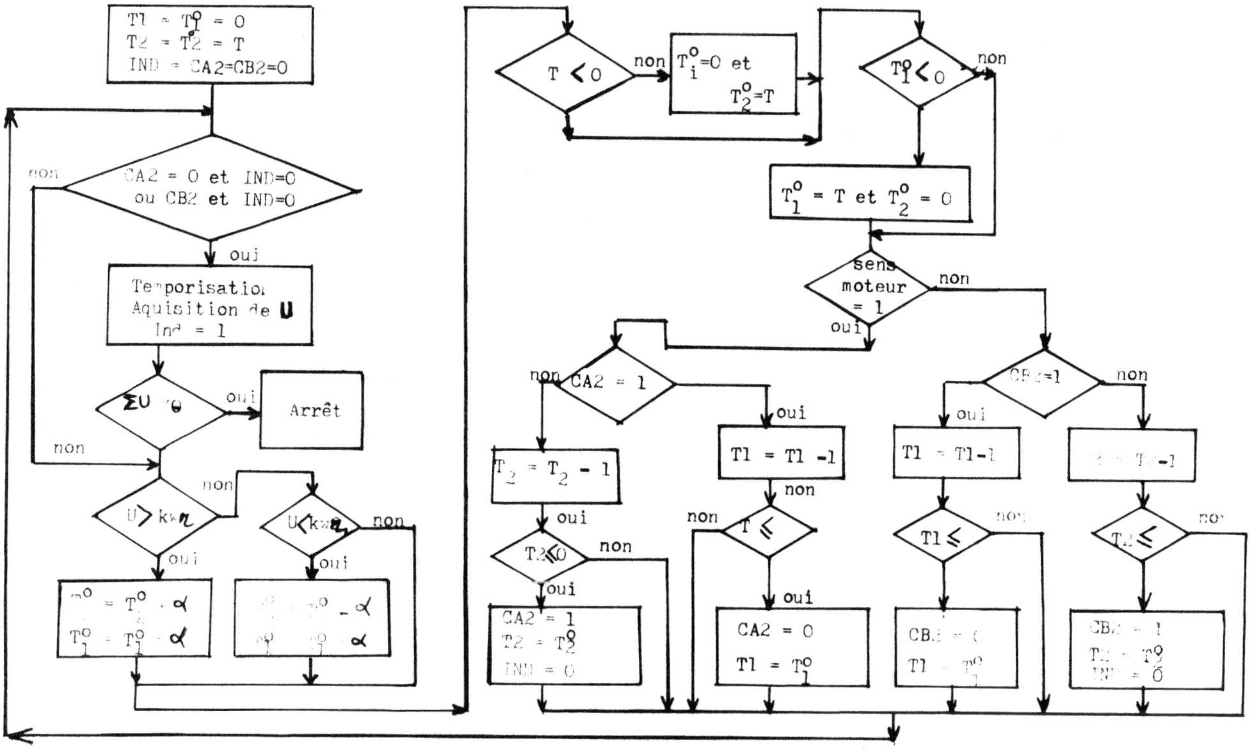

au précedent pour imposer un couple donné, une fois la rotation effectuée.

Un programme permet d'ouvrir et fermer la pince à vitesse maximale, d'un angle donné et d'imposer une force de serrage donnée.

Pour prendre une série d'objets identiques, on ajuste l'angle de fermeture et la force de serrage experimentalement, par approximation successives, en changeant deux octets de RAM, a partir du clavier.

On peut combiner ces programmes pour réaliser des sequences telles que: préhension - temporisation - rotation -couple de rotation constant.

12 - Bibliographie

/1/ - D.C. Motors, Speed Control, Servo-Systems
Engineering Handbook - Electro-Craft Corporation
1978

/2/ - R. CHAUPRADE
Commande Electronique des Moteurs à Courant Continu

/3/ - MEK 6800 D2 - Evaluation Manuel - Motorola Corporation

/4/ - M. Demerle./ J. Fromont
Regulation Numerique Directe de Vitesse
1ére et 2éme parties.
Revue Jeumont - Schneider n° 30/31
pp. 3 - 6/8 - 15

/5/ - R. Koll
Understanding Modern Control
Control Engineering 1975 - April - pp. 52 - 54

PERFORMANCE EVALUATION OF SEVEN DEGREES OF FREEDOM ROBOTS

R.G. Fenton
Professor

B. Benhabib
Graduate Student

A. Goldenberg
Associate Professor

University of Toronto
Department of Mechanical Engineering
Toronto, Ontario M5S 1A4
CANADA

ABSTRACT

A seven degrees of freedom robot can perform a specified task in an infinite number of different ways. It is possible to define a criterion for optimum performance, and to determine joint displacements for the robot so as to optimize its performance according to the specified criteria. In this paper, the performance of such an optimally controlled seven degrees of freedom robot is evaluated using simulation, and it is compared with the performance of a six degrees of freedom robot. It is found that the performance of the seven degrees of freedom robot is quite superior to that of the six degrees of freedom robot.

INTRODUCTION

A robot consists of a sequence of mechanical links connected generally by revolute or prismatic joints to form an open kinematic chain. The number of links and the type of joints connecting the links determine the degrees of freedom (DOF) of the robot. At least six DOF are needed to enable the robot to move its end effector into a specified position with given orientation in three dimensional space. A robot having only five DOF can only move with severely restricted end effector orientation. On the other hand, a robot with more than six DOF has a considerable flexibility in its operation [1].

A six DOF robot can reach any position in task space (a specified position with given orientation) in a small number of unique configuration. For example, the IBM 7565 robot with three prismatic and three revolute joints can reach a position in task space in one configuration only. The PUMA-600 robot, however, with six revolute joints can generally reach any position in task space in four (symmetric) configurations. Any of these configurations can be specified by a VAL-language control program as: LEFTY-ABOVE, LEFTY-BELOW, RIGHTY-ABOVE and RIGHTY-BELOW. The LEFTY and RIGHTY commands change the robot configuration to resemble a human's left or right arm respectively. The ABOVE and BELOW commands change the robot configuration so that the "elbow" of the robot is pointed up or down. A seven DOF robot, on the other hand, can move into any feasible task space position in an infinite number of different configurations.

The operation of a seven DOF robot can be controlled by selecting the best of the infinite number of available configurations each time the robot moves in task space. This selection can be done by using a specified optimality criterion. A seven DOF robot, whose motion is controlled with a computer program using an optimality criterion, is expected to perform its functions better than a comparable six degrees of freedom robot [2]. In this paper, the performance of a six DOF robot is examined and compared with that of a seven DOF robot.

ROBOT CONFIGURATION

For the purpose of this study, we consider a six DOF robot having six revolute joints, similar to the PUMA-600 robot, see Figure 1. One possibility to increase the degrees of freedom of this robot is to replace one of its links by two links. For example, link 2 may be replaced by two links. The orientation of the revolute joint axis connecting the two new links is, of course, an important parameter. If this joint axis is held parallel to joint axis 2, a seven DOF robot is obtained as shown in Figure 2. The work envelopes of the two robots shown on Figures 1 and 2, are identical if $a_0 = b_0$ and $a_1 + a_2 + a_3 = b_1 + b_2$.

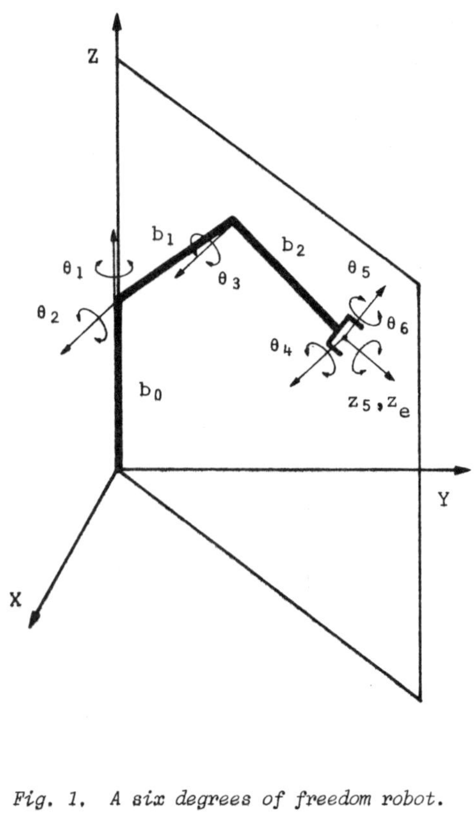

Fig. 1. A six degrees of freedom robot.

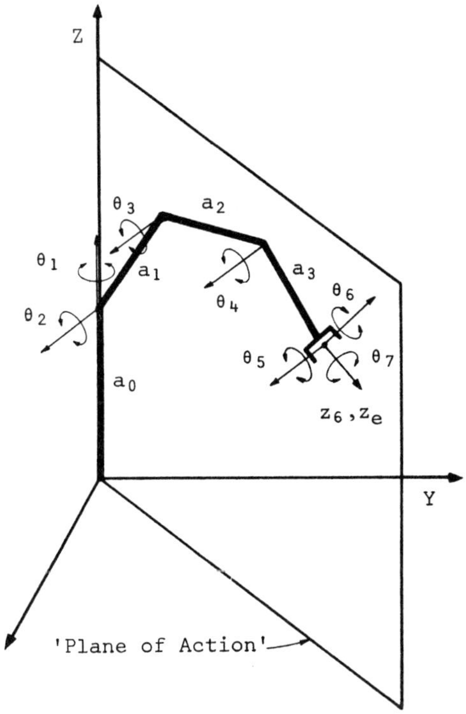

Fig. 2. A seven degrees of freedom robot.

PERFORMANCE INDEX

The relative performance of two robots will be determined on the basis of a specified performance index. Any performance characteristics (energy consumption, maximum joint torque, time required for motion, maximum distortion, or a weighted average combination of these quantities) can be used for defining a performance index. For the present study, the performance index is the average time required for the robot to move from one task space position to the next. This performance index is related to the speed the robot can perform its functions or to its productivity.

In order to evaluate the performance index of the robot, first the velocity profile of each joint has to be determined. The maximum angular velocity, acceleration and deceleration of each joint are specified, therefore, the velocity-time relation for a joint becomes as indicated on Figure 3. The joint begins its motion with maximum acceleration and moves with this acceleration until the maximum velocity is reached. Motion ends with maximum deceleration.

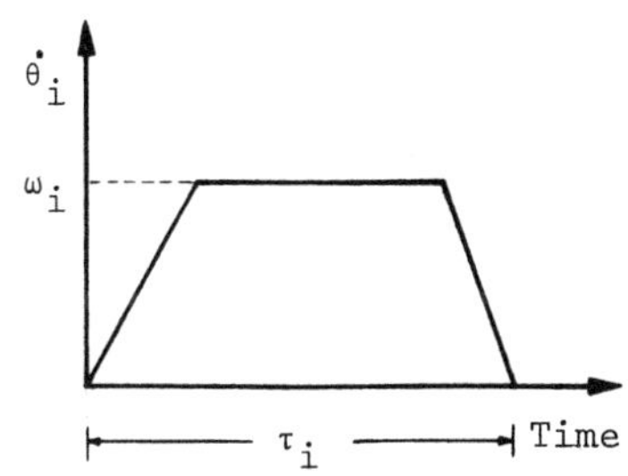

Fig. 3. Joint velocity profile.

The motion time for joint "i" is

$$\tau_i = \begin{cases} \sqrt{2\,\Delta\theta_i \left[\frac{1}{\alpha_{a,i}} + \frac{1}{\alpha_{d,i}}\right]} & \text{if } \Delta\theta_i \le \frac{\omega_i^2}{2}\left[\frac{1}{\alpha_{a,i}} + \frac{1}{\alpha_{d,i}}\right] \\ \frac{\Delta\theta_i}{\omega_i} + \frac{\omega_i}{2}\left[\frac{1}{\alpha_{a,i}} + \frac{1}{\alpha_{d,i}}\right] & \text{otherwise} \end{cases} \tag{1}$$

where $\alpha_{a,i}$ and $\alpha_{d,i}$ are the maximum angular acceleration and deceleration, ω_i is the maximum angular velocity, and $\Delta\theta_i$ is the required angular displacement of joint "i".

The robot generally performs a sequence of N motion steps starting at task space position P_0 and ending at position P_N. The time a six DOF robot needs to complete motion step "j" moving from task space position P_{j-1} to task space position P_j is

$$t'_j = \max[\tau_i;\ i = 1,6]_j \tag{2}$$

For the seven DOF robot the time required for this motion is

$$t''_j = \max[\tau_i(\theta_2^*);\ i = 1,7]_j \tag{3}$$

where θ_2^* defines the optimum configuration of the robot to perform motion step "j". The value of θ_2^* is determined using the specified optimally criterion as follows

$$\max[\tau_i(\theta_2);\ i = 1,7] = \text{minimum} \tag{4}$$

Solving equation (4) requires the use of a one-dimensional optimization search routine. Therefore, the motion time of the seven DOF robot becomes

$$t''_j = \min\{\max[\tau_i(\theta_2);\ i = 1,7]\}_j \tag{5}$$

The performance index for the six DOF robot is defined as

$$I' = \frac{1}{N}\sum_{j=1}^{N} \max[\tau_i;\ i = 1,6)_j \tag{6}$$

and for the seven DOF robot it becomes

$$I'' = \frac{1}{N}\sum_{j=1}^{N} \min\{\max[\tau_i(\theta_2);\ i = 1,7]\}_j \tag{7}$$

If N is large and the task space positions, P_j, are randomly generated, I'/I" can accurately quantify the relative performance of two robots on the basis of the selected criterion(minimum motion time).

PERFORMANCE INDEX EVALUATION

The performance index can be determined using Monte-Carlo simulation. Task space positions are determined with the help of a random number generator. Each task space position must be within the work envelope of the robots. The next step in the calculation is to determine the joint coordinates, $(\theta_i)_j$, corresponding to task space position P_j. This calculation is done for the six DOF robot on the basis of the Denavit-Hartenberg method [3,4]. The computation for the seven DOF robot is somewhat different.

The seven DOF robot of Figure 2 has seven joint variables. Angle θ_1 is the "body angle", it defines the plane containing links 2 to 4. Joint angles 1 to 4 define the position, and joint angles 5 to 7 the orientation of the end effector. If the end effector position is specified as $(x,y,z)_j$, the required joint angles can be computed from the following equations:

$$\left.\begin{aligned} \theta_1 &= \tan^{-1}\frac{y}{x} \\ \theta_4 &= \cos^{-1}\frac{A^2 + B^2 - a_2^2 - a_3^2}{2a_2a_3} \\ \theta_3 &= \tan^{-1}\frac{B}{A} - \sin^{-1}\frac{a_3\sin\theta_4}{\sqrt{A^2 + B^2}} \end{aligned}\right\} \quad (8)$$

$$\text{where } A = \sqrt{x^2 + y^2}\,\sin\theta_2 + (z - a_0)\cos\theta_2 - a_1$$

$$B = \sqrt{x^2 + y^2}\,\cos\theta_2 - (z - a_0)\sin\theta_2$$

Angle θ_2 is a free variable. A solution set $(\theta_1,\theta_3,\theta_4)_j$ exists for any feasible combination of end effector position, $(x,y,z)_j$, and θ_2. The angles θ_5 to θ_7 are computed in the same way as angles θ_4 to θ_6 are determined for the six degrees of freedom robot. For motion step "j", as the robot moves from task space position P_{j-1} to P_j, the required angular displacement increments can be determined. These quantities are functions of the free variable, θ_2^*

$$(\Delta\theta_i)_j = (\theta_i)_j - (\theta_i)_{j-1} = [\Delta\theta_i(\theta_2^*)]_j \quad (9)$$

and the optimum configuration of the robot represented by the value θ_2^* is computed by solving equation (4). This sequence of computation is continued until the performance index of equation (7) is evaluated.

RESULTS

The performance indices of the six and seven degrees of freedom robots of Figures 1 and 2 were determined with computer simulation using an 800 steps motion sequence. Velocity and acceleration constraints are listed in Table 1.

		Six DOF Robot	Seven DOF Robot
$\omega_i =$	2.0 1/s	for i = 1,3	for i = 1,4
	7.0 1/s	for i = 4,6	for i = 5,7
$\alpha_{a,i} = \alpha_{d,i} =$	25 $1/s^2$	for i = 1,3	for i = 1,4
	80 $1/s^2$	for i = 4,6	for i = 5,7

Table 1. Maximum joint angular velocities and accelerations.

The results of the Monte-Carlo simulation is as follows:

$$\frac{I'}{I''} = \frac{.996}{.703} = 1.417$$

indicating that the seven DOF robot is quite superior in performance. A histogram of the results, see Figure 4, shows that in 92% of the 800 steps of motion the seven DOF robot is superior to the six DOF robot. However, it was also found that in a small number of motion steps the six DOF robot quite unexpectedly performed better than the seven DOF robot,

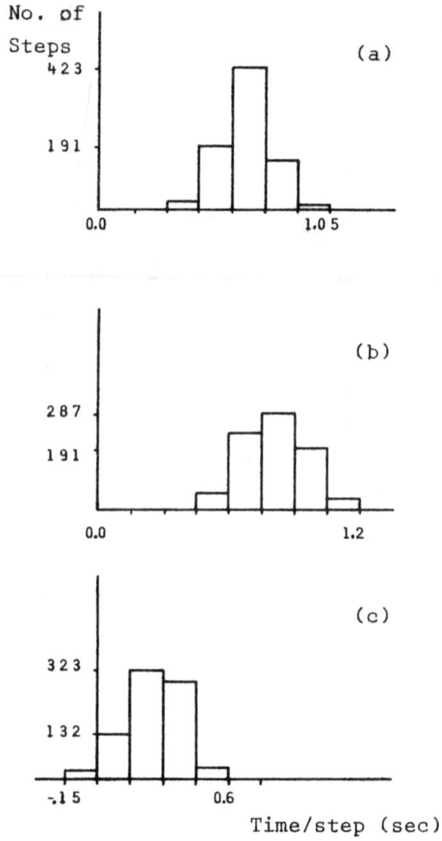

Fig. 4. A sequence of 800 steps of motion;

a) seven DOF robot $a_1 = a_2 = a_3 = 1.0$ average time/step = 0.703 sec.

b) six DOF robot $b_1 = b_2 = 1.5$ average time/step = 0.996 sec.

c) time difference (6DOF-7DOF) average time diff. = 0.294 sec. (44 equal time steps)

It should be noted that the angular displacement increment for the seven DOF robot is a function of the new task space position as well as the joint positions, $(\theta_i)_{j-1}$ at the end of the previous motion step. Values of $(\theta_i)_{j-1}$ were, however, obtained as a result of an optimization process performed in the previous step. Therefore, values of $(\Delta\theta_i)_j$ are dependent on the motion history of the robot, in addition to the new task space position. If both robots would start their motion each time from the same configuration, the results should indicate that the seven DOF robot is always superior to the six DOF robot.

In order to eliminate the effect of motion history on the performance index, the Monte-Carlo simulation performed previously was repeated for the same 800 motion steps, however, the starting position of each motion step was adjusted to $\theta_i = 0$. This simulation, therefore, does not represent a proper sequence of motion steps, it represents 800 individual steps of motion. Results of the computation again clearly indicate that the seven DOF robot is superior

$$\frac{I'}{I''} = \frac{0.777}{0.538} = 1.444$$

A histogram of the detailed results, see Figure 5, shows that the seven DOF robot is superior to the six DOF robot for each of the 800 random steps of motion performed.

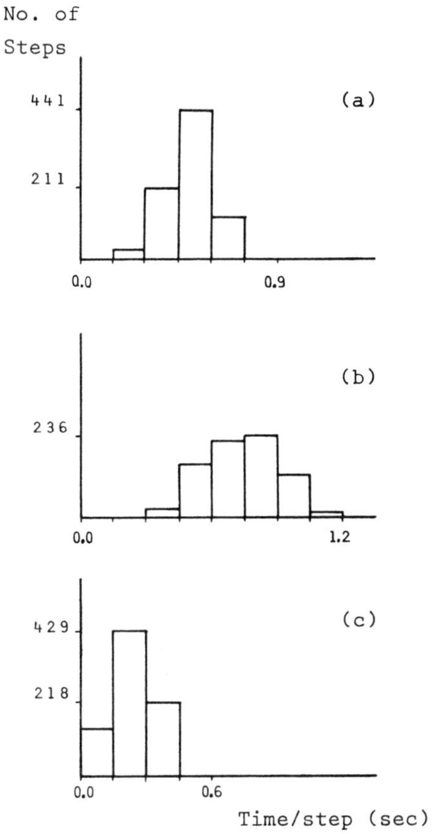

Fig. 5. 800 individual steps of motion;

a) seven DOF robot $a_1 = a_2 = a_3 = 1.0$ average time/step = 0.538 sec.

b) six DOF robot $b_1 = b_2 = 1.5$ average time/step = 0.777 sec.

c) time difference = (6DOF-7DOF) average time diff. = 0.239 sec. (62 equal time steps)

A six DOF robot may be considered to belong to the subset of a seven DOF robot if by holding one of the joint variables constant the seven DOF robot becomes identical to the six DOF robot. The six DOF robot of Figure 1 does not belong to a subset of the seven DOF robot of Figure 2 unless its dimensions are changed to $b_1 = 2$ and $b_2 = 1$. A six DOF robot with dimensions $b_1 = b_2 = 1.0$ would also belong to the subset of the seven DOF robot with dimensions $a_1 = 1$, $a_2 = 0.5$ and $a_3 = 1.5$. The relative performance of these two sets of robots were also analysed using computer simulation, and the computed results shown on Figures 6 and 7 again clearly indicate that the seven DOF robot is considerably superior in its performance to the six DOF robots.

CONCLUSIONS

The purpose of this study was to determine the relative performance of six and seven DOF robots. A specific robot with seven DOF was structured and its kinematic model was determined. An optimality criterion was established to control the motion of the seven DOF robot. Performance indices were defined and evaluated using computer simulation for several sets of six and seven DOF robots having identical work envelopes.

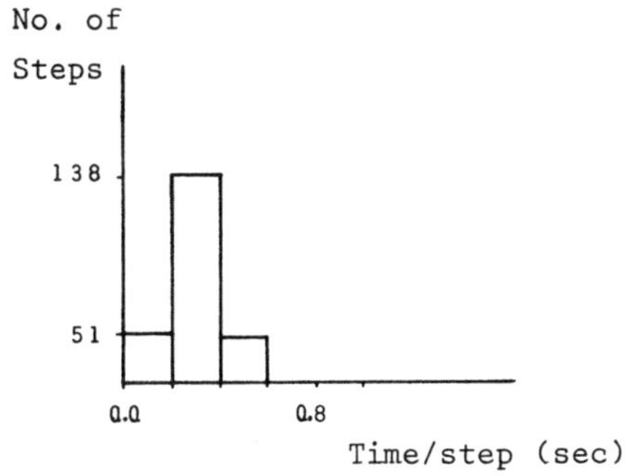

Fig. 6. 200 individual steps of motion;

seven DOF robot $a_1 = a_2 = a_3 = 1.0$
six DOF robot $b_1 = 2.0$ $b_2 = 1.0$
average time difference = 0.293 sec.
(6 equal time steps)

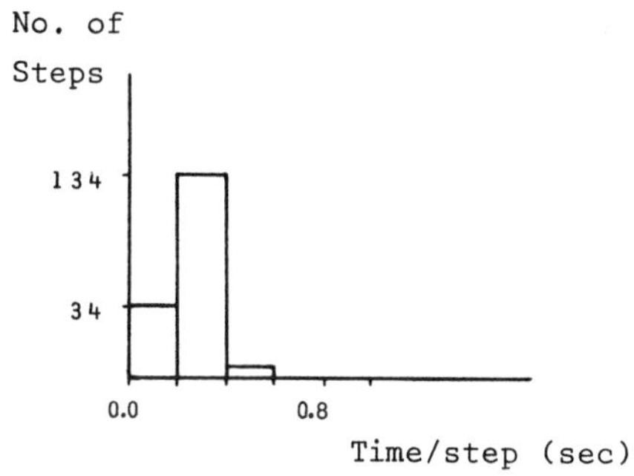

Fig. 7. 200 individual steps of motion;

seven DOF robot $a_1 = 1.0$ $a_2 = 0.5$ $a_3 = 1.5$
six DOF robot $b_1 = b_2 = 1.5$
average time difference = 0.254 sec.
(6 equal time steps)

It was found that

i) a seven DOF robot is superior in performance to that of a six DOF robot for any number of individual motion steps.

ii) if several motion steps are performed in a sequence the six DOF robot may be superior in performance to the seven DOF robot for a small number of these steps, but for the majority of steps, and for the sequence as a whole it is always the seven DOF robot which performs better.

iii) the difference in performance of a six and seven DOF robots can be substantial, representing as much as 40% improvement in favour of the seven DOF robot.

REFERENCES

(1) H. Hanafusa, "Analysis and Control of Articulated Robot Arms with Redundancy", Technical Report, Automation Research Laboratory, Kyoto University.

(2) M. Bermati, P. Morasso and V. Tagliasco, "The Inverse Kinematic Problem of Anthropomorphic Manipulator Arms", Trans. of A.S.M.E. J. of Dynamic Syst. Y. 104, March 1982, pp. 110-113.

(3) J. Denavit and R.S. Hartenberg, "A Kinematic Notation for Lower-Pair Mechanisms Based on Matrices", Trans. of A.S.M.E. J. of Appl. Mech., June, 1955, pp. 215-221.

(4) R.P. Paul, "Kinematic Control Equations for Simple Manipulators", Trans. of IEEE. Vol. 11, June 1981, pp. 449-455.

CARTESIAN PATH PLANNING BY GENERAL POLYNOMIAL INTERPOLATION

H.Bruhm
European Space Agency
ESTEC TPC Section
NL-2200 AG Noordwijk zh
The Netherlands

E. Ersü
Institute of Control Engineering
Section Control Systems Theory
Technical University of Darmstadt
D-6100 Darmstadt,FRG

ABSTRACT

A Cartesian path planning algorithm for robot arms is presented, in which the path is comprised of individual segments of different types. Each segment is distinguished by the shape of the curve and the development of the end-effector's orientation. Straight lines, circular and parabolic arcs and splines are available for path description. The orientation of the end-effector may be chosen to be constant with respect to the task (e.g. to a workpiece) or with respect to the path. Alternatively roll-pitch-yaw angles may be interpolated linearly between pre-specified initial and final orientations. To determine free parameters, a specific number of set points must be given for each segment.

Transition from one segment to the next can take place with interruption of motion at the transition point or continuously, as one chooses. In the latter case, a polynomial establishing a smooth connection between the two segments is inserted automatically. The length of the connecting segment is adapted to the speed of motion according to a heuristic method.

The programmed path may be followed with an arbitrary speed profile. Furthermore, the desired speed value may be modified during runtime, depending on sensory data. Standard speed profiles with smooth accelerations and decelerations have been implemented.

1. INTRODUCTION

With today's microprocessors Cartesian path planning for robot arms can be done on-line and at reasonable cost. The fundamental advantages of path planning in Cartesian coordinates over path planning in joint coordinates have been explained by Paul [1] and recently in a paper by Brooks [2]. Paul [1] introduced a simple interpolation scheme in which the path is made up of line segments with smooth transitions. An interpolation scheme proposed by Stute and Keppeler in [3] offers line segments and circular arcs as elements for path construction. The orientation of the robot hand is interpolated linearly in both references.

In the case of the interpolation scheme to be introduced here, the combination of various algorithms for the interpolation of position and orientation yields 11 different types of segments. Smooth transitions are possible between any two types of segments. A speed profile can be prescribed independently of the path specification. The path planning has thus a degree of versatility which is expected to meet all practical requirements.

2. THE PRINCIPLE OF TWO-STAGE INTERPOLATION

If the task of path interpolation is completely assigned to the Cartesian level of the robot control system, the achievable sampling rate of the servo controllers will be determined by the frequency at which solutions of the inverse kinematic problems can be obtained. In order to evade this limitation, a subordinate interpolation in joint coordinates was provided. Thus, even very quick arms requiring servo rates in the 1 KHz order of magnitude can be controlled in Cartesian coordinates in real time. This increase in speed can be achieved with an insignificant loss of tracking accuracy.

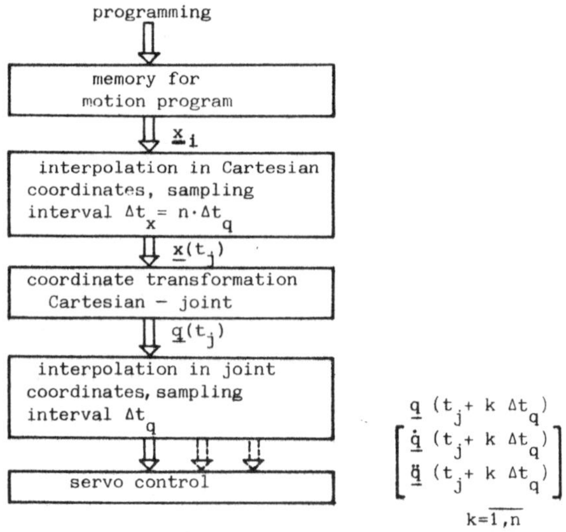

Fig.1: Architecture of path control system with two-stage interpolation (feedback control loops not considered)

Figure 1 illustrates how the two interpolation algorithms and the coordinate transformation act together. Cartesian interpolation is done between the points $\underline{x}_i$ specified in the teaching phase, and produces intermediate Cartesian points $\underline{x}(t_j)$ in time intervals Δt_x. For each point $\underline{x}(t_j)$ the coordinate transformation algorithm calculates the corresponding vector of joint coordinates $\underline{q}(t_j)$. In the second stage of interpolation additional set points $\underline{q}(t_j + k\Delta t_q)$ are produced at the desired servo rate between each pair of $\underline{q}(t_j)$ and $\underline{q}(t_{j+1})$. Time derivatives $\dot{\underline{q}}(t_j + k\Delta t_q)$ and $\ddot{\underline{q}}(t_j + k\Delta t_q)$ can be provided with little additional cost, if they are needed by the servo controllers.

The interpolation in joint coordinates was mentioned only to provide the reader with a better understanding of the system architecture. The following discussion will be confined to the problems of Cartesian interpolation.

3. PATH SPECIFICATION

3.1 Description of End-Effector Position and Orientation

In order to give a mathematical description of robot arm movements, we define a coordinate frame which maintains a fixed spatial relationship to the end-effector (called "hand" for convenience) (cf. Paul [1]). In fig. 2 the hand frame's axial directions are given by the unity vectors $\vec{i}_h$, $\vec{j}_h$ and $\vec{k}_h$. The motion of the hand is now determined by the motion of the hand frame with respect to a reference frame, the so-called task frame (vectors $\vec{i}_r$, $\vec{j}_r$ and $\vec{k}_r$ in fig. 2). The task frame can be fixed in space or moving, depending on the type of task (fixed or moving, cf. Brooks [2]).

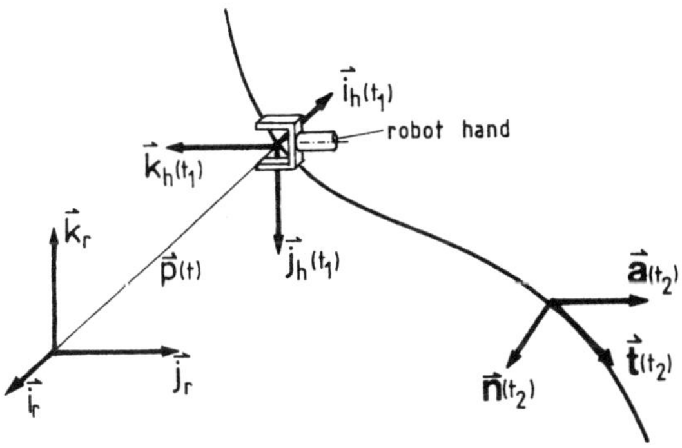

Fig.2: Coordinate frames used in the description of robot motions

The relative motion of the hand frame with respect to the task frame can be regarded as a trajectory in the six-dimensional space of generalized Cartesian coordinates

$$\underline{x}'(t) = (\underline{p}'(t), \underline{\phi}'(t)) \tag{1}$$

where $\underline{p}$, the vector from the origin of the task frame to the origin of the hand frame, describes the hand's position, and $\underline{\phi}$ its orientation in terms of roll, pitch and yaw (RPY) angles. The prime denotes the transpose.

The coordinate frame constituted by the tangent, normal and approach vectors to a certain point of the path ($\vec{t}$-$\vec{n}$-$\vec{a}$ frame), which is also shown in fig. 2, can be used as reference for the orientation of the hand alternatively to the task frame (see next chapter). It is of course only defined for curved paths.

3.2 Types of Path Segments

Every robot motion program consists of a sequence of individual path segment specifications. The type of each segment is designated by the type of the algorithms applied for the interpolation of hand position and orientation. The permissible combinations have been collated in figure 3.

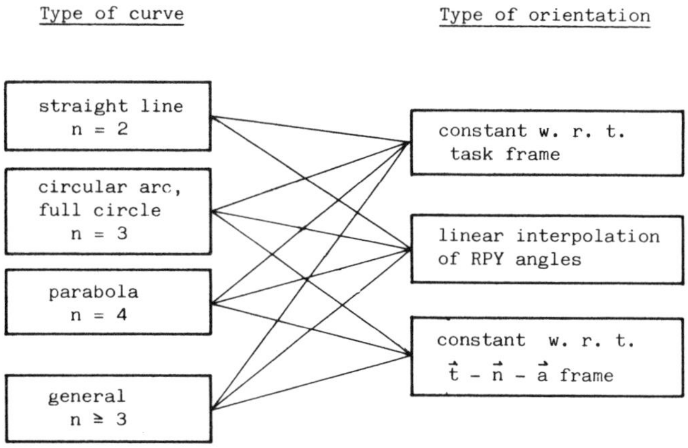

Fig.3: Permissible combinations of interpolation algorithms (n: number of set points to be specified)

Motion programs may be specified by means of teach-in, alphanumeric input from a keyboard or by direct data transfer from CAD. In the case of teach-in, the method which is still used most frequently, the number of set points to be stored for each segment depends on the type of position interpolation. The appropriate numbers are indicated in fig. 3.

4. EQUATIONS OF INTERPOLATION

4.1 Interpolation of Position

The interpolation of position uses a parameter u which sweeps the closed interval [0,1] for every segment or, in the case of general curves, for every sub-segment, i.e. between every two subsequent set points specified in the motion program. All types of position interpolation can formally be described by

$$\underline{p}(u) = \underline{K}\,\underline{u}(u) \tag{2}$$

where $\underline{K}$ is a matrix of constant coefficients and $\underline{u}(u)$ a vector of functions of the parameter u. Note that the time-dependance of u is yet unspecified. The dimensions of $\underline{K}$ and the meanings of $\underline{u}(u)$ are tabulated hereafter for all types of curves.

Type of curve	$\{\underline{K}\}$ =	$\underline{u}'(u)$ =
straight line	3 x 2	$(1,u)$
circular arc	3 x 3	$(1,\cos\phi_m u,\sin\phi_m u)$
parabola	3 x 3	$(1,u,u^2)$
general curve	3 x 6	$(1,u,u^2,u^3,u^4,u^5)$

ϕ_m is the central angle of the circular arc. General curves are quintic splines of Hermitian type. For an easy to understand treatise on technical applications of spline theory see Späth [4].

4.2 Interpolation of Orientation

For the purpose of synchronisation of translational and rotational motions, the parameter u is used for interpolation of both position and orientation.

In the cases of constant orientation with respect to the task frame and of linear interpolation of RPY angles, the equations describing the interpolation of orientation read:

$$\underline{\phi}(u) = \underline{\phi}_a \quad \text{and} \tag{3}$$

$$\underline{\phi}(u) = \underline{\phi}_a + (\underline{\phi}_e - \underline{\phi}_a)\,\frac{u-u_a}{\Delta u} \tag{4}$$

$$u \in [u_a, u_e], \qquad \Delta u = u_e - u_a$$

ϕ_a and ϕ_e are the vectors of RPY angles at the beginning and at the end of the subsegment under consideration, u_a and u_e the corresponding values of the parameter u.

In the case of constant orientation with respect to the $\vec{t}$-$\vec{n}$-$\vec{a}$ frame, the orientation of the hand is described by the matrix equation

$$\underline{R}_h(u) = \underline{R}_b(u) \cdot {}^b\underline{R}_h \tag{5}$$

$$\text{with} \quad \underline{R}_b(u) = (\underline{t}(u), \underline{n}(u), \underline{a}(u))$$

$$\text{and} \quad {}^b\underline{R}_h = \underline{R}_b^{-1}(0) \cdot \underline{R}_h(0)$$

The matrices denote the following relative rotations:

$\underline{R}_h$: rotation of hand relative to task frame

$\underline{R}_b$: rotation of $\vec{t}$-$\vec{n}$-$\vec{a}$ frame relative to task frame

${}^b\underline{R}_h$: rotation of hand relative to $\vec{t}$-$\vec{n}$-$\vec{a}$ frame

The RPY angles $\underline{\phi}(u)$ can be calculated from $\underline{R}_h(u)$.

5. PROGRAM EXECUTION WITH PRESCRIBED SPEED PROFILE

In many cases a given path will have to be traced by the robot with a prescribed speed profile v(t) which can be shaped to provide, for example, gentle accelerations and decelerations or to meet the requirements of a certain machining technology like arc welding or grinding. Such requirements can be fulfilled by reparametrisation of the equations of interpolation according to the differential equation

$$\dot{u}(t) = \frac{v(t)}{\left\| \frac{d\underline{p}(\xi)}{d\xi} \right\|_{\xi=u(t)}} \tag{6}$$

This differential equation will be integrated numerically, because an analytical solution u(t) can only be obtained for a few types of curves. Suitable procedures for numeric integration can be found in Törnig [5]. A theoretical study performed in [7] shows that the error caused by singularities in eq. (6), which can be found on spline segments, is negligible. Closed loop Cartesian speed control is feasible if reparametrisation is done in real time.

6. SEGMENT TRANSITION

Transition from one path segment to the next normally implies a sudden change of tangent direction which cannot be effected with finite acceleration at a speed greater than zero. The path planning algorithm presented in this paper offers two different solutions to this problem. Both methods permit segment transitions with steady velocity and acceleration.

With the first method, called segment transition with intermediate stop, the speed is reduced to zero in the segment intersection point. Thus the inward-bound segment ends in a deceleration part and the outward-bound segment begins with an acceleration part.

With the second method, called segment transition via connecting segment, a smooth connecting segment is inserted between two points P_a on the inward-bound segment and P_e on the outward-bound segment (see fig. 4). The connecting segments are established by a polynomial interpolation of 5th degree which is very similar to the spline interpolation used for general curves. Their length is adapted to the speed of motion by appropriate selection of end points P_a and P_e.

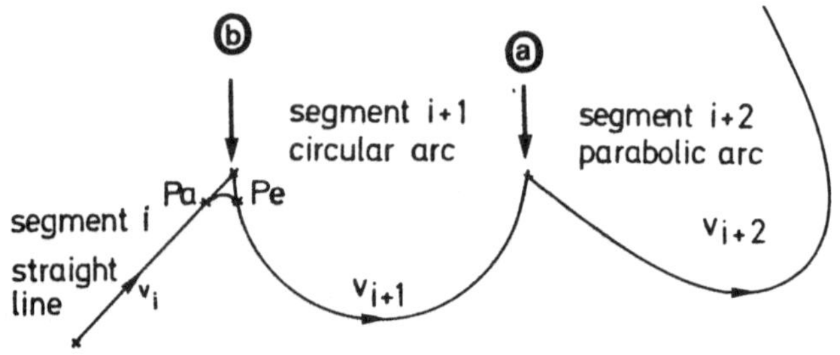

Fig.4: a) Segment transition with intermediate stop
b) Segment transition via connecting segment

7. STANDARD SPEED PROFILES

A motion appears to be smooth only if its speed profile is steady and steadily differentiable. Therefore a standard speed profile was implemented in the path planning algorithm, which is made up of intervals with constant speed connected by polynamials of degree 3. Two typical examples are presented in fig. 5, one for segment transition with intermediate stop and one one for segment transition via connecting segment.

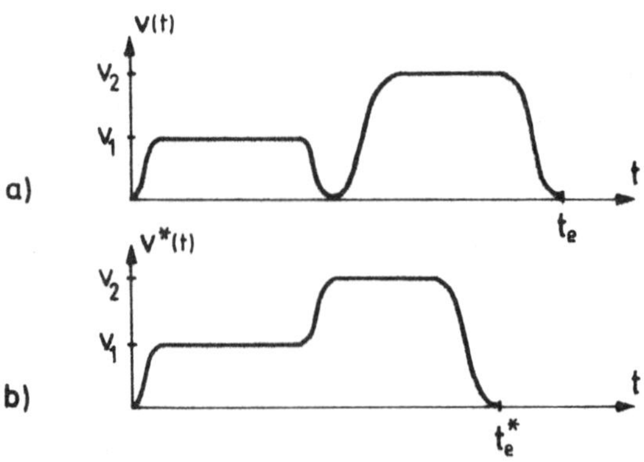

Fig.5: Standard speed profiles for a path consisting of two segments with desired speeds v_1 and v_2.

a) Segment transition with intermediate stop
b) Segment transition via connecting segment

8. CONCLUSION

The path planning algorithm presented in this paper incorporates the exact mathematical representation of those curves that are most frequently used in mechanical engineering. All "unusual" curves can be approximated by spline interpolation between a reasonable number of set points. It is therefore justified to say that the mathematical foundations of a "genuine" continuous path (CP) control system for robots have been supplied. A detailed mathematical treatment of the associated problems, including for example analytical expressions for the maximum acceleration within a path segment, can be found in [7]. The algorithms have been implemented on a digital computer PDP 11/34 for off-line testing.

The main advantage of this path control system compared to others which approximate continuous path motions by a sequence of very short point-to-point movements, appears as an enormous reduction of the memory space occupied by (complex) motion programs. The possibility to prescribe arbitrary speed profiles for the execution of programs must also be mentioned in this connection.

Continuous path control has been customary in the field of NC machine tools for many years. The industrial application of genuine CP robot control is, for the present, being delayed mainly because the servo controllers of today's robots are not designed for CP operation. The type of controller which is required, must have the capability to produce a motion between two points in joint coordinate space with well-defined joint rates, taking into account the dynamic coupling between joints, which is very essential for high speed operation. Joint control systems that can cope with these dynamic problems have already been proposed in literature, e.g. by Patzelt in [6]. The next generation of industrial robots is therefore very likely to exhibit the desired control features.

REFERENCES

[1] R. Paul
Manipulator Cartesian Path Control
IEEE Trans. SMC, Vol. 9, No. 11, Nov. 1979.
pp. 702 - 711

[2] T.L. Brooks
The Mathematics of Supervisory Computer Control of Manipulators
Journal of Dynamic Systems, Measurement and Control, Vol. 104, Dec. 1982, pp. 323 - 330

[3] G. Stute, M. Keppeler
Führungsgrößenerzeugung für Handhabungssysteme
wt, Zeitschrift für Industrielle Fertigung, No. 71, 1981
pp. 147 - 151

[4] H. Späth
Spline - Algorithmen zur Konstruktion glatter Kurven und Flächen
R. Oldenbourg Verlag, München, Wien, 1978

[5] W. Törnig
Numerische Mathematik für Ingenieure und Physiker, Band 2.
Springer Verlag, Berlin, Heidelberg, New York, 1979.

[6] H. Steusloff (Hrsg.)
Wege zu sehr fortgeschrittenen Handhabungssystemen
Springer Verlag, Berlin, Heidelberg, New York, 1980.

[7] H. Bruhm
Development of Suitable Algorithms for Cartesian Path Planning for Industrial Robots
Diplomarbeit at the Institute of Control Engineering, TH Darmstadt, directed by E. Ersü (in German)
Darmstadt, 1982.

AN APPROACH VIA ROBUST LINEAR MULTIVARIABLE SYSTEM DESIGN METHODS TO NONLINEAR MANIPULATOR CONTROL

E. Ersü and W. Neddermeyer

Institute of Control Engineering
Section Control Systems Theory
Technical University of Darmstadt
D-6100 Darmstadt, FRG

Abstract. In manipulator systems very good quality dynamic performance is difficult to achieve by conventional control methods because of both the inherent geometric nonlinearities of these systems and the dependence of the system dynamics on the characteristics of manipulated objects. The general statement made by many papers published on this topic is that it is not possible to design linear control systems which will yield uniformly high system performance for a wide range of manipulator tasks. With the recent advances in the field of linear multivariable system design, adjusted to great plant uncertainties, however, this is only partly true. In this paper a variable structure control law using a linear multivariable feedback system design method for plants with large parameter uncertainties will be presented which results in manipulator trajectories with dynamic characteristics which are nearly invariant in spite of system nonlinearities and the characteristics of the manipulated object.

The basic ideas underlying the approach are: Subdividing the working space of the manipulator into subspaces where in each subspace the nonlinear system dynamics can be modelled by a corresponding set of linear differential equations with parameter variations within the range of the subspace, design of a robust multivariable controller for each subspace which copes with the parameter variations.

A simulation example with a two degree of freedom system shows the efficiency of the proposed method.

I. INTRODUCTION

Manipulator dynamics present from the theoretical point of view a very interesting case due to the various nonlinearities and the on-line parameter changes. Besides the nonlinear dynamic couplings such as e.g. Coriolis or centrifugal forces on-line applications show parameter changes such as e.q. hand load and friction changes. Various different methods are proposed in the linterature to design a controller for this highly complex multivariable system e.g. self-tuning control [4], nonlinear decoupling method [5] or the method of inverse system with linear feedback control [6]. The main disadvantage of most of these ideas is the on-line computational effort which causes problems for high speed manipulators.

The method introduced in this paper works with a robust variable structure control by which each controller designed by linearizing the plant equations is evaluated off-line for a specific subspace of the manipulator's global workspace. Its special feature is that the controller for each subspace is a linear robust one which can cope with nonlinearities, parameter changes and modelling errors in the corresponding subspace.

The paper is organized in five sections. Section II. introduces the general control concept being applicable to manipulators and other nonlinear system also. In section III. the robust control design used in the control concept is explained. Simulation examples are discussed in section IV. Section V concludes the paper with an outlook and some remarks regarding the practical implementation problem.

II. THE CONTROL CONCEPT

As there is no unique theory to design control systems for nonlinear plants a conventional way is via linearization. Usually the present methods propose a linearization in one operating point, which takes into account some working conditions. Thus if the plant moves away from this working conditions, the linearized system equations are not valid for the actual operating point due to the deviations from the linearization point. This problem can be overcome if the control system can cope with changes of the nonlinear system due to the variations in the workspace.
A basic idea in this direction is, that the operating point variations can be interpreted in general as parameter changes in the linearized system equations. Thus design of a nonlinear control for the nonlinear plant can be reduced to a desgn of linear control which tolerates parameter variations or uncertainities respectively in a specified sense. Such linear control design methods are called robust methods which become popular and modern in control theory in the last years. One of the most interesting robust methods is that of Horowitz in the frequency domain for single-input single-output systems [1], which he and his co-workers extended to multivariable systems [2], [3] later on. The nonlinear control concept in this paper combines these ideas of robust design with the variable structure controller design.

The mathematical considerations are as follows:

Given a nonlinear system

$$\dot{\underline{y}} = \underline{f}(\underline{y},\underline{u},\underline{p}) \qquad \text{(II.1.)}$$

with

$\underline{u} \in R^{n_u}$ input vector

$\underline{y} \in R^{n_y}$ output vector

$\underline{p} \in R^{n_p}$ scalar parameter vector

where $\underline{p}$ represents the possible changes in system's parameters e.g. changing hand load of a manipulator.

Defining the global workspace of the system (II.1.) by

$$W_G: = \{\underline{y}/\underline{y}_{lb}{}^{1)} \le \underline{y} \le \underline{y}_{ub}\} \quad \text{(II.2.)}$$

and subdividing W_G into subspaces W^i with

$$W_G = \bigcup_{i=1}^{n} W^i \quad \text{(II.3.)}$$

with

$$W^i = \{\underline{y}/\underline{y}^i_{lb} \le \underline{y} \le \underline{y}^i_{ub}\} \quad \text{(II.4.)}$$

the system II.1. can be represented in each subspace W^i by a set P^i of linearized system equations

$$P^i = \{P^i(s, \underline{a}^{ij}) \;/\; P^i(s, \underline{a}^{ij}) = L\{S^j\} \wedge \underline{y}^j \in W^i;\; j = 1, \dots k\} \quad \text{(II.5.)}$$

with $L\{\ \}$ the Laplace operator

S^j the linearized system equations in state space representation at the operating point $\underline{y}^j$

$P^i(s, \underline{a}^{ij})$ the $(n_y \times n_u)$ transfer function matrix of the linearized system in subspace W^i at the set point

$\underline{a}^{ij}$ Scalar vector of parameters of the transfer function matrix.

The scalar k is the number of different fixed parameter set points $\underline{y}^j$, and its choice depends on the admissible error ε between the real nonlinear system and the linearized one measured by

$$\| \underline{y}(t)_{nonlinear} - \underline{y}(t)_{linear} \| < \varepsilon \quad \text{(II.6.)}$$

for all t, u(t) and

$$y(0)_{nonlinear} = y(0)_{linear}$$

By the above descripted linearization procedure the set of linear transfer function matrices

$$\bigwedge_{i=1}^{n} P_i \quad \text{(II.7.)}$$

will represent the system (II.1.) with an accuracy determined by ε.

Some practical considerations have to be made:

i. The choice of the single subspaces W^i or the corresponding upper and lower bounds is practically dependent on the degree of nonlinearities in system equations (II.1.). A rule of thumb can be that the structure of each $P^i(s, \underline{a}^{ij})$ are similar in each subspace W^i.

ii. In practice the error ε can be measured only with great amount of simulation efforts, thus k will only be estimated to fulfill the eq. (II.6.).

iii. Another practical assumption to be made is that $n_y = n_u$.

Following this linearized nonlinear system description a variable structure controller can be designed by robust methods. For each subspace W^i a robust controller C^i can be calculated with a prescribed closed-loop behaviour for all parameter variations $\underline{a}^{ij}$; j=1, ..., k, so that C^i will be switched on if the system operates in W^i.

The whole design procedure can be summarized as follows:

i. Linearize the nonlinear system at n•k set points and obtain $P^l(s, \underline{a}^l)$, l=1,...,n•k,

ii. Determine each workspace W^i, i=1,...,n and the corresponding $P^i(s, \underline{a}^{ij})$, i=1,...,n; j=1,...,k, depending on the charakteristics of each $P^l(s, \underline{a}^l)$, e.g. pole-zero-configuration or order of the transfer function.

iii. Evaluate C^i for each W^i by the robust control design method (s. Section III.).

In this context of variable structure controller it must be pointed out that problems can arise during switching from C^i to C^{i+1} due to different gains and structures of each controller C^i which will result in discontinuous control signals. In practice effects of such signals on manipulator hardware should be taken into account by some smoothing methods.

III. ROBUST CONTROL DESIGN METHOD

Following the control concept described above one will obtain in each subspace W^i a n x n ($n_y = n_u = n$) system which is defined in Fig. III.1, with the matrices:

Plant P^i, compensation G^i and prefilter F^i.
The plant will be specified in V-canonical form as $P_i = [p^i_{uv}] = (I - \Lambda^i\,\Theta^i)^{-1}\,\Theta^i$, with $\Lambda^i(s)$ a diagonal matrix of transfer functions and $\Theta^i(s)$ a matrix with nondiagonal elements only. For the compensation $G^i = [g^i_{uv}]$ a diagonal matrix suffices as well as for the prefilter-matrix $F^i = [f^i_{uv}]$. All matrices are n x n, with transfer functions als elements.

E.g. every transfer function of the plant $P^i_{uv}(s, \underline{a}^{ij}_{uv})$ depends on a parameter vector $\underline{a}_{uv}$, that varies within the bounds

$$\underline{\varepsilon}^i_{1uv},\ \underline{\varepsilon}^i_{2uv}$$

$$\text{with} \quad \underline{c}^i_{1uv} \le \underline{a}^{ij}_{uv} \quad \underline{c}^i_{2uv}. \quad \text{(III.1.)}$$

The goal of the design process is to ensure that the closed loop transfer function matrix $T^i = [t^i_{uv}(s)]$ in Fig. III.1 satisfy conditions of the form $A^i_{uv}(\omega) \le |t^i_{uv}(j\omega)| \le B^i_{uv}(j\omega)$. In our case one wants all non diagonal elements of T^i to be small. For such a case, denoted as 'basically non-interacting', $A^i_{uv} = \emptyset$, for $u \neq v$, and the corresponding f^i_{uv} are also to be taken zero.

The design philosophy is to convert the MIMO problem into a number of single-loop problems, with parameter uncertainty, external disturbances, and performance tolerances derived from the original problem.
(For simplicity the high index i will be neglected in the following.)

The equivalent single-loop problem:
For the sake of completeness, the derivation, which is quite short without the proof, of the equivalent single-loop problems, is presented. In Fig. III.1 one can see that,

$$\underline{y} = PG(F\underline{w} - \underline{y}) \quad \text{(III.2.)}$$

and transformed

$$(P^{-1} + G)\,\underline{y} = G\,F\underline{w}. \quad \text{(III.3.)}$$

Substitute $T\underline{w}$ for $\underline{y}$ in the last equation, giving

$$(P^{-1} + G)\,T\underline{w} = G\,F\underline{w} \quad \text{(III.4.)}$$

Let the elements of P^{-1} be denoted by p^1_{ui} (note that $\hat{p}_{uu} = 1/\lambda_{uu}$ and $\hat{p}_{ui} = -\Theta_{ui}$ for $u \neq i$) and let $r_v \neq \emptyset$ and $r_i \equiv \emptyset$, $i \neq v$ then the u-th element of (III.3.) is

$$\sum_{i=1}^{n} (\hat{p}_{ui} + g_{ui})\,T_{iv} = \sum_{i=1}^{n} g_{ui}\,f_{iv} \quad \text{(III.5.)}$$

and with the special condition, that G and F are diagonal one can write t_{uv} with $u \neq v$ as

---

1) lb: lower bound
ub: upper bound

$$t_{uv} = \frac{-d_{uv}/\hat{p}_{uu}}{1 + g_{uu}/\hat{p}_{uu}} \quad \text{(III.6.)}$$

and t_{uu} as

$$t_{uu} = \frac{(1/\hat{p}_{uu})\, g_{uu}\, f_{uu} - d_{uu}/\hat{p}_{uu}}{1 + g_{uu}/\hat{p}_{uu}} \quad \text{(III.7.)}$$

with

$$d_{uv} = \sum_{i=u}^{n} \hat{p}_{ui}\, t_{iv}.$$

Now, it is easily seen that the above equations define n single loop design tasks, with the equivalent plants $q_{uu} = \hat{p}_{uu}^{-1}$ and n disturbance inputs for each loop.

One can devide the disturbance inputs in (n-1) "non-diagonale-disturbances" (NDD) which are $- d_{uv}\, q_{uu}$ ($v \neq u$) and one "diagonal disturbance" (DD), that is $- d_{uu}\, q_{uu}$.

However, for every single loop we have a design constraint, that can be written as

$$A_{uu} \le \left| \frac{f_{uu}\, g_{uu}\, q_{uu} - d_{uu}\, q_{uu}}{1 + g_{uu}\, q_{uu}} \right| \le B_{uu} \quad \text{(III.8.)}$$

In order to separate the DD-term we obtain

$$A_{uu} \le \left| \frac{f_{uu}\, g_{uu}\, q_{uu}}{1 + g_{uu}\, q_{uu}} \right| \pm \left| \frac{d_{uu}\, q_{uu}}{1 + g_{uu}\, q_{uu}} \right| \le B_{uu} \quad \text{(III.9.)}$$

with

$$a_{uu} \le \left| \frac{f_{uu}\, g_{uu}\, q_{uu}}{1 + g_{uu}\, q_{uu}} \right| \le b_{uu} \quad \text{(III.10)}$$

and

$$c_{uu} \ge \left| \frac{d_{uu}\, g_{uu}}{1 + g_{uu}\, q_{uu}} \right| \quad \text{(III.11)}$$

with

$$A_{uu} = a_{uu} - c_{uu} \qquad B_{uu} = b_{uu} + c_{uu}. \quad \text{(III.12)}$$

The above equations define two kinds of constraints on the n-th single loop compensator design.

$$\text{I.} \quad \Delta \left| \frac{g_{uu}\, q_{uu}}{1 + g_{uu}\, q_{uu}} \right| \le \frac{b_{uu}}{a_{uu}} \quad \text{(III.13)}$$

$$\text{II.} \quad \left| \frac{d_{uu}\, q_{uu}}{1 + g_{uu}\, q_{uu}} \right| \le c_{uu} \ \text{(DD-term)} \quad \text{(III.14)}$$

For each loop a third kind of constraints is defined by the (n-1) NDD-terms, which are

$$\text{III.} \quad \left| \frac{d_{uv}\, q_{uv}}{1 + g_{uu}\, q_{uu}} \right| \le B_{uv} \ \text{(NDD-terms)} \quad \text{(III.15)}$$

In the design process the compensators G_{uu} have to be choosen such that I, II and III are fulfilled.

Single loop design process is as follows:
The first step is to choose a parameter set a_{uv}^{n} as nominal. In the second step, with the help of a Nichols chart, with its loci of constant $|L/1+L|$ values, the first kind of constraints can be compiled into bounds, which define forbidden regions for the locus of the nominal open loop $R_{uu}^{n} = g_{uu}\, q_{uu}^{n}$ for a number of selected frequency points w_i.

The modified Nichols chart with loci of constant $(1+L)^{-1}$ is useful to calculate forbidden regions for II and III.

The compensator g_{uu} must be choosen in such a way, that the nominal open loop l_{uu} doesn't enter into the forbidden regions and for minimum controller gain coincides with the bounds.

For stability of the MIMO problem the open-loop frequency responses l_{uv}^{n} have to fulfill the Nyquist-criterion.

For choosing the regulators g_{uu} we have defined a vector performance index that contains the above developed boundaries and other terms.

The index is used to optimize the parameter of g_{uu} in an interactive procedure.

After the compensator design is done a prefilter is easily calculated using (III.10).

Some concluding remarks are to be made:

i. In general a controller can cope with parameter variations and disturbances by high feedback gains assuming that they don't cause stability problems. In the case of the above discussed single loop robust design method one obtains a controller with minimum gain to hold the design specifications.

ii. In the proposed design method the controller gain will surely depend on the subspace specification. For larger subspaces one will obtain higher gains, and so the hardware system will need higher input energy.

IV. SIMULATION EXAMPLE

Process Model

The nonlinear dynamics of a manipulator with a series of n one-degree-of-freedom joints can be formulated via the Lagrangian dynamics for each joint either revolute or prismatic by

$$F_i = G_i(\underline{q}) + \sum_{j=1}^{n} J_{ij}(\underline{q})\, \ddot{q}_i + \sum_{j=1}^{n} \sum_{k=1}^{n} C_{ijk} \cdot \dot{q}_j \dot{q}_k + F_{R_i}(\dot{q}_i) \quad \text{(IV.1.)}$$

where F_i is the generalized force either a torgue for a revolute joint or a force for a prismatic one, q_i as the joint variable being either an angle or a distance, G_i represents the gravity term, J_{ij} the intertial terms, C_{ijk} the Coriolis and centrifugal terms and F_{R_i} the friction term.

For a two-degrees-of-freedom manipulator considered in this paper as an example and shown in Fig. IV.1 its dynamics are represented by

$$F_1(t) = (m_a+m_1)\cdot\ddot{r}(t) - [(m_a+m_1)\cdot r(t) - \frac{m_a-1}{2}]\dot{\phi}^2(t) + f_R(\dot{r}) \quad \text{(IV.2.)}$$

$$M(t) = [K+(m_a+m_1)r^2(t) - m_a l\, r(t)]\ddot{\phi}(t) + [(m_a+m_1)r(t) - \frac{m_a l}{2}] \cdot 2\dot{r}(t)\dot{\phi}(t) + f_R(\dot{l}) \quad \text{(IV.3.)}$$

with

$$K = \frac{m^* r^{*2}}{2} + \frac{m_a l^2}{3}.$$

The friction terms $f_R(\dot{r})$ and $f_R(\dot{\phi})$ were approximated by linear terms; these are

$$f_R(\dot{r}) = -a\dot{r} \quad \text{and} \quad \phi(\dot{\phi}) = -b\phi.$$

The parameters m^*, m_a, l and r^* have fixed values:

$$m^* = 14\ \text{kg} \qquad m_a = 5\ \text{kg} \quad \text{(IV.4.)}$$
$$l = 1{,}5\ \text{m} \qquad r^* = 0{,}2\ \text{m}.$$

In order to get a good real-world approximation the simulations are carried out with the parameter variations for hand load and friction changes:

The global workspace was given by

$$W_G: \quad 0{,}8\ m \le r \le 1{,}3\ m \qquad (IV.5.)$$
$$0 \le \phi \le 1{,}57\ rad$$
$$-1\ m/s \le \dot{r} \le 1\ m/s$$
$$-0{,}7\ rad/s \le \dot{\phi} \le 0{,}7\ rad/s$$

Controller Design

The first task for the design of the variable structure controller is to determine the bounds of each workspace. In the case of the example at hand the nonlinearities in the system dynamic equations are caused by the translational joint variable r and the velocities of the joint variables $\dot{r}$ and $\dot{\phi}$.

Thus each workspace is defined by

$$W^i = \{r \wedge \dot{r} \wedge \dot{\phi} \ / \ r^*_{lb} \le r \le r_{ub}; \dot{r}_{lb} \le \dot{r} \le \dot{r}_{ub}; \dot{\phi}_{lb} \le \dot{\phi} \le \dot{\phi}_{ub}\} \qquad (IV.6.)$$

The linearization of the nonlinear equations at significant points of the $\dot{r}$-$\dot{\phi}$-space (see marked points in Fig. IV.2) was carried out and the pole-zero-configuration for each linearization point was analysed. The linearized systems vary from stable and minimumphase to unstable and nonminimumphase systems. For the demonstration the worst case of the shaded area in Fig. IV.2 is chosen where the system has unstable and nonminimumphase properties. The corresponding workspace W^1 is defined by

$$W^1: \quad 0{,}8\ m \le r \le 1{,}3\ m \qquad (IV.7.)$$
$$-1\ m/s \le \dot{r} \le -0{,}3\ m/s$$
$$-0{,}7\ rad/s \le \dot{\phi} \le 0{,}7\ rad/s$$

Besides the parameter variations $\underline{a}^{1j}$ in W^1 due to nonlinear dynamics on-line load mass and friction changes are considered for the controller design.

The simulation results in the following demonstrate the performance of the controller inside and outside of the specified workspace W^1 with also changes in load mass and friction. Simulations with changing workspaces and switching controllers are now under work.

Simulation Results

Fig. IV.3 and IV.4 illustrate the case where the system is nearly in the workspace W^1 the angle velocity is for a short moment outside of the specified bounds. The step responses for the set point input $r=1m$, $\phi=-10^o$ of the simulated nonlinear system meets very well the design specifications.

In case of Fig. IV.3 the simulation was done with the load mass $m_l=0{,}5$ kg and the friction parameters a=3, b=3. In Fig. IV.4 these parameters were chosen as

$$m_l = 1{,}5\ kg \quad a = 4 \quad b = 4.$$

Despite these changes the step responses don't vary from each other which is a result of robust control design method.

Fig. IV.5 and Fig. IV.6 show a simulation example with parameter changes as in Fig. IV.3 or Fig. IV.4 respectively outside of the workspace W^1 due to the positive velocity of the translational joint $\dot{r}$. The used controller was the same as in Fig. IV.3 and Fig. IV.4 and the set point was given us $r=1{,}1m$, $\phi=10^o$.

The simulation shows nearly the same behaviour of the sytem as in the above mentioned cases, but one have to note a small overshooting in the r and ϕ trajectories. In the next two figures IV.7 and IV.8 a further simulation example is illustrated. It can be seen that the trajectories are outside of the workspace W^1 over a wide range.

Due to this fact the trajectories of r and ϕ have an unacceptable overshooting. This demonstrates, that a change of the controller is necessary as assumed in the overall control concept.

V. CONCLUSIONS

A variable structure controller concept is introduced which can cope with the nonlinear manipulator dynamics especially with the dynamic couplings. Subdividing the whole manipulator workspace in subspaces in which the nonlinear system is approximated by a set of linearized system equations with parameter variations the concept utilizes for each subspace a linear multivariable controller designed by a robust method to cope with the parameter variations in the corresponding subspace. Additionally the concept allows consideration of other possible on-line parameter changes such as different load masses and friction changes.

In this paper the concept is demonstrated by an simulation example of a 2 d.o.f. manipulator arm. In this example only one subspace is defined and the good performance of the designed controller is shown inside and outside of this subspace. In the ongoing work simulations are planned for adjucent subspaces to find out the effects of variable structure controller such as discontinous trajectories. In the long range the design will be extended to a 6 d.o.f. manipulator arm. In this context it must be pointed out that the computational effort during the design of the controller will be very high, and it is only possible by a huge software package as it was used for the above simulation example [8].

IV. REFERENCES

[1] I. Horowitz, M. Sidi: Synthesis of feedback systems with large plant ignorance for prescribed time domain tolerances. Int. J. Control, 1972, Vol. 16, No. 2

[2] I. Horowitz: Quantitative synthesis of uncertain multiple input-output feedback systems. Int. J. Control, 1979, Vol. 30, No. 1.

[3] I. Horowitz, D. Loecher: Design of a 3x3 multivariable feedback system with large plant uncertainty. Int. J. Control, 1981, Vol. 33, No. 4.

[4] Koivo, A.J., Paul, R.P.: Manipulator with self-tuning controller. Proc. Int. Conf. on Cybernetics and Society, Boston, MA, USA, 8-10 Oct. 80

[5] E. Freund: Input Isolation and Group Decoupling of Nonlinear Systems. Proc. of IACC, San Fransisco, CA, USA, June 1977

[6] W. Paztelt, W.: A Robot Position Control Algorithm for the Grip Onto An Accelerated Conveyor Belt. Proc. 12th Int. Conf. on Industrial Robots, Paris, France, June 9th-11th, 1982

[7] H. Kranz: Untersuchung von Reglersynthesemöglichkeiten für nichtlineare Mehrgrößensysteme am Beispiel eines Industrieroboters. Diplomarbeit am Fachgebiet Regelsystemtheorie der TH Darmstadt (directed by E. Ersü and W. Neddermeyer)

[8] W. Carls: Entwicklung eines Programmpakets zum quantitiven Entwurf von Mehrgrößensystemen nach Horowitz Studienarbeit am Fachgebiet Regelsystemtheorie der TH Darmstadt (directed by W. Neddermeyer and A. Gräser)

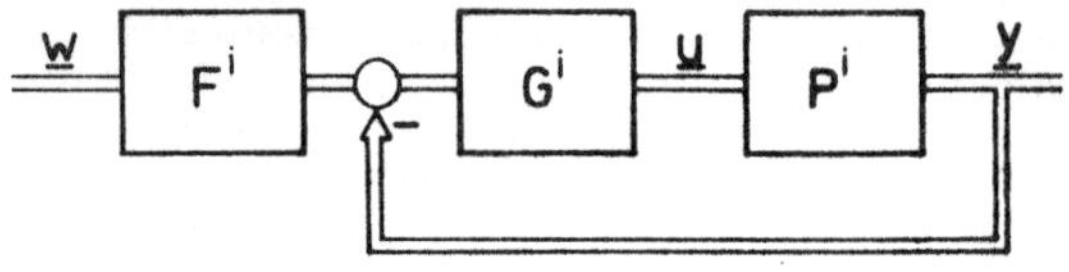

Figure III.1

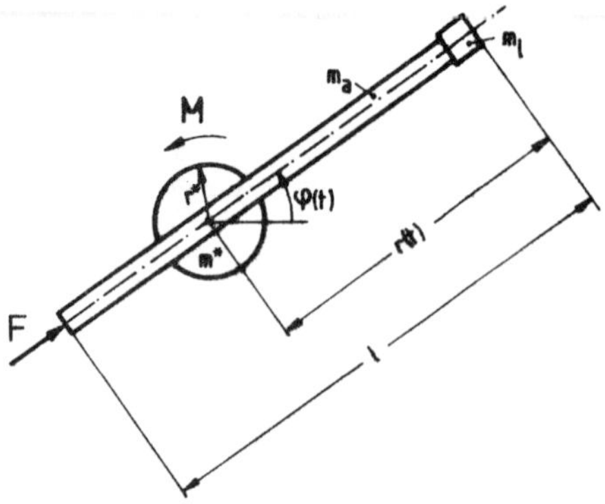

Figure IV.1

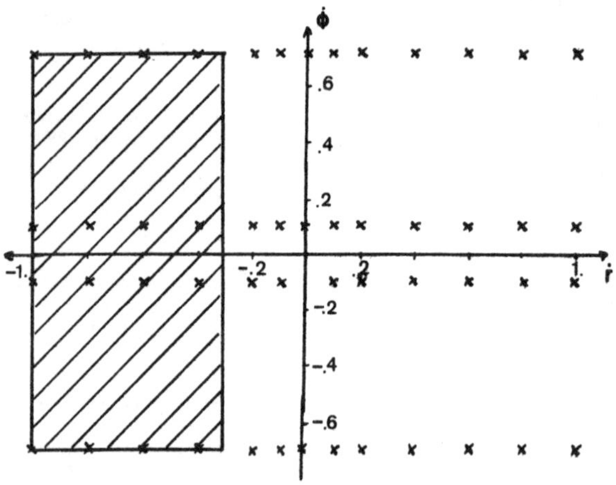

Figure IV.2

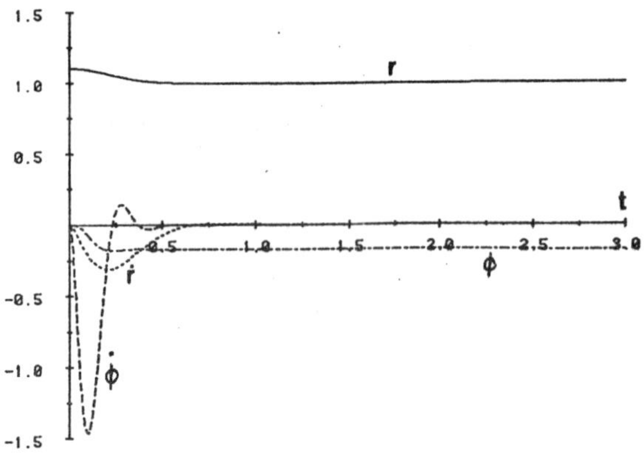

Figure IV.3

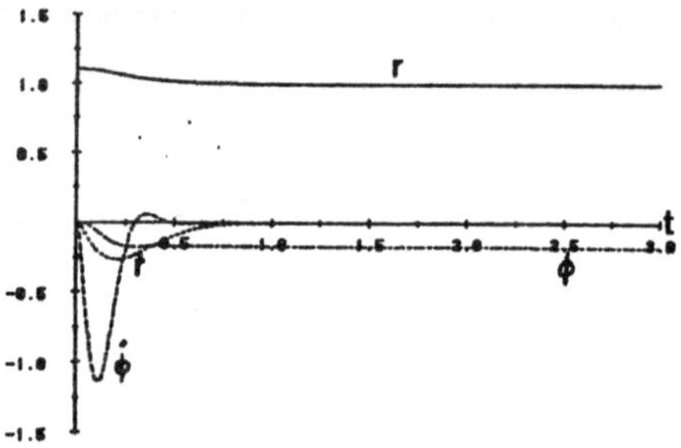

Figure IV.4

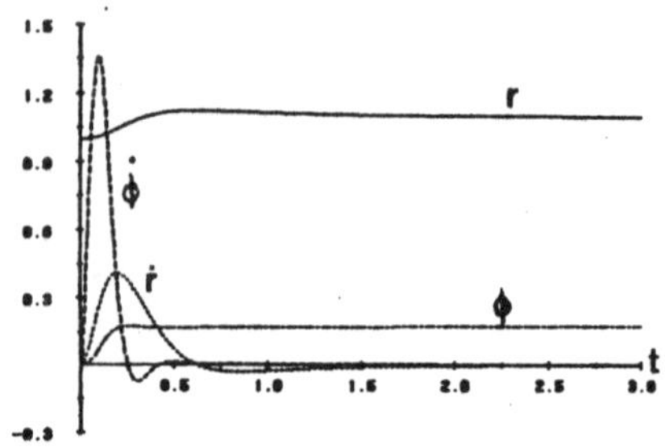

Figure IV.5

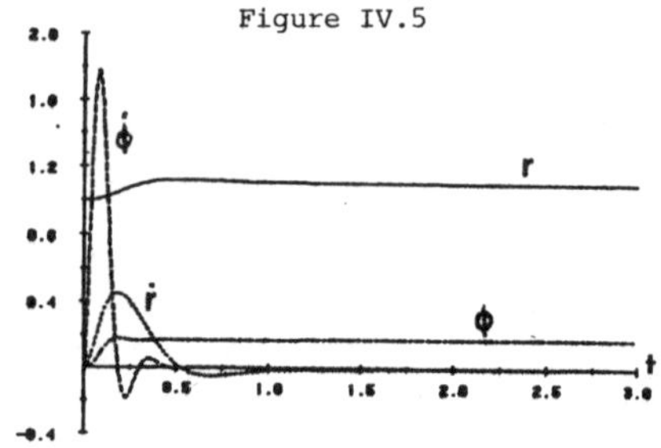

Figure IV.6

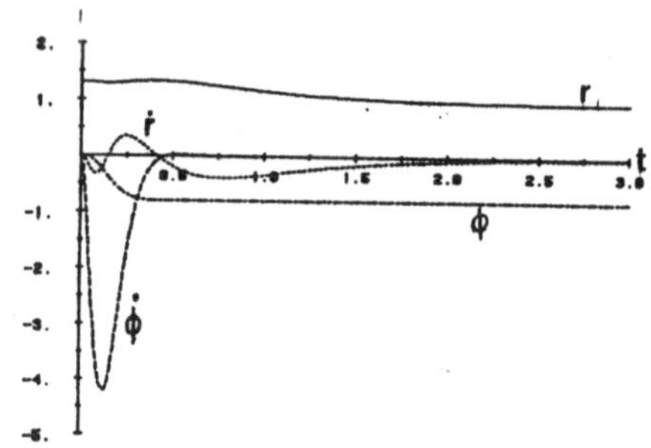

Figure IV.7

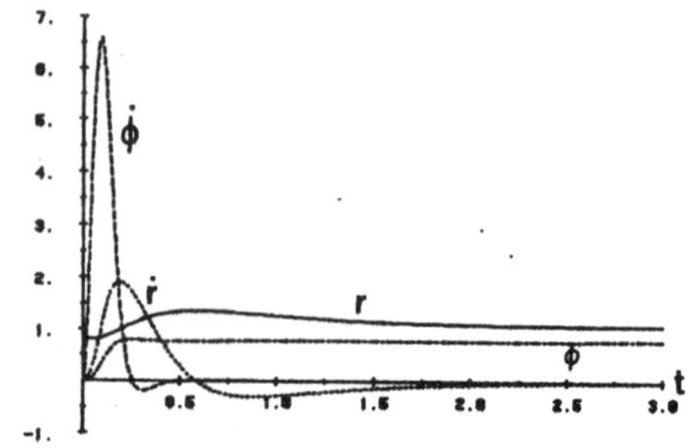

Figure IV.8

MODULAR ROBOT SYSTEMS

Moore P R, Weston R H,
Thatcher T W, Gascoigne J D

Department of Engineering Production
Loughborough University of Technology
United Kingdom

ABSTRACT

This paper provides an overview of currently available modular robot systems and considers the design and development of a new family of single degree of freedom mechanical units and associated control system modules which are physically linked, via standard faceplate interfaces, and electrically coupled, via standard power and data buses, to form a multi-axis group or groups which function to handle or process components. The family of modules described are designed to have far wider application than conventional pedestal mounted industrial robots and contain control system elements which function to assist industrial users in distributing "intelligent" mechanical hardware throughout manufacturing systems.

INTRODUCTION

The 1980's promise to be the decade in which flexible automation schemes, including robotic systems, find wide acceptance and application in industry. The reasons for introducing robots are diverse. Potential benefits from their implementation can be economic cost effectiveness, environmental safety, product/process development, and manufacturing flexibility. The degree and scale of investment in introducing robot systems can range from the purchase of a relatively low cost pick and place robot to the installation of an extremely high cost flexible manufacturing system (FMS).

A popular concept of a robot is that of a complex machine which can provide considerable dexterity and flexibility in automating complex tasks which are normally accomplished by manual operatives. Many industrial robots conform to this concept and can be associated with significant capital cost, particularly if complex sensors such as vision systems are required. However, robots will play just as important a role in performing simple handling and processing tasks which are commonplace in manufacturing industries. Similarly, FMS is considered by many to be synonymous with large scale factory automation schemes. However, a generation of industrial robots and support systems are now emerging which will facilitate the distribution of manipulator and computer controls throughout manufacturing plants in a piecemeal fashion (1). In such distributed automation schemes, manipulators ranging from the simple to complex will be linked via computer based communication systems to serve and process components within an integrated control scheme.

Universal pedestal mounted industrial robots can suffer two major shortcomings:

(i) a high degree of sophistication, and hence high capital cost, is usually necessary to ensure flexibility so that often a robot has many redundant capabilities in relation to the task requirements;

and

(ii) a fixed mechanical structure which may not be ideally suited to individual tasks, but may allow a range of tasks to be performed in a sub-optimal manner. For example, the kinematic requirements of a press loading robot differ considerably from those of an arc welding robot.

For the above reasons a wide variety of industrial robots have been manufactured to meet demand and yet there are many application areas for which there is no industrial robot of suitable mechanical form, eg single axis articulation. An alternative approach is to utilise modular elements from a family of modules designed to allow users to construct a robot with specific kinematic functions and control system requirements. This paper describes the elements of such a modular robot system which is being designed and engineered within a Science and Engineering Research Council funded research programme involving Loughborough University of Technology (LUT) and Martonair Ltd, UK.

It should be stated that the term "robot" will be used in the widest context embracing all categories of workhandling systems, manipulators and robots. This includes the simple end-stop arm and the multi-degree of freedom contouring robot.

ACTUATORS FOR INDUSTRIAL ROBOTS

Robotics is a discipline that embraces many technologies. In designing and marketing a modular robotic system it is necessary to consider many factors relating to each of these technologies. The choice of actuator is one such factor which is extremely important as this choice has significant implications with regard to mechanical design, control system design, dynamics and kinematics, application areas for which such systems are suitable, and cost.

Robot actuators usually fall into one of two general categories, Electrical and Fluidic (2), which can be considered to contain the actuator types described by Table 1. However, it is not uncommon for robots to use hybrid drives comprising two or more actuator types. For example, the Remek PAM, which has combined pneumatic and stepper motor driven axes, and the Dainichi Sykes range which uses pneumatic balancing in conjunction with low power dc motors

ELECTRIC ACTUATORS	(i)	DC MOTORS
	(ii)	AC MOTORS
	(iii)	STEPPER MOTORS
FLUIDIC ACTUATORS	(i)	HYDRAULIC CYLINDERS AND MOTORS
	(ii)	PNEUMATIC CYLINDERS AND MOTORS

TABLE 1 : ROBOT ACTUATORS

Electric Actuators

Electric actuators for robots are usually motors of the dc type although there are some examples of commercial industrial robots using ac motor or stepper motor drives. Different excitation techniques produce various selected performances although generally all electric motors are fairly low torque high speed devices. The literature is rich with technical appraisals of electric motors and their application in servo systems (3).

In servo applications, typical dc motors offer superior torque and power characteristics when compared to ac motors. Similarly ac motors compare favourably with stepper motors which supply only modest torques and mechanical power (750W). Electric motors are usually used in conjunction with some form of gearing to produce better torque characteristics. Obviously any necessary mechanical gearing introduces associated problems such as backlash, stiction, friction, and increased system inertia. However, taking these factors into account an increasing trend is apparent towards the use of electric actuators as robot drives. The dc motor is fast becoming the most popular robot actuator often being utilised in preference to fluidic actuators particularly where high accuracy positioning is required. Stepper motors will continue to find application where fine positioning is required at low speeds or over a short traverse.

Fluidic Actuators

Fluidic actuators require some form of power source, normally an electrical motor to generate the fluid power. In hydraulic systems this is normally a small electrical pump that stores hydraulic power in an accumulator, the power requirement of the actuator determining the size of the accumulator. In the pneumatic system an electrical compressor normally supplies a reservoir that is large enough to cater for the demands of a whole manufacturing area. In this way pneumatic actuators can simply tap into the pneumatic supply line at any relevant point in the system. Hydraulic drives have a disadvantage in that it is not normally practical to site the actuator remotely with respect to its power source.

Generally, fluid actuators exhibit a higher power to weight ratio than equivalent electric actuators. In particular, as high working pressures (100 bar or more) can be employed, hydraulic actuators are ideal for producing extremely large output powers (4). Correspondingly pneumatic actuators are more suited to medium power applications normally operating at a maximum pressure of 10 bar.

The Application of Hydraulic Actuators

Hydraulic actuators have been used since the introduction of the first commercial industrial robots for high, medium, and low power applications, particularly where servo controlled positioning of loads and/or tools is required. For high power applications they remain the state of the art for the point to point positioning and/or contouring of large workpieces. However, as stated above, for low to medium power applications electric dc motors are tending to supercede their hydraulic counterparts. However, in some applications it is most convenient to use hydraulic actuators with only end stop control as the high working pressures allow very small actuators to be used (a significant advantage in end-effector design).

The Application of Pneumatic Actuators

Pneumatic actuators have found widespread use as drive elements for limited function robots providing pick and place motion to pre-set mechanical end-stops. At present the largest sector of the robot market (in terms of volume of units) is served by such pneumatically actuated robots which demonstrate relatively high traverse speeds, intrinsic safety in many hazardous environments, and relatively low cost.

However, for applications requiring point to point positioning of a workpiece and/or tools, a number of problems (resulting from the use of air as a working fluid) are to be found if pneumatic actuators are used. In particular, the compressibility of air and the highly non-linear dynamics of such drive systems has severely limited their use. Various researchers have analysed the use of pneumatic actuators within position and velocity servomechanism (5,6,7,8) but only now is it possible to introduce at a realistic cost, a measure of compensation for the effects of compressibility and non-linearities through the application of microelectronics and appropriate software compensation algorithms. The use of a pneumatic actuator as the drive element for closed-loop position and velocity controls would have significant potential benefit particularly in regard to capital cost if similar accuracy and dynamic response, to that demonstrated by hydraulic drives, is achieved.

MODULAR ROBOT SYSTEMS CURRENTLY AVAILABLE

Before describing the features of the Loughborough University/Martonair Ltd modular systems work, it is necessary to review commercially available modular robots and their associated control systems. The majority of the present generation of modular robots have been designed and manufactured in Europe. Table II summarises the major features of such systems.

Bosch Automation (West Germany)

At present, Bosch Automation offer one of the most comprehensive range of modular equipment which can be used to construct user defined robot systems. A user can purchase catalogue units, or modules from Bosch to form complete production installations to meet his specific needs including conveying systems, manual/automatic workstations, pallet systems, pick and place robots with pneumatic actuation, and dc electric motor driven robots with servo controls. In respect of their modular robotic devices, Bosch offer both rotational and translational single axis modules which range in size and lifting capabilities. Both dc electric motor and pneumatic driven modules are available (typical characteristics of an electric translational module being: maximum velocity 1.0 ms $^{-1}$; accuracy +0.1 mm; repeatability +0.05 mm; with variable payload dependant upon module size) which can be configured into multi-axis workhandling equipment and fitted with catalogue wrist and gripper units.

Bosch also market control system components which enable sequence control of multi-axis structures. Within the control system range of modules, units can be purchased to provide servo control of both position and velocity for dc motor driven axes and/or to achieve end stop control of pneumatically actuated axes. Although some of the advanced control system features available with the present generation of pedestal robots are not available to the user of Bosch modular robot systems it should be emphasised that far greater mechanical flexibility can be achieved.

Electrolux/Asea (Sweden)

The Asea MHU range of robots are pneumatically actuated to adjustable mechanical stops. The range comprises three sizes of system with each system being pedestal mounted to provide cylindrical co-ordinate articulation (maximum payload 15 Kg). The systems are modular in respect of construction methods and flexibility allowing more than one manipulator arm to be mounted on the pedestal at any one time.

Eracond (Japan)

Eracond offer a range of low cost arms and grippers which are pneumatically actuated to mechanical end stops. These arms and grippers are designed to handle small components and to be integrated into a larger workhandling system. Eracond are one of the few manufacturers to offer a reasonably comprehensive range of robot end effectors (many robot manufacturers consider the end effector or gripper to require individual tooling design dependent upon the application).

Felss (West Germany)

The Felss range of pneumatically actuated handling modules includes rotary and translational modules which, in conjunction with a gantry element, can be used to handle modest payloads (up to 15 Kg). Modules from the Felss range allow particular flexibility in constructing machine loading equipment but can also be utilised in general automation systems.

Fibrotakt/Fibromanta (West Germany)

Fibromanta offer an extensive range of modular workhandling units with either hydraulic or pneumatic actuation. Modules from this range can be selected to handle payloads of 4 to 1700 Kg with maximum working speeds within the range 0.1 to 1.15 ms^{-1}. Fibromanta usually supply control systems to provide motion to mechanical end stops but can provide hydraulic units with servo controlled point to point motion.

H H Freudenberg (West Germany)

Freudenberg offer a range of either pneumatically or hydraulically actuated rotary and translational modular units for workhandling. Control systems are normally supplied to provide axis motion to mechanical end stops.

INA Automation (United Kingdom)

INA Automation market a range of pneumatically actuated rotary and translational modules generally aimed at handling relatively light components (2 to 5 Kg). Their modular units include a general purpose designed gantry unit. INA also offer special purpose built automation equipment.

Menziken (Switzerland)

Menziken offer a range of rotary and translational modular units with pneumatic actuation. The largest modules are designed for payloads in excess of 40 Kg, but the majority of the range is designed to move smaller payloads (5 to 10 Kg) quickly.

These units represent one of the truly modular workhandling systems presently available and are ideally suitable for inclusion within specialised automation systems.

Sciaky/Roboflex (France)

Sciaky produce an extensive range of hydraulically actuated modules which are designed to manipulate large payloads over a considerable distance (up to 10 m). These modules form one of the most comprehensive ranges of hydraulically actuated modular systems, and workhandling equipment can be configured to provide as many degrees of freedom as dictated by the application. Sciaky modular units are supplied with closed-loop controls providing point to point axis motion.

Robomat AG (West Germany)

Robomat supply a range of pneumatically actuated modules demonstrating end-stop motion to mechanical stops. Both rotary and translational modules are available which are designed to handle small payloads of less than 1 Kg.

Robotic Systems (United Kingdom)

Robotic systems produce a range of modular workhandling units driven by electrical printed circuit motors. The modules all demonstrate rotary motion which limits the possible axis configurations. The modules are supplied with closed loop position controls and are fitted with electromagnetic brakes to increase the kinematic stiffness of multi-axis systems configured from the modules. Payloads of up to 45 Kg can be handled.

A NEW MODULAR ROBOT SYSTEM

The aim of collaborative research and product development, involving Loughborough University of Technology and Martonair Ltd, UK has been to design and develop an autonomous family of modular mechanical and control system elements to allow industrial users maximum flexibility in constructing robotic devices. To achieve these objectives it is clearly necessary to produce modules which can accomplish the simplest workhandling or processing task, at a realistic cost, but also allow a large number of modules to be combined to accomplish complex tasks.

The brief appraisal of commercially available modular systems presented above serves particularly to illustrate the need for an autonomous family of control system modules. A user may wish to integrate the functioning of a complete sector of plant involving the co-ordinated working of a large number of mechanical modules with existing production machines. A user should be able to form an integrated control scheme to achieve his desired control and monitoring requirements without the need for extensive in-house development and testing. Our appraisal of existing modular systems has also served to highlight the importance attached to choice of actuator and its effect on the cost and characteristics of associated mechanical modules. Pneumatically actuated modules are relatively low cost but at present all commercial modular systems employing pneumatic actuation provide only end stop motion. That is movement to mechanical stops. Many component processing, handling and assembly operations can be performed by employing a number of axes moving to preset mechanical stops but clearly a considerable increase in flexibility and range of application is accomplished by providing servo control of position for as many modules as the application dictates. Particularly for batch production with components of variable size.

The foregoing observations coupled with the commercial interests of Martonair Ltd served to form the following programme of research and development, which began in late 1981:

(i) to appraise existing modular systems;

(ii) to design and test single axis mechanical modules with pneumatic actuation;

(iii) to design and construct microprocessor-based controls to achieve servo control of single axis pneumatically driven modules by providing software compensation for the effects of compressibility and non-linearities;

(iv) to design and test a heirarchy of control system modules designed to allow the integration of axis units with other production equipment and systems attempting to maximise flexibility but minimise implementation and reconfiguration costs;

(v) to design and construct prototype mechanical and control system modules, evaluating their characteristics in host applications;

(vi) to perform an application study to consider industrial applications over and above those served by conventional pedestal robots;

(vii) to extend and modify the range of modular units as dictated by the findings of the applications study. Such extensions possibly including electric motor driven mechanical modules.

In summary the choice of pneumatic drive systems for first generation mechanical units was made for the following reasons:

(a) substantial economic advantages should be apparent,

(b) pneumatic systems are generally available and acceptable within manufacturing plants,

(c) pneumatic drives lend themselves to the design of modular elements,

(d) pneumatic end stop devices form the largest single sector of robotic devices used in industry,

(e) the collaborating company, Martonair Ltd, have considerable expertise and a ready market (both in the UK and Europe) in the application of industrial pneumatic systems.

MECHANICAL HARDWARE SYSTEM DESIGN

A family of mechanical modules will be developed based on existing prototype single degree of freedom units under test. These modules will be used to construct manipulators with kinematic structure and degrees of freedom as determined by the specific application. Each module will contain its own pneumatic actuator and feedback system. Both end stop and servo controlled modules will be developed, the servo controlled units having an integral microprocessor-based controller. Modules will be joined mechanically at standard jointing interfaces and coupled electrically through standard power and communications buses.

By employing a modular mechanical structure it is possible to construct a wide variety of axis configurations. A number of such axis groups could be formed to function together within conventional automation systems with integration achieved using a supervisory computer control module.

The characteristics of the mechanical modules necessarily determines the characteristics of axis groups. Any axis configuration must have adequate structural stiffness, both static and dynamic. Clearly the analysis of robot dynamics and the design of control system modules becomes very complex when considerable variations in mechanical structure are possible. Figures 1 to 7 illustrate the major features of one family of mechanical modules proposed. Figure 8 illustrates some of the axis configurations possible with such a family of modules.

In summary, the major features of servo controlled modules are:

(i) one degree of freedom, either rotational or translational,

(ii) pneumatic drive system,

(iii) encoder position transducer,

(iv) lock mechanism,

(v) individual axis microprocessor controller with standard interface bus to supervisory modules,

(vi) standardised faceplate location,

(vii) direct drive on all modules to alleviate problems associated with mechanical transmissions,

(viii) rotary modules would preferably have vane actuation to minimise size requirements,

(ix) axis inertia to be minimised by reducing the mass of moving elements while maintaining high mechanical stiffness.

CONTROL SYSTEM ARCHITECTURE

The proposed hardware structure is outlined in Figure 9, which illustrates the hierarchy of controls. The assignment of a single chip micro computer to each axis should not increase the price of a module unacceptably and removes the burden of realtime position control from the supervisory controller. By using such an approach the flexibility of the mechanical elements is further enhanced. At present, both individual axis controllers and supervisory computer controls have been developed in prototype form using Texas Instruments 9900 family microprocessor elements. However, single chip microcomputer controllers will be utilised in the near future to implement single axis control strategy based on the 9900 prototype designs.

Software for the supervisory controller has been designed using a Mascot top down design approach and is presently being implemented in Texas Instruments MPP Pascal. This design approach has been taken to allow supervisory system software appropriate to the application concerned, to be configured at manufacture and reconfigured by a user as his control systems requirements change. This supervisory software has been designed to supply commands to servo controlled axes via a standard data link.

Considerable effort is being made to provide sequence programming aids including teach, edit, and textural programming software to assist users unfamiliar with computer systems. Furthermore, particular attention has been paid to the need for the integrated sequence programming of more than one group of axis modules.

CONCLUSIONS

The introduction of new families of modular robots and associated control equipment will accelerate the introduction of distributed automation schemes throughout manufacturing industries. This paper has considered the features of a proposed family of both mechanical and control system modules which are based on prototype designs presently being evaluated. The mechanical modules comprise single degree of freedom units with standard jointing interfaces so that an industrial user or system builder will construct a multi-axis modular group or groups to form a robot system with kinematic structure suited to his particular application. To improve cost effectiveness, first generation mechanical modules will be pneumatically actuated but will be supplied with a microprocessor-based servo-controlled axis positioning system and mechanical locking mechanism to increase system stiffness. Supervisory control modules are also being constructed which will function to allow industrial users to integrate the sequencing of robot axis groups with other production equipment. It is anticipated that our family of mechanical modules complemented by comprehensive control system modules should find wide application in all realms of materials handling.

REFERENCES

(1) Riley, F "The use of modular flexible assembly systems as a half way path between special design and robots" 3rd Int Conf on Assembly Automation. May 1982

(2) Diaz, R G "Robotic actuators: a technology assessment" Int Comp Tech Conf 1980 on Advances in Comp. Tech. pp 225-232.

(3) Canfield, E B "Electromechanical Control Systems and Devices" John Wiley & Sons Inc, New York, pp 142-263 1965.

(4) Burrows, C R "Fluid Power Servomechanisms" Van Nostrand Reinhold Company, London, 1972.

(5) Shearer, J L "Non-linear analogue study of a high pressure servo mechanism" Trans. A S M E, 79 p 465 (1957).

(6) McCloy, D and Martin H R "Control of Fluid Power: Analysis and Design" 2nd (Revised) Ed., John Wiles & Sons Inc, New York pp 337-379, 1980.

(7) Drazan, P J and Thomas P B "Simulation of an Electro-pneumatic Manipulator System". Proc 5th Int Fluid Power Symp. Sept. 1978.

(8) Weston, R H, Miles, K E, Moore P R, Gascoigne J D and Thatcher, T W "Digitally Controlled Modular Work Handling Systems" Proc 1st Grantees Conference, SERC Robotics Initiative, Univ. of Birmingham, Sept 1982.

Features / Manufacturer	Pneumatic	Hydraulic	Electric	End Stop	Servo-control	Rotation	Translation	Modular control system
Bosch	X		X	X	X	X	X	X
Electrolux	X			X		X	X	
Eracond	X			X		X	X	
Felss	X			X		X	X	
Fibrotakt	X	X		X	X	X	X	
Freudenberg	X	X		X		X	X	
I.N.A.	X			X		X	X	
Menziken	X			X		X	X	
Sciaky		X			X	X	X	X
Robomat								
Robotic Systems			X		X	X		

Modular Robot Manufacturers

TABLE II

Figure (1) Rotary Base Unit

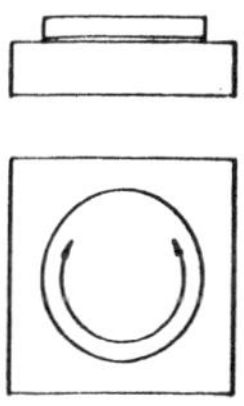

Torque 28 kpm @ 10 bar.
Speed 180°s^{-1}

Rotation 270°

Figure (2) Rotary Intermediate Unit

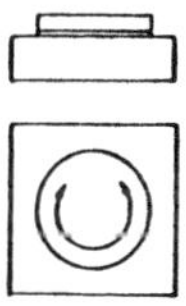

Torque 7 kpm @ 10 bar
Speed 90°s^{-1}

Rotation 270°

Figure (3) Vertical Unit

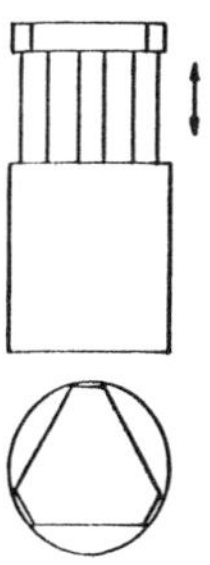

Stroke 500 mm
Piston ∅ 50 mm
Speed 300 mms

Figure (4) Gantry Unit

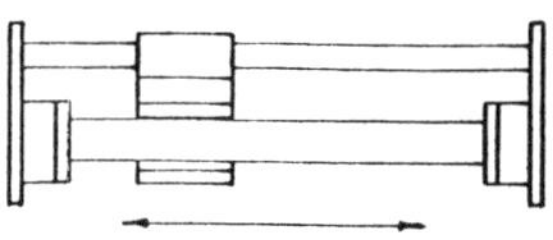

Stroke 500-1500 mm
Piston ∅ 63.5 mm
Speed 600 mms^{-1}

Figure (5) Horizontal Units

(A)

Stroke 300 mm
Piston ∅ 38 mm
Speed 400 mms^{-1}

(B)

Stroke 600 mm
Piston ∅ 50 mm
Speed 400 mms^{-1}

Figure (6) Wrist Units

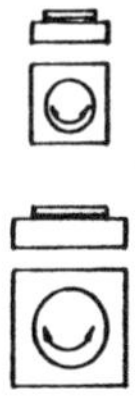

(A) Rotation 180°
End Stop only
Speed 180°s^{-1}

(B) Rotation 180°
End Stop only
Speed 180°s^{-1}

Figure (7) Wrist Translation Unit

(A)

Stroke 100 mm
Piston ∅ 38 mm
End Stop only
Speed 600 mm s^{-1}

(B)

Stroke 50 mm
Piston ∅ 25 mm
End Stop only
Speed 600 mm s^{-1}

Figure (8) Some Possible Axis Configurations

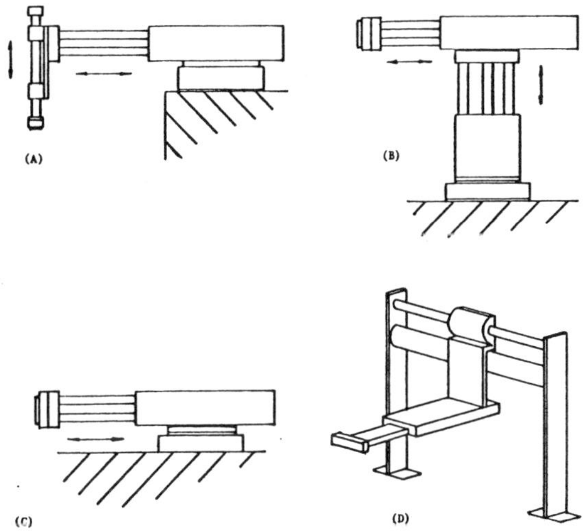

FIGURE (9) THE HARDWARE SCHEME

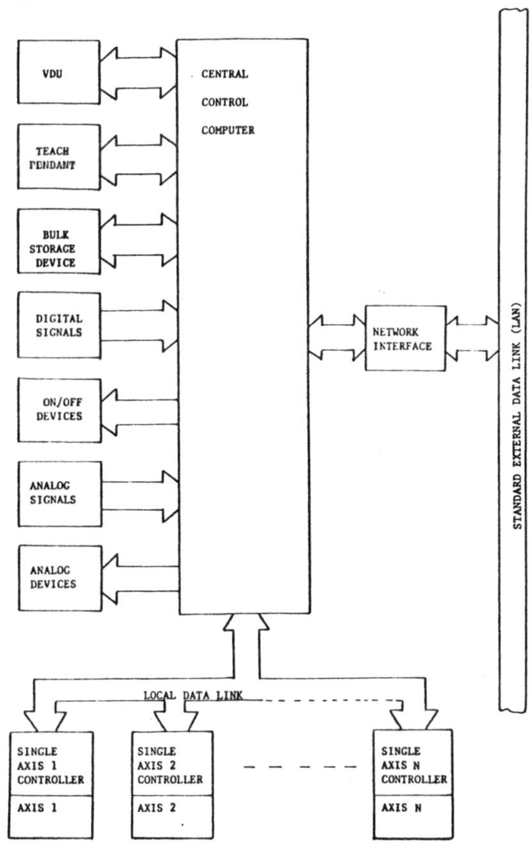

DESIGN OF A GRAB MANIPULATOR SYSTEM FOR PEAT HANDLING

Jaakko Häggman
Pekka Ranta Cons. Co
Jyväskylä, FINLAND

Heikki N. Koivo
Tampere University of Technology
Tampere Finland

Pekka Ranta
Pekka Ranta Cons. Co
Jyväskylä, FINLAND

Abstract. The paper discusses the automation of a grap manipulator system for peat handling. The system consists of a grab, a bridge crane and a microprocessor control system. Both hardware and software aspects of the control system are investigated.

1. INTRODUCTION

Peat handling in thermal power stations requires high investment and maintenance costs. The grab manipulator presented in this paper represents new technology with which the costs can be reduced.

Combining a grab and a bridge crane together with a microprocessor control system an efficient and economical transport system can be designed. Such a system replaces conventional conveyor belts and storage silos. Storage building costs would be considerably reduced.

The paper presents a design for the instrumentation, hardware and software parts of a grab manipulator system. First, a general description of peat handling using the grab manipulator in a thermal power plant is given. A brief account of the required instrumentation is included in section 3. The nucleus of the grab manipulator system is the control system. The hardware elements in the control system are discussed in section 4. The control system specifications and the software design procedure are given in sections 5 and 6, respectively.

2. GENERAL DESCRIPTION OF PEAT HANDLING

2.1. General

Peat is transported to the power plant by trucks, which dump their loads into the input pockets (Fig. 1). Peat is transferred by the grab manipulator to the middle storage or, if needed, directly to the feed silo of the boiler. From the middle storage the peat is transferred to the feed silo of the boiler by the grab manipulator. From there it is moved by the screw conveyor to the input of the spout, through which it falls to the boiler. The peat is neither ground nor dried. It is fed directly to a fluidized bed combustion chamber. The grab manipulator system requires no conveyor belts nor storages unloading from the bottom.

A basic requirement in the grab manipulator system is that the feed silo always be filled with the right grade of peat above a certain minimum level. This level is determined by safety requirements. An explosion may occur if the feed silo is empty and an air pocket develops in the boiler.

The walls in the peat storage building are made of sheet metal. The support balks of the storage building create obstructions which have to be taken into account in the control system design.

2.2. Grab manipulator operation

Three different duty cycles of the grab manipulator can be visualized (Fig. 2):

1. Transfer of peat from the middle storage into the feed silo.
2. Transfer of peat from the input pocket to the middle storage.
3. Transfer of peat from the input pocket into the feed silo.

Since the size of the grab manipulator is 6 m^3, several cycles are required to empty the input pocket. Rough calculations indicate that a typical grab manipulator has to operate 9...10 hours emptying the input pocket to satisfy the daily consumption. The truck deliveries can be arranged during regular working hours at about one hour intervals. To supply the weekend consumption of peat to the feed silo, the grab manipulator must operate another four hours. Since no operator is needed for the grab manipulator system, part of this work is carried out during the evening.

2.3. Locating peat in storage

The peat level in the storages is kept as flat as possible. For this reason the storages are divided into delivery and fetch positions. Introducing an xyz-coordinate system for storages, delivery and fetch address tables can be formed which can be programmed into computer memory.

2.4. Level vectors

To combine different movements of the grab manipulator smoothly, the peat level in the storages has to be known. When the grab is driven above the middle storage, the vertical movement can be combined with the horizontal movement, if it is known that the grab does not hit the peat surface too early. For this purpose level vectors for each storage are defined in the computer memory. Each element of these vectors contains the level of a certain delivery and fetch location of the storage. The level vectors can then be used to schedule the grab manipulator in such a fashion that in the middle storages peat is delivered to a place where the least amount of it exists and fetched from a place where most of it is found.

3. INSTRUMENTATION

A short description of the instrumentation is given in this section. Special emphasis is given to the position and velocity sensing of the grab manipulator.

The bridge crane is moved using shunt motors both in transfer movements and in lifting and lowering. The continuous velocity control of transfer movements is realized by inverters. These are also used for motor braking - separate discbrakes are not installed. A pole-changing motor having two different lift speeds is used as a lift motor.

The movement of the grab manipulator is detected by pulse sensors attached to the bridge grane. Both the distance travelled by the grab and its speed are measured with the pulse sensor. Position information is obtained with the aid of coordinate registers by adding or subtracting from the old coordinate values the distance indicated by the number of measured pulses. The speed measurement is arranged by counting the number of pulses during the measurement interval.

Distorted position information due to disturbances or loss of pulses is corrected by installing correcting sensors (flip-flops) in the middle points of x- and y-coordinates.

A scale is attached to the system to weigh the amounts of peat transported and to observe the system operation. The scale is installed at the point where the grab is attached to the steel cable.

A four-jawed grab is used, because it has proven to be better in practice for handling peat and wood chips. The jaws of the grab are governed hydraulically.

A capacitive level limiter is installed in the feed silo of the boiler.

4. MICROCOMPUTER SYSTEM

The control computer has to satisfy the following specifications:

- The control computer should act as a nucleus for the manless grab manipulator system. It communicates with the grab manipulator and the operator via the interface cards and peripheral equipment.
- The control computer must have mass memory, where the peat flow information can be stored and retrieved.

After careful consideration the system structure, shown in Fig. 3, was chosen. It consists of the following:

- A single board computer which has a 6809 CPU with 2 k of RAM, 16 + 4 parallel I/O parts, an RS 232C serial bus interface, 3 16-bit counter/timers, automatic restart, and a DMA possibility,
- 3 RAM modules having 8102 x 8 bit static RAM,
- an 8192 x 8 bit static CMOS RAM with battery back up on the board,
- 24 k EPROM,
- a serial bus interface module,
- an A/D conversion module, which has an 8-channel multiplexer,
- a D/A conversion module, which has 4 individual D/A converters,
- 2 parallel interface modules with 8 optoseparated inputs and outputs.

The status of the grab manipulator system is monitored with a control terminal. The keyboard of the terminal is used to change the address and peat level tables and to acknowledge alarms.

5. CONTROL SYSTEM SPECIFICATIONS

The basic control requirements for the grab manipulator are the correct timing of the feed silo filling operation and the input pocket emptying operation. The system must be able to function without an operator, although manual operation is also allowed. A video display terminal is stationed in the control room, so that process information can be displayed to ensure the correct operation of the grab manipulator system. The system includes a semi-automatic mode which turns on in alarm situations. Alarm situations occur when the normal operation of the grab manipulator is interrupted. The grab manipulator can be controlled in the semi-automatic mode manually from the control room instead of climbing up to the car of the bridge crane.

The system must be able to keep track of the time of the day and the amount of peat that has come to the plant during the day and has been taken to the feed silo. This information is used in billing and in calculating the efficiency of the power plant.

The above goals depend on the plant. The grab manipulator system has been designed to be more generally applicable. This has been achieved by using certain tables. The tables include address tables and peat level tables. The optimal motion of the grab manipulator to the target point is realized with obstruction tables.

6. SOFTWARE DESIGN

The structure of the software was studied first. The structure is divided into parallel processes, data structures, and an operating system. Only the software related to the control of the grab manipulator is desribed here.

A separate operating system was not needed. The reason for not using a real-time operating system is that relatively simple system software is all that is needed for the grab manipulator. Process interrupts can be used to handle tasks that appear to be simultaneous.

Part of the data structure was presented in subsections 2.3 and 2.4. The modular program structure is shown in Fig. 4. The modules in Fig. 4 are organized in hierarchical level.

Three basic program design methods were used (McGowan and Kelly, 1975): TOP/DOWN method, modularity, and structured programming. The TOP/DOWN method helps both in design and documentation. The system functions are first represented with simple flow diagrams at the highest level. The representation then sharpens at lower levels.

Modularity is used so that independent activities in the system operation are separated into blocks, which can be called when needed. Modules are formed so that to be used a module need only refer to submodules (modules on a lower hierarchical level). There is no need to refer to modules on the same level. Modules are given starting values and they return the result. There are four hierarchical levels of modules as shown in Fig. 4. Detailed desription of modules can be found in Häggman, Koivo and Ranta (1983).

Structured programming makes the block diagrams easily readable. The details of the block diagrams are not given in this paper. In program testing both BOTTOM-UP and TOP-DOWN testing are used.

7. CONCLUSION

The paper has presented the design of a grab manipulator system for peat handling in a thermal power station. The system consists of a grab, a bridge crane, and a microprocessor control system.

The heart of the grab manipulator system is the microcomputer. The software design of the control system was explained. Structured programming, modularity and the TOP-DOWN method were used. The most important program objective was to minimize the drive times of the grab manipulator. This is accomplished by having the program construct well-planned grab trajectories, which allow close movement to obstructions.

The grab manipulator system presented offers the following advantages over the conventional conveyor-belt transport systems:

- It is more reliable than conventional systems, which contain many individual components increasing the fail-

ure probability. It has been estimated that about 80 % of the service failures in a peat power plant are due to the fuel handling system.
- Energy usage of a grab manipulator system is smaller than in conventional systems.
- The wear of conventional conveyor belts is higher than the wear of a grab manipulator.
- The grab-manipulator system with its storage requires less spatial area and is less of an eyesore than the conventional set-up.

The grab manipulator system presented is also applicable in other situations for transferring solid materials. The size of grab and its speed are the limiting factors of the system applicability.

REFERENCES

Häggman, J., Koivo, H.N., and P. Ranta (1983). Design of a grab manipulator system for peat handling. Internal report. Tampere University of Tech. Control Eng., Tampere, Finland.

McGowan, C., and J. Kelly (1975). Top-Down Structured Programming Techniques. McGraw-Hill, New York.

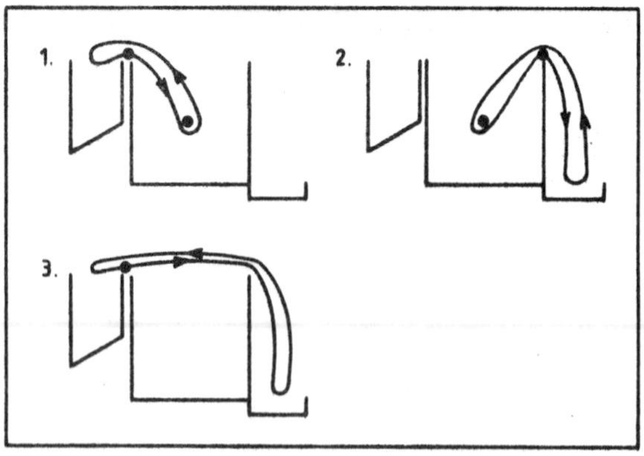

Fig. 2. Duty cycles of the grab manipulator.

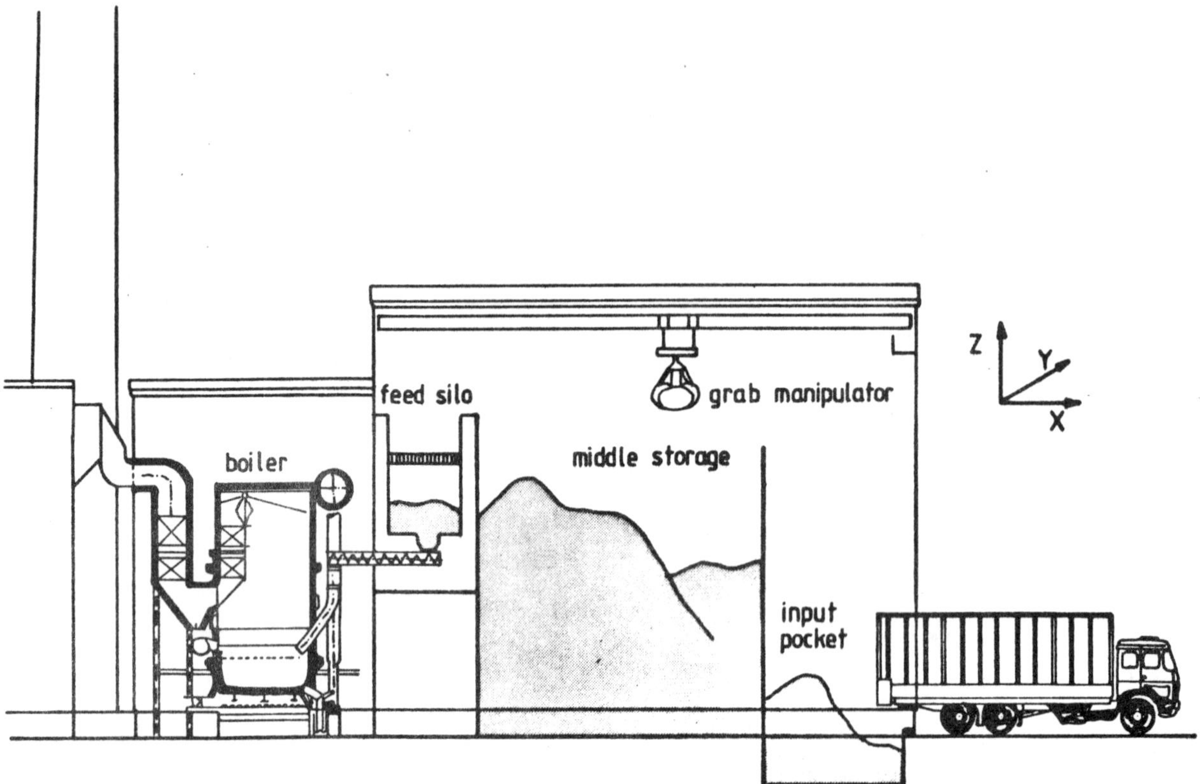

Fig. 1. Peat handling in a thermal power plant.

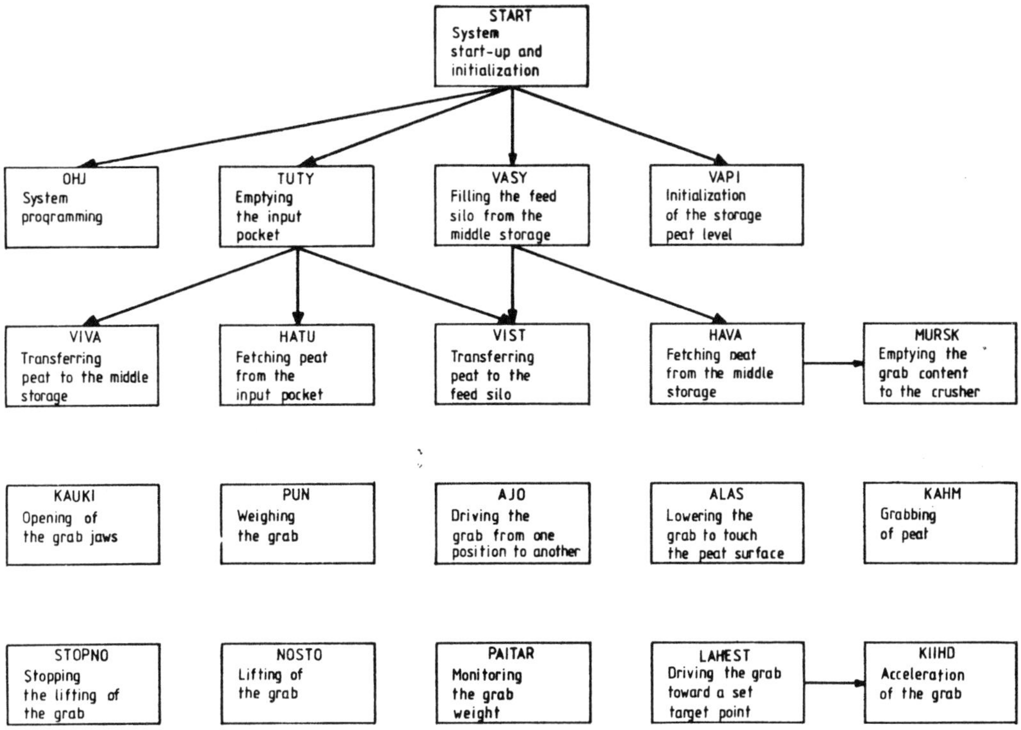

Fig. 4. The modular program structure of the grab manipulator control system. The arrow indicates reference to submodules. At the two lowest levels the arrows are not drawn for clarity's sake.

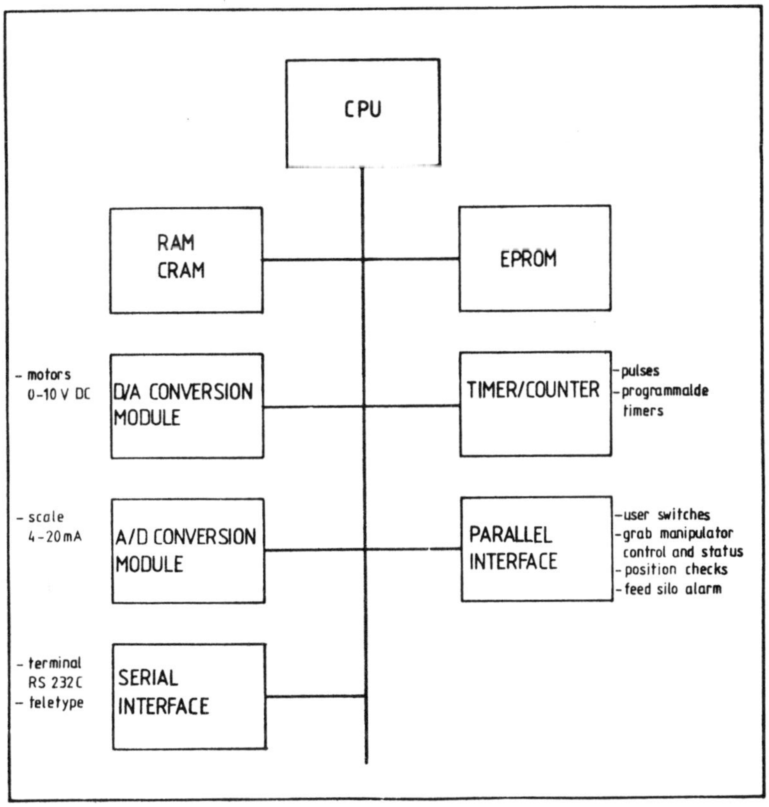

Fig. 3. The structure of the microcomputer system for controlling the grab manipulator.